Advanced Structured Materials

Volume 112

Series Editors

Andreas Öchsner, Faculty of Mechanical Engineering, Esslingen University of Applied Sciences, Esslingen, Germany
Lucas F. M. da Silva, Department of Mechanical Engineering, Faculty of Engineering, University of Porto, Porto, Portugal
Holm Altenbach, Faculty of Mechanical Engineering, Otto-von-Guericke-Universität Magdeburg, Magdeburg, Sachsen-Anhalt, Germany

Common engineering materials reach in many applications their limits and new developments are required to fulfil increasing demands on engineering materials. The performance of materials can be increased by combining different materials to achieve better properties than a single constituent or by shaping the material or constituents in a specific structure. The interaction between material and structure may arise on different length scales, such as micro-, meso- or macroscale, and offers possible applications in quite diverse fields.

This book series addresses the fundamental relationship between materials and their structure on the overall properties (e.g. mechanical, thermal, chemical or magnetic etc) and applications.

The topics of *Advanced Structured Materials* include but are not limited to

- classical fibre-reinforced composites (e.g. glass, carbon or Aramid reinforced plastics)
- metal matrix composites (MMCs)
- micro porous composites
- micro channel materials
- multilayered materials
- cellular materials (e.g., metallic or polymer foams, sponges, hollow sphere structures)
- porous materials
- truss structures
- nanocomposite materials
- biomaterials
- nanoporous metals
- concrete
- coated materials
- smart materials

Advanced Structured Materials is indexed in Google Scholar and Scopus.

More information about this series at http://www.springer.com/series/8611

Konstantin Naumenko ·
Holm Altenbach

Modeling High Temperature Materials Behavior for Structural Analysis

Part II. Solution Procedures and Structural Analysis Examples

Springer

Konstantin Naumenko
Fakultät für Maschinenbau
Institut für Mechanik
Otto-von-Guericke-Universität
Magdeburg, Germany

Holm Altenbach
Fakultät für Maschinenbau
Institut für Mechanik
Otto-von-Guericke-Universität
Magdeburg, Germany

ISSN 1869-8433 ISSN 1869-8441 (electronic)
Advanced Structured Materials
ISBN 978-3-030-20383-2 ISBN 978-3-030-20381-8 (eBook)
https://doi.org/10.1007/978-3-030-20381-8

This Springer imprint is published by the registered company Springer Nature Switzerland AG
The registered company address is: Gewerbestrasse 11, 6330 Cham, Switzerland

Preface

Many structural components operate at high temperature and mechanical loadings over a long period of time. Examples include components of power plants, chemical refineries and heat engines. Long-term strength analysis, design and life-time assessments require to take into account inelastic deformation and damage processes. The main subject of "modeling materials behavior at high temperature for structural analysis" is to develop methods for the analysis of inelastic processes in components, such as time-dependent changes of stress and strain states and damage evolution up to the critical stage of rupture.

The scope of this book is related to the fields "creep mechanics" (Betten, 2008; Hyde et al, 2013; Naumenko and Altenbach, 2007; Odqvist, 1981), "continuum creep and damage mechanics" (Hayhurst, 2001; Murakami, 2012), "mechanics of high-temperature plasticity" (Ilschner, 1973) or in a more broad sense to "behavior of materials and structures at high temperature". In the first part of the book Naumenko and Altenbach (2016) basic equations of continuum mechanics and constitutive models describing the mechanical behavior of structural materials under multi-axial stress states are introduced. This, second part is devoted to the application of structural mechanics models, such as beams, plates and shells, as well as the solution procedures of nonlinear initial-boundary value problems. These subjects have become traditional since the pioneering texts written in 1950s by Prager (1959) and 1960s by Odqvist and Hult (1962); Hult (1966), and Rabotnov (1969), among others. These classical books provide a first collection of solutions to plasticity and creep problems for elementary structures such as rods, beams and circular plates based on the simple constitutive models like the Norton-Bailey equation. The monographs of Penny and Mariott (1995) (first edition in 1971) and Viswanathan (1989) concentrate on robust methods and empirical relationships which are useful for the design procedures. The books of Kraus (1980), Malinin (1981) and Boyle and Spence (1983), published in 1980s, introduce advanced constitutive models with internal state variables and apply numerical methods for the structural analysis.

The aim of modeling and simulation is to reflect basic features of inelastic behavior in structures including the development of inelastic deformations, relaxation and redistribution of stresses as well as the local reduction of material strength. A model should be able to account for material deterioration processes in order to predict long term structural behavior and to analyze critical failure zones. Structural analysis usually requires the following steps:

1. Assumptions must be made with regard to the geometry of the structure, types of loading and heating as well as kinematical constraints.
2. A suitable structural mechanics model must be applied based on the assumptions concerning kinematics of deformations, types of internal forces (moments) and related balance equations.
3. A reliable constitutive model must be formulated to reflect inelastic deformations and processes like hardening/recovery and damage.
4. A mathematical model of the structural behavior (initial-boundary value problem) must be formulated including the material independent equations, constitutive and evolution equations as well as initial and boundary conditions.
5. Numerical solution procedures to solve non-linear initial-boundary value problems must be developed.
6. The verification of the applied models must be performed including the structural mechanics model, the constitutive model, the mathematical model as well as the numerical methods and algorithms.

The first two steps are common within continuum mechanics and engineering mechanics. Here, mathematical models of idealized solids and structures are developed and investigated. Examples include the models of three-dimensional solids, beams, rods, plates and shells. The idealizations are related to the continuum hypothesis, cross section assumptions, etc. The above models were originally developed within the theory of linear elasticity, e.g. Hahn (1985); Timoshenko (1953). In creep mechanics they are applied together with constitutive and evolution equations describing idealized inelastic behavior, e.g. steady-state creep (Boyle and Spence, 1983; Hult, 1966; Kraus, 1980; Malinin, 1981; Odqvist, 1974). As mentioned in the first part (Naumenko and Altenbach, 2016), many structural materials exhibit non-classical phenomena such as different inelastic strain rates under tension and compression, stress state dependence of tertiary creep, damage induced anisotropy, etc. Consideration of such effects may require various extensions of available structural mechanics models. For example, the concept of the stress free (neutral) plane widely used in the theory of beams and plates becomes invalid if the material shows different creep rates under tension and compression. In this book we discuss the applicability of classical and refined models of beams, plates and shells to the inelastic analysis. Based on several examples we examine the accuracy of cross section assumptions for displacement and stress fields.

The mathematical model of an inelastic structure is the initial-boundary value problem (IBVP) which usually includes partial differential equations describing kinematics of deformation and balance of forces, ordinary differential equations describing inelastic processes as well as initial and boundary conditions. The numerical solution can be organized as follows, e.g. Boyle and Spence (1983). For given values of the inelastic strain tensor and internal state variables at a fixed time the boundary value problem (BVP) is solved. Here direct variational methods, e.g.

the Ritz method, the Galerkin method, the finite element method are usually applied. In addition, a time step procedure is required to integrate constitutive and evolution equations. In this book various methods are reviewed and discussed with respect to their efficiency and numerical accuracy.

In recent years the finite element method has become the widely accepted tool for structural analysis. The advantage of the finite element method is the possibility to model and analyze engineering structures with complex geometries, various types of loadings and boundary conditions. General purpose finite element codes like ABAQUS or ANSYS were developed to solve various problems in solid mechanics. In application to the inelastic analysis one should take into account that a general purpose constitutive equation which allows to reflect the whole set of creep and damage processes in structural materials over a wide range of loading and temperature conditions is not available at present. Therefore, a specific constitutive model with selected internal state variables, special types of stress and temperature functions as well as material constants identified from available experimental data should be incorporated into the commercial finite element code by writing a user-defined material subroutine. In this book the ABAQUS and ANSYS finite element codes are applied to the numerical analysis. In order to consider damage processes the user-defined subroutines are developed and implemented. The subroutines serve to utilize constitutive and evolution equations with damage state variables. In addition, they allow the postprocessing of damage, i.e. the creation of contour plots visualizing damage distributions.

An important question in any structural analysis is that on reliability of the applied models, numerical methods and obtained results. The reliability assessment may require the following verification steps:

- *Verification of developed finite element subroutines.* To assess that the subroutines are correctly coded and implemented, results of finite element computations must be compared with reference solutions of benchmark problems. Several benchmark problems have been proposed in Becker et al (2002) based on an in-house finite element code. In this book several closed form and semi-analytical solutions of steady-state creep in elementary structures are presented. To extend these solutions to the primary and tertiary creep ranges the Ritz and the time step methods will be applied. The advantage of these problems is the possibility to obtain reference solutions without a finite element discretization. Furthermore, they allow to verify finite element subroutines over a wide range of finite element types including beam, shell and solid type elements.
- *Verification of applied numerical methods.* Here the problems of the suitable finite element type, the mesh density, the time step size and the time step control must be analyzed. They are of particular importance in creep damage related simulations. In this book these problems are discussed based on numerical tests and by comparison with reference solutions.
- *Verification of constitutive and structural mechanics models.* This step requires tests of model structural components and the corresponding numerical analysis by the use of the developed techniques. Examples of experimental studies on creep in structures include beams, transversely loaded plates (Boyle and Spence,

1983), thin-walled tubes under internal pressure (Koundy et al, 1997; Krieg, 1999), pressure vessels (Eggeler et al, 1994; Fessler and Hyde, 1994), circumferentially notched bars (Hayhurst, 1994).

In Chapt. 1 basics of inelastic structural analysis are introduced starting with elementary examples for bars and bar systems. Main features of inelastic structural responses including creep, relaxation, stress redistribution under various stress-controlled and displacement-controlled loading profiles are illustrated. Governing equations a for two-bar system are introduced and initial value problems for one-dimensional inelastic stress analysis are formulated. Closed-form solutions for a two-bar system under different types of material behavior are presented. Furthermore, examples for numerical time-step methods are introduced. They include one-step explicit and implicit time integration schemes.

Chapter 2 gives a summary of governing mechanical equations to describe inelastic behavior in three-dimensional solids. The set of equations includes material independent equations, constitutive and evolution equations, as well as the initial and boundary. The formulated initial-boundary value problem can be solved by numerical methods. Explicit and implicit time integration methods are presented to analyze three-dimensional solids. With time-step procedures, linearized boundary value problems should be solved within time and/or iteration steps. Variational formulations and direct variational methods are discussed in detail. In addition, time-scales methods are presented in order to formulate efficient solution procedures for components subjected to cyclic loadings.

Chapter 3 collects examples of inelastic structural analysis for beams. Both the classical Bernoulli-Euler beam theory and the first order shear deformation theory are introduced. Governing equations and variational formulations for inelastic analysis are presented. Closed-form solutions and approximate analytical solutions are derived for beams from materials that exhibit power law creep and stress regime dependent creep. Creep-damage constitutive models are applied to illustrate basic features of stress redistribution and damage evolution in beams. The reference solutions for various problems obtained by the Ritz method are applied to verify user-defined creep-damage material subroutines, implemented inside general purpose finite element codes.

Chapter 4 presents examples of inelastic structural analysis for plane stress and plane stress problems. Elementary structures including a pressurized thick cylinder, a rotating disc, and a plate with a circular hole are analyzed. Classical solutions with the power law creep constitutive equations as well as solutions with stress regime dependent inelastic behavior are presented.

Chapter 5 gives an overview of modeling approaches for plates and shells in the inelastic range and presents examples of inelastic analysis of thin-walled structures. Governing equations of the first order shear deformation theory of plates are discussed in detail. An emphasis is placed on the direct formulation of inelastic constitutive laws. For circular plates, numerical solutions of steady-state creep are presented. Advanced constitutive models with internal state variables, such as the damage parameter require the use of advanced plate theories to consider edge effects. An example of a rectangular plate with different types of boundary conditions is presented to illustrate edge effects. Finally, an example for a thin-walled pipe subjected to the internal pressure and the bending moment is discussed.

Several chapters of this book have grown out of our lectures and lecture notes on fundamentals of continuum mechanics, mechanics of materials and inelastic structural analysis for graduate level students and PhD students held at the Martin-Luther-Universität Halle-Wittenberg, Otto-von-Guericke-Universität Magdeburg, Fraunhofer Institut für Werkstoffmechanik (Halle/Saale), Politechnico Milano, Nagoya University, Politechnika Lubelska and National Technical University "Kharkiv Polytechnical Institute", among others. Many results presented originate from scientific and academic exchange projects. We wish to acknowledge financial support from the German Research Foundation (DFG), German Academic Exchange Service (DAAD), the State Saxony-Anhalt, and European Commission (ERASMUS). This book partly includes structural analysis examples published in the monograph *Modeling of Creep for Structural Analysis* (Naumenko and Altenbach, 2007). Many additional examples of elementary structural analysis, which can be solved in a class by standard numerical methods, are introduced.

We would like to acknowledge Professors J. Betten, J. Boyle, O. T. Bruhns, E. Gariboldi, T. Hyde, R. Kienzler, Z. L. Kowalewski, O. K. Morachkowski and N. Ohno for many fruitful discussions which stimulated our research in mechanics of inelastic material behavior. For the careful reading of the manuscript we thank Dr. Johanna Eisenträger. We would like to thank Dr. Christoph Baumann from Springer Publisher for the assistance and support during preparing the book.

Magdeburg
Spring 2019

Konstantin Naumenko
Holm Altenbach

References

Becker AA, Hyde TH, Sun W, Andersson P (2002) Benchmarks for finite element analysis of creep continuum damage mechanics. Comp Mat Sci 25:34 – 41

Betten J (2008) Creep Mechanics, 3rd edn. Springer, Berlin

Boyle JT, Spence J (1983) Stress Analysis for Creep. Butterworth, London

Eggeler G, Ramteke A, Coleman M, Chew B, Peter G, Burblies A, Hald J, Jefferey C, Rantala J, deWitte M, Mohrmann R (1994) Analysis of creep in a welded P91 pressure vessel. International Journal of Pressure Vessels and Piping 60:237 – 257

Fessler H, Hyde TH (1994) The use of model materials to simulate creep behavior. The Journal of Strain Analysis for Engineering Design 29(3):193 – 200

Hahn HG (1985) Elastizitätstheorie. B.G. Teubner, Stuttgart

Hayhurst DR (1994) The use of continuum damage mechanics in creep analysis for design. The Journal of Strain Analysis for Engineering Design 25(3):233 – 241

Hayhurst DR (2001) Computational continuum damage mechanics: its use in the prediction of creep fracture in structures - past, present and future. In: Murakami S, Ohno N (eds) IUTAM Symposium on Creep in Structures, Kluwer, Dordrecht, pp 175 – 188

Hult JA (1966) Creep in Engineering Structures. Blaisdell Publishing Company, Waltham

Hyde T, Sun W, Hyde C (2013) Applied Creep Mechanics. McGraw-Hill Education

Ilschner B (1973) Hochtemperatur-Plastizität. Springer, Berlin et al.
Koundy V, Forgeron T, Naour FL (1997) Modeling of multiaxial creep behavior for incoloy 800 tubes under internal pressure. Trans ASME J Pressure Vessel & Technology 119:313 – 318
Kraus H (1980) Creep Analysis. John Wiley & Sons, New York
Krieg R (1999) Reactor Pressure Vessel Under Severe Accident Loading. Final Report of EU-Project Contract FI4S-CT95-0002. Tech. rep., Forschungszentrum Karlsruhe, Karlsruhe
Malinin NN (1981) Raschet na polzuchest' konstrukcionnykh elementov (Creep Calculations of Structural Elements, in Russ.). Mashinostroenie, Moskva
Murakami S (2012) Continuum Damage Mechanics: A Continuum Mechanics Approach to the Analysis of Damage and Fracture. Solid Mechanics and Its Applications, Springer
Naumenko K, Altenbach H (2007) Modelling of Creep for Structural Analysis. Springer, Berlin et al.
Naumenko K, Altenbach H (2016) Modeling High Temperature Materials Behavior for Structural Analysis: Part I: Continuum Mechanics Foundations and Constitutive Models, Advanced Structured Materials, vol 28. Springer
Odqvist FKG (1974) Mathematical Theory of Creep and Creep Rupture. Oxford University Press, Oxford
Odqvist FKG (1981) Historical survay of the development of creep mechanics from its beginnings in the last century to 1970. In: Ponter ARS, Hayhurst DR (eds) Creep in Structures, Springer, Berlin, pp 1 – 12
Odqvist FKG, Hult J (1962) Kriechfestigkeit metallischer Werkstoffe. Springer, Berlin u.a.
Penny RK, Mariott DL (1995) Design for Creep. Chapman & Hall, London
Prager W (1959) An Introduction to Plasticity. Addison-Wesley Publishing Company
Rabotnov YN (1969) Creep Problems in Structural Members. North-Holland, Amsterdam
Timoshenko SP (1953) History of Strength of Materials. Dover Publ., New York
Viswanathan R (1989) Damage Mechanisms and Life Assessment of High Temperature Components. ASM international

Contents

About the Authors

Prof. Dr.-Ing. habil. Konstantin Naumenko, Member of the International Association of Applied Mathematics and Mechanics. Employment history includes positions at Otto-von-Guericke-University Magdeburg and at Martin-Luther-University Halle-Wittenberg, both Germany. Graduated from Kharkiv Polytechnic Institute in 1989 (diploma in Dynamics and Strength of Machines). Defended PhD in 1996 at Otto-von-Guericke-University Magdeburg, habilitation in 2006 at Martin-Luther-University Halle-Wittenberg.

Areas of scientific interest:

- Inelastic structural analysis.
- Damage and fracture mechanics.
- Advanced constitutive models for high-temperature materials.

Author/Co-author/Co-editor of 5 Books (textbooks/monographs/proceedings) and 115 peer-reviewed scientific papers.

Prof. Dr.-Ing. habil. Dr. h. c. mult Holm Altenbach, Member of the International Association of Applied Mathematics and Mechanics, and the International Research Center on Mathematics and Mechanics of Complex Systems (M&MoCS), Italy. Employment history includes positions at Otto-von-Guericke-University Magdeburg and at Martin-Luther-University Halle-Wittenberg, both Germany. Graduated from Leningrad Polytechnic Institute in 1980 (diploma in Dynamics and Strength of Machines). Defended PhD in 1983, awarded Doctor of technical sciences in 1987, both at the same Institute.

Present position: Full Professor in Engineering Mechanics at the Otto-von-Guericke-University, Faculty of Mechanical Engineering, Institute of Mechanics (since 2011), acting director of the Institute of Mechanics since 2015.

Areas of scientific interest:

- General theory of elastic and inelastic plates and shells.
- Creep and damage mechanics.
- Strength theories.
- Nano- and micromechanics.

Author/Co-author/Co-editor of 60 Books (textbooks/monographs/proceedings), appr. 380 scientific papers (among them 250 peer-reviewed) and 500 scientific lectures. Managing Editor (2004 to 2014) and Editor-in-Chief (2005 – to date) of the Journal of Applied Mathematics and Mechanics (ZAMM) – the oldest journal in Mechanics in Germany (founded by Richard von Mises in 1921), Advisory Editor of the journal "Continuum Mechanics and Thermodynamics" since 2011, Associate Editor of the journal "Mechanics of Composites" (Riga) since 2014, Co-Editor of the Springer Series "Advanced Structured Materials" since 2010.

Awards: 1992 Krupp-Award (Alexander von Humboldt-Foundation), 2000 Best paper of the year Journal of Strain Analysis for Engineering Design, 2003 Gold Medal of the Faculty of Mechanical Engineering, Politechnika Lubelska, Lublin, Poland, 2004 Semko-Medal of the National Technical University Kharkov, Ukraine, 2007 Doctor honoris causa, National Technical University Kharkov, Ukraine, 2011 Fellow of the Japanese Society for the Promotion of Science, 2014 Doctor honoris causa, University Constanta, Romania, 2016 Doctor honoris causa, Vekua-Institute, Tbilisi, Georgia, 2018 Alexander von Humboldt Award (Poland).

Chapter 1
Bars and Bar Systems

Bars are structural members that support uni-axial tensile or compressive loadings. Chapter 1 presents elementary examples for bars and bar systems and gives an introduction to inelastic stress analysis. Basic features of inelastic structural responses including creep, relaxation, stress redistribution and others are discussed. In Sect. 1.1 governing equations a for two-bar system are introduced and initial value problems for one-dimensional inelastic stress analysis are formulated. Section 1.2 presents elementary solutions for stresses in linear thermo-elastic bars subjected to non-uniform heating. Closed-form solutions for a two-bar system under assumption of linear viscous material behavior are presented in Sect. 1.3. Various force-controlled and displacement-controlled loading profiles are discussed. In addition to analytical solutions, examples for numerical time-step methods are introduced. They include one-step explicit and implicit time integration methods. Results are compared with closed form solutions to conclude on numerical accuracy and stability. Section 1.3 gives an overview of uni-axial constitutive models describing idealized non-linear inelastic behavior. Hardening, softening and damage processes are neglected to make the analysis transparent. Solutions are presented for different loading paths illustrating stress-range dependent creep, creep recovery, relaxation and tensile behaviors. Finally time-step methods are discussed to show basic features of numerical analysis for non-linear inelasticity problems.

1.1 Governing Equations for Two-bar System

An elementary example of engineering mechanics is the two-bar system introduced in Naumenko and Altenbach (2016, Sect. 1.2.2). Let us consider two pipes, which are rigidly connected as shown in Fig. 1.1a. For the sake of simplicity let us assume that the diameter of each pipe is much smaller than the length such that the stress state is uni-axial and a two-bar model can be used to analyze the structural behavior. The governing equations for the analysis of the considered two-bar system can be summarized as follows

K. Naumenko and H. Altenbach, *Modeling High Temperature Materials Behavior for Structural Analysis*, Advanced Structured Materials 112,
https://doi.org/10.1007/978-3-030-20381-8_1

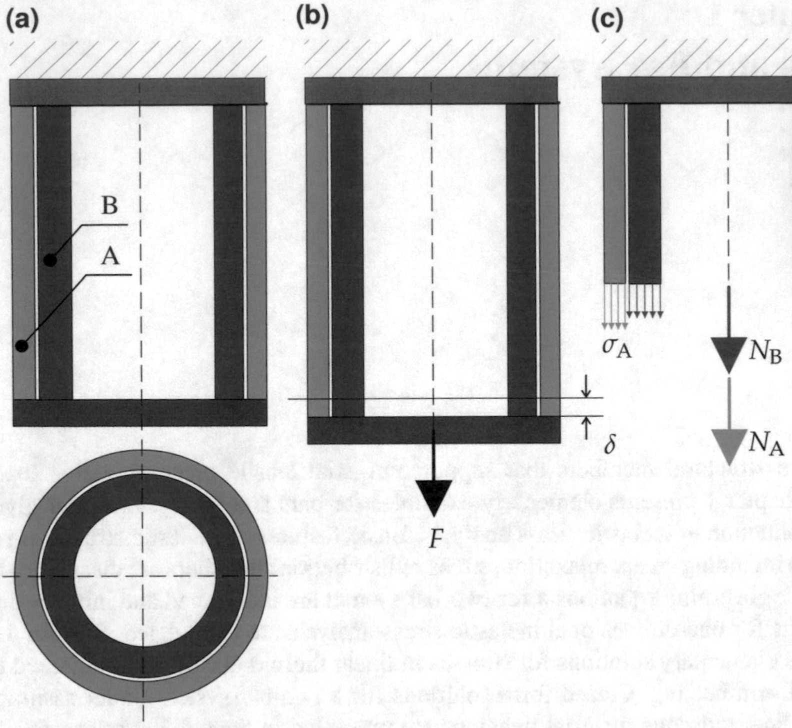

Fig. 1.1 Two-bar system. **(a)** Geometry, **(b)** loading, **(c)** free body diagram

– kinematical equations
  – strain-displacement relation

$$\varepsilon_\mathrm{A} = \frac{\delta_\mathrm{A}}{l}, \quad \varepsilon_\mathrm{B} = \frac{\delta_\mathrm{B}}{l}, \tag{1.1.1}$$

  where l is the length of each pipe in the reference state, ε_i with $i = \mathrm{A}, \mathrm{B}$ are strains and δ_i are displacements
  – compatibility condition

$$\delta_\mathrm{A} = \delta_\mathrm{B} = \delta \quad \Rightarrow \quad \varepsilon_\mathrm{A} = \varepsilon_\mathrm{B} = \varepsilon = \frac{\delta}{l}, \tag{1.1.2}$$

  where δ is the displacement of the two-bar system, Fig. 1.1b.

– equilibrium condition

$$N_\mathrm{A} + N_\mathrm{B} = F \quad \Rightarrow \quad \sigma_\mathrm{A} A_\mathrm{A} + \sigma_\mathrm{B} A_\mathrm{B} = F, \tag{1.1.3}$$

where N_i are internal forces in bars, σ_i are the corresponding stresses, Fig. 1.1c, and A_i are cross-section areas. For the sake of brevity we assume $A_\mathrm{A} = A_\mathrm{B} = A$

such that

$$\sigma_{\rm A} + \sigma_{\rm B} = \frac{F}{A}$$

- constitutive equations

$$\varepsilon_{\rm A} = \varepsilon_{\rm A}^{\rm el} + \varepsilon_{\rm A}^{\rm th} + \varepsilon_{\rm A}^{\rm pl}, \quad \varepsilon_{\rm B} = \varepsilon_{\rm B}^{\rm el} + \varepsilon_{\rm B}^{\rm th} + \varepsilon_{\rm B}^{\rm pl}, \tag{1.1.4}$$

where $\varepsilon_i^{\rm el}, \varepsilon_i^{\rm th}, \varepsilon_i^{\rm pl}$ are elastic, thermal and inelastic strains, respectively. Let us assume that the bars A and B have the same elastic and thermal material properties. The elastic strains are then defined as follows

$$\varepsilon_{\rm A}^{\rm el} = \frac{\sigma_{\rm A}}{E}, \quad \varepsilon_{\rm B}^{\rm el} = \frac{\sigma_{\rm B}}{E}, \tag{1.1.5}$$

where E is the Young's modulus. For the thermal strains we assume

$$\begin{aligned} \varepsilon_{\rm A}^{\rm th} &= \alpha_T \Delta T_{\rm A}, \quad \Delta T_{\rm A} = T_{\rm A} - T_0, \\ \varepsilon_{\rm B}^{\rm th} &= \alpha_T \Delta T_{\rm B}, \quad \Delta T_{\rm B} = T_{\rm B} - T_0, \end{aligned} \tag{1.1.6}$$

where α_T is the coefficient of thermal expansion, which is assumed the same in both bars, T_i are absolute temperatures in the bars and T_0 is the reference temperature. For the inelastic strains the following rate equations are specified, see Naumenko and Altenbach (2016, Sect. 3.3)

$$\begin{aligned} \dot{\varepsilon}_{\rm A}^{\rm pl} &= a_{\rm A} \exp\left(-\frac{Q_{\rm A}}{RT_{\rm A}}\right) f\left(\frac{|\sigma_{\rm A}|}{\sigma_{0\rm A}}\right) \operatorname{sgn}(\sigma_{\rm A}), \\ \dot{\varepsilon}_{\rm B}^{\rm pl} &= a_{\rm B} \exp\left(-\frac{Q_{\rm B}}{RT_{\rm B}}\right) f\left(\frac{|\sigma_{\rm B}|}{\sigma_{0\rm B}}\right) \operatorname{sgn}(\sigma_{\rm B}) \end{aligned} \tag{1.1.7}$$

where $f(x)$ is the stress function and a_i, Q_i, $i = {\rm A}, {\rm B}$ are parameters to be identified from experimental data and R is the universal gas constant. In general σ_{0i} can be temperature-dependent. In this chapter we assume that σ_{0i} are constants. Equations (1.1.7) describe "ideal" inelastic material behavior by neglecting hardening processes. A widely used stress function is the power law

$$f(x) = x^n, \tag{1.1.8}$$

where n is assumed constant. Further examples for stress functions will be discussed in Sect. 1.4. For $n = 1$, Eqs (1.1.7) and (1.1.8) provide the model of a linear viscous fluid. For moderate values of n, say for $3 \leq n \leq 10$, Eqs (1.1.7) and (1.1.8) describe the power law creep regime which is widely discussed in creep mechanics. For large values of n, the inelastic strain rate sensitivity according to Eqs (1.1.7) becomes negligible. For $n \to \infty$, the constitutive equations of rate-independent plasticity (St. Venant model) with the yield stresses σ_{0i} follow from Eqs (1.1.7) and (1.1.8).

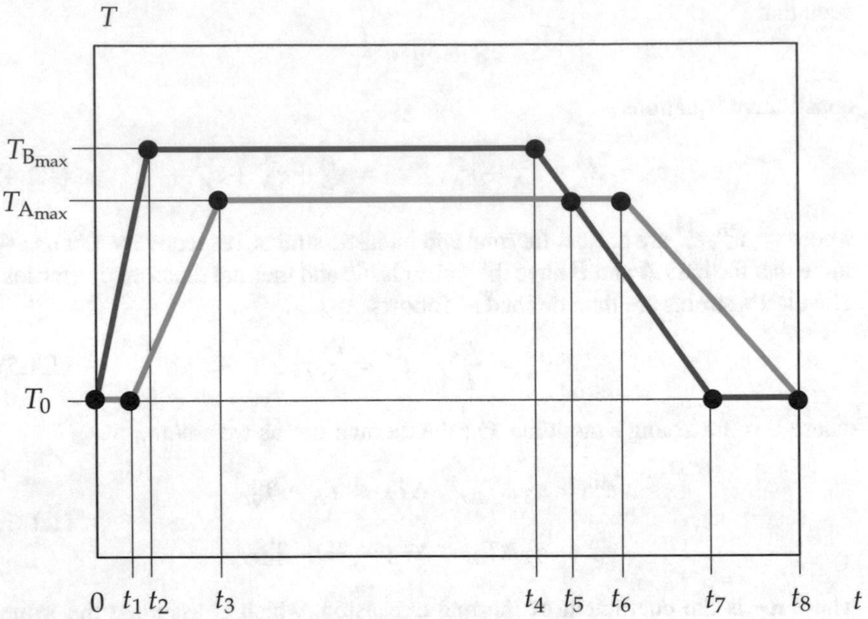

Fig. 1.2 Profiles of absolute temperature in bars A and B

1.2 Thermo-elasticity with Temperature Changes

Let us consider the two-bar system subjected to the non-uniform heating with the temperature profiles in bars shown in Fig. 1.2. The absolute temperature of pipe B increases from T_0 up to $T_{B_{max}}$ during a time interval $0 - t_2$. As the heat flows towards the outer pipe A, the temperature T_A increases with a delay during the time interval $t_1 - t_2$. The greatest difference in temperatures is assumed at the time point t_2. This is sometimes called "upshock" (Skelton, 2003), the pipe B is subjected to the compression with the maximum stress magnitude. During the time interval $t_2 - t_3$ the temperature difference between the pipes decreases while the absolute temperature of pipe A increases. The time interval $t_3 - t_4$ is the steady operation period. The time interval $t_4 - t_7$ is the cool-down stage, the steam temperature and the temperature of the pipe B decrease. During the time interval $t_5 - t_7$ the temperature difference increases again. However, the temperature of pipe B is now lower than the temperature of pipe A. During this "downshock" stage, pipe B is subjected to tension with the maximum stress value at the time point t_7. The external force is zero and the internal forces in the bars arise due to non-uniform heating only. For the sake of brevity we assume that the Young's modulus E and the coefficient of thermal expansion α_T are constant within the considered temperature interval $T_0 - T_{B_{max}}$. Let us analyze the stress and strain states in the pipes during the whole thermal cycle. With $F = 0$ the obvious equilibrium condition for the forces (1.1.3) yields $N_A = -N_B$. Since the cross-section areas are assumed the same, the relation between the stresses is $\sigma_{\mathrm{A}} = -\sigma_{\mathrm{B}}$. Neglecting inelastic strains the constitutive equations for the bars (1.1.4) – (1.1.7) can be given as follows

$$\varepsilon_{\mathrm{A}} = \frac{\sigma_{\mathrm{A}}}{E} + \alpha_T \Delta T_{\mathrm{A}}, \quad \varepsilon_{\mathrm{B}} = \frac{\sigma_{\mathrm{B}}}{E} + \alpha_T \Delta T_{\mathrm{B}}, \tag{1.2.9}$$

Subtracting Eq. $(1.2.9)_2$ from $(1.2.9)_1$ and taking into account the compatibility condition (1.1.2), the following equation for the stress in the bar B can be obtained

$$\sigma_{\mathrm{B}} = \frac{E\alpha_T}{2}(T_{\mathrm{A}} - T_{\mathrm{B}}) \tag{1.2.10}$$

If the inelastic strains are negligible, then the stress in the bar B is related to the difference in absolute temperatures between the bars A and B. As a result the bar B is subjected to compression during warm-up stage and tension during the cool-down stage. For the temperature profile shown in Fig. 1.2 the stress σ_{B} takes the minimum compressive value at t_2, the maximum tensile value at t_7 and zero values at t_5 as well as at the beginning and the end of the cycle.

Figure 1.3 illustrates qualitatively the stress value in the pipe B as a function of time. The results are normalized with the minimum stress value during the warm-up stage

$$\sigma_{\mathrm{B_{min}}} = \sigma_{\mathrm{B}}(t_2)$$

Two stress peaks are observed - the compressive one during the warm up stage with the maximum temperature difference $T_{\mathrm{B}} - T_{\mathrm{A}}$ at the time point t_2 and the tensile peak during the cool-down stage with the minimum temperature difference at the time point t_7. At t_5 the stresses in the bars take zero values since $T_{\mathrm{B}} = T_{\mathrm{A}}$. After adding Eqs $(1.2.9)_1$ to $(1.2.9)_2$, the mean strain of the bar system can be computed as follows

$$\varepsilon = \frac{\alpha_T}{2}(\Delta T_{\mathrm{A}} + \Delta T_{\mathrm{B}}) \tag{1.2.11}$$

1.3 Linear Viscous Behavior

Linear viscous behavior can be observed for many materials only within specific stress and temperature ranges. The aim of this section is to illustrate basic ideas of numerical time step procedures. The assumption of linear viscous behavior is advantageous since closed-form analytical solutions can be derived. Furthermore, stability and accuracy of the numerical solution procedure can be analyzed explicitly.

Let us assume that the absolute temperatures in the bars are constant. With $n = 1$ the constitutive Eqs (1.1.7) take the following form

$$\dot{\varepsilon}^{\mathrm{pl}}_{\mathrm{A}} = \dot{\varepsilon}_{0\mathrm{A}} \frac{\sigma_{\mathrm{A}}}{\sigma_{0\mathrm{A}}}, \quad \dot{\varepsilon}^{\mathrm{pl}}_{\mathrm{B}} = \dot{\varepsilon}_{0\mathrm{B}} \frac{\sigma_{\mathrm{B}}}{\sigma_{0\mathrm{B}}}, \tag{1.3.12}$$

where

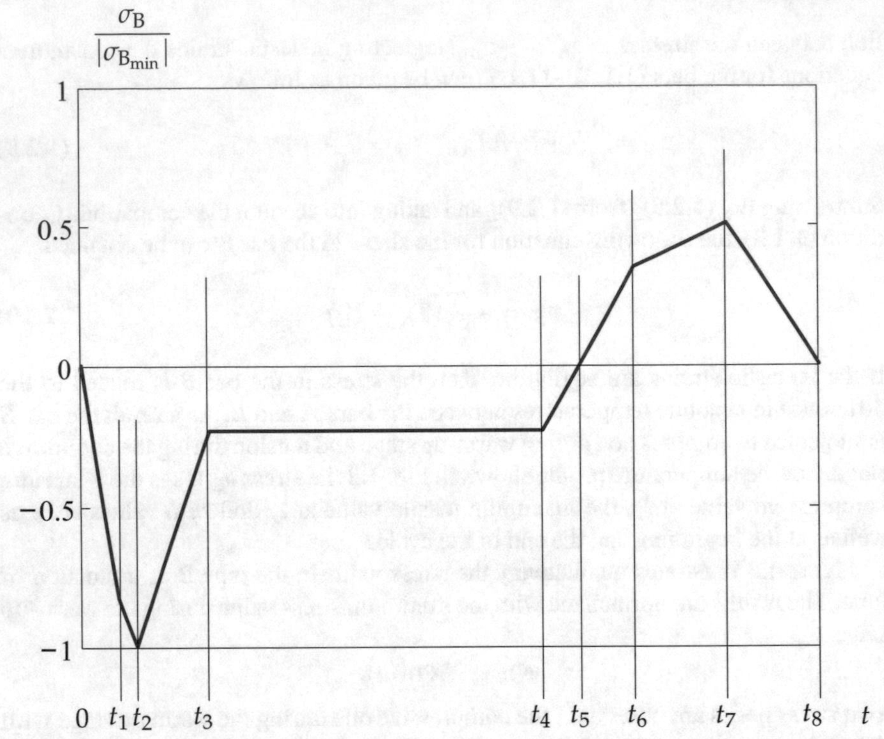

Fig. 1.3 Normalized stress in the pipe B vs. time

$$\dot{\varepsilon}_{0A} = a_A \exp\left(-\frac{Q_A}{RT_A}\right), \quad \dot{\varepsilon}_{0B} = a_B \exp\left(-\frac{Q_B}{RT_B}\right) \tag{1.3.13}$$

Introducing the viscosities

$$\mu_A = \frac{\sigma_{0A}}{\dot{\varepsilon}_{0A}}, \quad \mu_B = \frac{\sigma_{0B}}{\dot{\varepsilon}_{0B}},$$

the constitutive equations for the inelastic strain rates are

$$\dot{\varepsilon}^{pl}_A = \frac{\sigma_A}{\mu_A}, \quad \dot{\varepsilon}^{pl}_B = \frac{\sigma_B}{\mu_B} \tag{1.3.14}$$

Taking the time derivative of Eqs (1.1.4) and considering Eqs (1.1.5) and (1.3.14) the following differential equations for the stresses in the bars can be obtained

$$\dot{\sigma}_A + \lambda_A \sigma_A = E\dot{\varepsilon}_A, \quad \dot{\sigma}_B + \lambda_B \sigma_B = E\dot{\varepsilon}_B, \quad \lambda_A = \frac{E}{\mu_A}, \quad \lambda_B = \frac{E}{\mu_B} \tag{1.3.15}$$

The general solutions to Eqs (1.3.15) are

$$\sigma_i(t) = E\exp(-\lambda_i t)\int \exp(\lambda_i t)\dot{\varepsilon}_i(t)\mathrm{d}t + C_i\exp(-\lambda_i t), \qquad (1.3.16)$$

where C_i are integration constants. With the initial conditions $\sigma_i(0) = \sigma_{i_0}$ the general solution can also be given as follows

$$\sigma_i(t) = \sigma_{i_0}\exp(-\lambda_i t) + \int\limits_0^t \exp[-\lambda_i(t-\tau)]\dot{\varepsilon}_i(\tau)\mathrm{d}\tau \qquad (1.3.17)$$

1.3.1 Displacement-controlled Loading Paths

The displacement of the rigid bar δ is given as a function of time. The force F required to move the bar system and the stresses in the bars have to be computed. First assume that the displacement is a linear function of time such that

$$\delta(t) = v_\delta t,$$

where v_δ is the velocity. From Eqs (1.1.2) the strains are computed as follows

$$\varepsilon_i = \varepsilon = \frac{v_\delta}{l}t$$

The strain rates are

$$\dot{\varepsilon}_i = \dot{\varepsilon} = \frac{v_\delta}{l}$$

The general solution (1.3.16) is now specified as follows

$$\sigma_i(t) = \mu_i\dot{\varepsilon} + C_i\exp(-\lambda_i t), \quad i = \mathrm{A},\mathrm{B}, \qquad (1.3.18)$$

Assuming that the bars are stress-free in the reference state, i.e. $\sigma_i(0) = 0$, the solution (1.3.18) takes the form

$$\sigma_i(t) = \mu_i\dot{\varepsilon}[1-\exp(-\lambda_i t)] \qquad (1.3.19)$$

With $\varepsilon = \dot{\varepsilon}t$ the stresses can be specified as functions of the strain and the strain rate

$$\sigma_i(\varepsilon,\dot{\varepsilon}) = \mu_i\dot{\varepsilon}\left[1-\exp\left(-\frac{E\varepsilon}{\mu_i\dot{\varepsilon}}\right)\right] \qquad (1.3.20)$$

For small strain values such that $E\varepsilon \ll \mu_i\dot{\varepsilon}$, the solution (1.3.20) yields stress values within the linear-elastic range

$$\sigma_i = E\varepsilon$$

On the other hand with increasing strain the solutions approach the steady-state asymptotic values, i.e.

$$\sigma_i \to \sigma_{\mathrm{ss}_i}, \quad \text{for} \quad \frac{E\varepsilon}{\mu_i \dot{\varepsilon}} \to \infty, \quad \Rightarrow \quad \sigma_{\mathrm{ss}_i} = \mu_i \dot{\varepsilon}$$

With the solutions for stresses in bars the force F can be computed from Eq. (1.1.3).

The next example deals with a periodic loading in the form

$$\varepsilon(t) = \varepsilon_{\mathrm{a}} \sin \omega t, \tag{1.3.21}$$

where ε_{a} is the stain amplitude and ω is the angular frequency. In this case

$$\int \exp(\lambda_i t)\dot{\varepsilon}_i(t)\mathrm{d}t = \frac{\varepsilon_{\mathrm{a}}\varsigma_i}{1+\varsigma_i^2}\exp(\lambda_i t)(\cos\omega t + \varsigma_i \sin\omega t), \quad \varsigma_i = \frac{\omega}{\lambda_i}$$

and the general solution (1.3.16) takes the following form

$$\sigma_i(t) = \frac{E\varepsilon_{\mathrm{a}}\varsigma_i}{1+\varsigma_i^2}(\cos\omega t + \varsigma_i \sin\omega t) + C_i \exp(-\lambda_i t) \tag{1.3.22}$$

Assuming the initial conditions $\sigma_i(0) = 0$ the solution is specified as follows

$$\sigma_i(t) = \frac{E\varepsilon_{\mathrm{a}}\varsigma_i}{1+\varsigma_i^2}[\cos\omega t + \varsigma_i \sin\omega t - \exp(-\lambda_i t)] \tag{1.3.23}$$

1.3.2 Force-controlled Loading Paths

The force F is assumed to be a function of time. To compute the stresses σ_i in the bars and the strain ε let us introduce the following new variables

$$\sigma_\Delta = \frac{\sigma_{\mathrm{A}} - \sigma_{\mathrm{B}}}{2}, \quad \sigma = \frac{\sigma_{\mathrm{A}} + \sigma_{\mathrm{B}}}{2} = \frac{F}{2A} \tag{1.3.24}$$

Subtracting Eq. $(1.3.15)_2$ from $(1.3.15)_1$ provides the following differential equation for the stress difference

$$\dot{\sigma}_\Delta + \lambda\sigma_\Delta = \lambda_\Delta \sigma, \tag{1.3.25}$$

where

$$\lambda = \frac{\lambda_{\mathrm{A}} + \lambda_{\mathrm{B}}}{2}, \quad \lambda_\Delta = \frac{\lambda_{\mathrm{B}} - \lambda_{\mathrm{A}}}{2}$$

are effective material properties of the two-bar system. By adding Eq. $(1.3.15)_1$ and $(1.3.15)_2$ the following equation for the strain rate can be obtained

$$\dot{\varepsilon} = \frac{\dot{\sigma}}{E} + \frac{1}{E}(\lambda\sigma - \lambda_\Delta\sigma_\Delta) \tag{1.3.26}$$

The general solution to Eq. (1.3.25) is

$$\sigma_\Delta(t) = \sigma_\Delta(0)\exp(-\lambda t) + \lambda_\Delta \int_0^t \exp[-\lambda(t-\tau)]\sigma(\tau)d\tau \tag{1.3.27}$$

As an example we consider a creep test of the two-bar system. In this case the structure is rapidly loaded within the elastic range such that for $t = 0$

$$\sigma_A(0) = \sigma_B(0) = \sigma = E\varepsilon(0) \quad \Rightarrow \quad \sigma_\Delta(0) = 0$$

Then the applied force F and consequently stress σ are kept constant. In this case the solution (1.3.27) reads

$$\sigma_\Delta(t) = \frac{\lambda_\Delta}{\lambda}\sigma[1 - \exp(-\lambda t)] \tag{1.3.28}$$

Since $T_B > T_A$ is assumed, it follows from Eqs (1.3.13) and (1.3.15) that $\lambda_B > \lambda_A$ and $\lambda_\Delta > 0$. In this case σ_Δ increases from zero towards the steady state value

$$\sigma_{\Delta_{ss}} = \frac{\lambda_\Delta}{\lambda}\sigma$$

As a result the stress in the bar A increases while the stress in the bar B decreases

$$\sigma_A = \sigma + \sigma_\Delta, \quad \sigma_B = \sigma - \sigma_\Delta$$

Equation (1.3.28) reveals that the rate of stress redistribution is determined by λ while its magnitude is related to λ_Δ. For the bars with the same creep properties $\lambda_\Delta = 0$ and the stresses remain constant.

The creep rate of the bar system can be computed from Eqs (1.3.26) and (1.3.28) as follows

$$\dot{\varepsilon}(t) = \frac{\sigma}{\lambda E}\left\{\lambda^2 - \lambda_\Delta^2[1 - \exp(-\lambda t)]\right\} \tag{1.3.29}$$

The initial (maximum) creep rate after the loading of the bar system is determined by

$$\dot{\varepsilon}_{\mathrm{in}} = \dot{\varepsilon}(0) = \frac{\lambda}{E}\sigma,$$

while the minimum creep rate in the steady state is

$$\dot{\varepsilon}_{\mathrm{ss}} = \frac{4\lambda_A\lambda_B}{\lambda E}\sigma$$

It is evident, that the primary creep of the two bar system is determined by the stress redistribution in the bars. If creep properties of the bars are the same, i.e. for the case $\lambda_\Delta = 0$, the creep rate does not change in time.

1.3.3 Time-step Methods

Systems of ordinary differential equations like (1.3.15) can be solved by numerical methods. Various approximate solution procedures to initial-value problems are classified and analyzed in textbooks of numerical mathematics, e.g. Butcher (2016); Hairer and Wanner (1996). Based on the two linear differential equations (1.3.15), let us illustrate basic ideas of several one-step time integration methods.

Within any numerical time step procedure the considered time interval $t_0 - t_n$ is broken into time steps $\Delta t_k = t_{k+1} - t_k, k = 0, 1, \ldots, n$ with discrete time points t_k. Approximate solutions to ordinary differential equations are evaluated in discrete time points. With regard to Eqs (1.3.15) the discrete solutions are specified as follows

$$\sigma_{\mathrm{A}_k} = \sigma_{\mathrm{A}}(t_k), \quad \sigma_{\mathrm{B}_k} = \sigma_{\mathrm{B}}(t_k), \quad k = 0, 1, \ldots, n$$

Assume that the strains $\varepsilon_{\mathrm{A}} = \varepsilon_{\mathrm{B}} = \varepsilon$ are given as functions of time. To compute the stresses let us integrate Eqs (1.3.15) within the time step Δt_k as follows

$$\begin{aligned}
\int\limits_{t_k}^{t_{k+1}} \dot{\sigma}_{\mathrm{A}} \mathrm{d}t &= -\lambda_{\mathrm{A}} \int\limits_{t_k}^{t_{k+1}} \sigma_{\mathrm{A}} \mathrm{d}t + E \int\limits_{t_k}^{t_{k+1}} \dot{\varepsilon} \mathrm{d}t, \\
\int\limits_{t_k}^{t_{k+1}} \dot{\sigma}_{\mathrm{B}} \mathrm{d}t &= -\lambda_{\mathrm{B}} \int\limits_{t_k}^{t_{k+1}} \sigma_{\mathrm{B}} \mathrm{d}t + E \int\limits_{t_k}^{t_{k+1}} \dot{\varepsilon} \mathrm{d}t
\end{aligned} \qquad (1.3.30)$$

After evaluating the integrals we obtain

$$\begin{aligned}
\Delta\sigma_{\mathrm{A}_k} &= -\lambda_{\mathrm{A}} \int\limits_{t_k}^{t_{k+1}} \sigma_{\mathrm{A}} \mathrm{d}t + E\Delta\varepsilon_k, \quad \Delta\sigma_{\mathrm{A}_k} = \sigma_{\mathrm{A}_{k+1}} - \sigma_{\mathrm{A}_k}, \\
\Delta\sigma_{\mathrm{B}_k} &= -\lambda_{\mathrm{B}} \int\limits_{t_k}^{t_{k+1}} \sigma_{\mathrm{B}} \mathrm{d}t + E\Delta\varepsilon_k, \quad \Delta\sigma_{\mathrm{B}_k} = \sigma_{\mathrm{B}_{k+1}} - \sigma_{\mathrm{B}_k}
\end{aligned} \qquad (1.3.31)$$

for the given $\Delta\varepsilon_k = \varepsilon_{k+1} - \varepsilon_k$. The integrals on the right-hand side of Eqs (1.3.31) are evaluated approximately.

1.3.3.1 Explicit Euler Method

One possibility to approximate the integrals is

$$\int\limits_{t_k}^{t_{k+1}} \sigma_{\mathrm{A}} \mathrm{d}t \approx \sigma_{\mathrm{A}_k} \Delta t_k, \qquad \int\limits_{t_k}^{t_{k+1}} \sigma_{\mathrm{B}} \mathrm{d}t \approx \sigma_{\mathrm{B}_k} \Delta t_k \qquad (1.3.32)$$

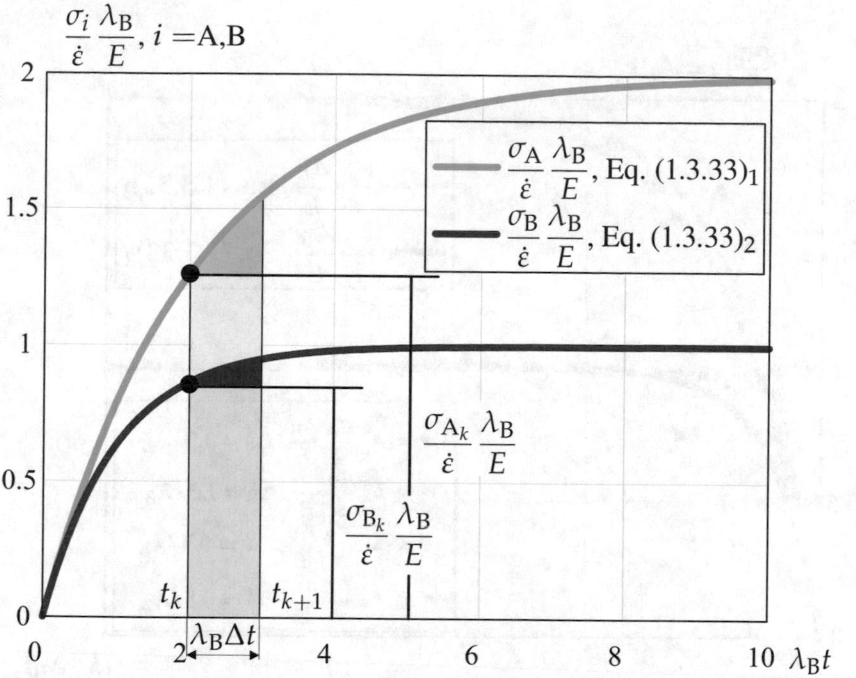

Fig. 1.4 Approximate evaluation of integrals based on solutions at time point t_k for $\lambda_B = 2\lambda_A$

Let us consider the displacement-controlled loading with a constant rate. The closed form solutions (1.3.19) can be given in a normalized form as follows

$$\frac{\lambda_B}{E\dot{\varepsilon}}\sigma_A(t) = \zeta[1 - \exp(-\lambda_B t/\zeta)], \quad \frac{\lambda_B}{E\dot{\varepsilon}}\sigma_B(t) = 1 - \exp(-\lambda_B t), \tag{1.3.33}$$

where

$$\zeta = \frac{\lambda_B}{\lambda_A}$$

Figure 1.4 illustrates the solutions (1.3.33) for the case $\zeta = 2$. The approximations of integrals (1.3.32) are illustrated by rectangular areas below the stress vs time curves based on solutions at time point t_k. The accuracy of these approximations depends on the time step size Δt_k. With (1.3.31) and (1.3.32) we obtain the following one-step time integration formulas for the evaluation of solutions

$$\begin{aligned} \sigma_{A_{k+1}} &= \sigma_{A_k}(1 - \lambda_A \Delta t_k) + E\Delta\varepsilon_k, \\ \sigma_{B_{k+1}} &= \sigma_{B_k}(1 - \lambda_B \Delta t_k) + E\Delta\varepsilon_k, \end{aligned} \tag{1.3.34}$$

where $\Delta\varepsilon_k = \dot{\varepsilon}\Delta t_k$. Starting with initial data, i.e. σ_{A_0} and σ_{B_0}, Eqs (1.3.34) allow the step-by-step evaluation of stress values at time step $k+1$ from known solutions at time step k. This approach is referred as explicit or forward Euler method. Let

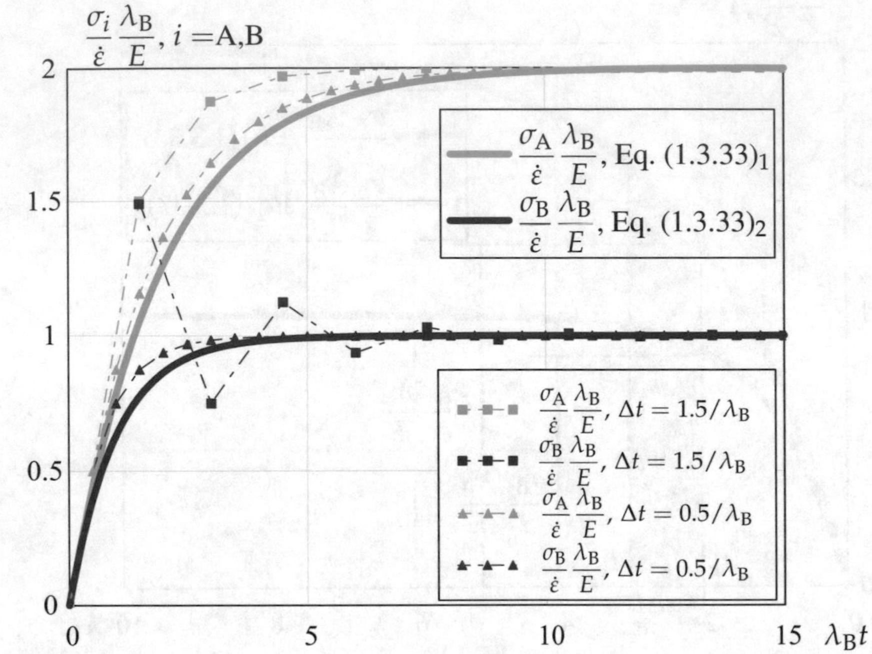

Fig. 1.5 Numerical solutions by explicit Euler method with different time steps for $\lambda_B = 2\lambda_A$

us specify the constant time step size $\Delta t_k = \Delta t$ such that the strain increments are constant, i.e. $\Delta\varepsilon_k = \dot{\varepsilon}\Delta t$. Figure 1.5 illustrates the approximate solutions by the explicit Euler method for the case $\lambda_B = 2\lambda_A$, $\sigma_{A_0} = \sigma_{B_0} = 0$ and different time steps. It can be observed that with a decrease of the time step size, the numerical solutions approach the solutions (1.3.33). For the time step size $\Delta t = 1.5/\lambda_B$ the solution for σ_A is stable while for σ_B the solution oscillates within the transition stage and converges towards the steady state stress value.

Let us analyze the stability of the explicit Euler method applied to the considered example. By induction one may derive the following series solutions from Eqs (1.3.34)

$$\sigma_{A_k} = E\dot{\varepsilon}\Delta t \sum_{j=0}^{k-1}(1 - \lambda_A\Delta t)^j, \quad \sigma_{B_k} = E\dot{\varepsilon}\Delta t \sum_{j=0}^{k-1}(1 - \lambda_B\Delta t)^j \tag{1.3.35}$$

Equations (1.3.35) are approximate solutions to the differential Eqs (1.3.15) for the stress values at the time step k. As the number k may be arbitrarily large, it is important to examine the convergence of the series solutions. This is provided if the following inequalities are satisfied

$$|1 - \lambda_A\Delta t| < 1, \quad |1 - \lambda_B\Delta t| < 1 \tag{1.3.36}$$

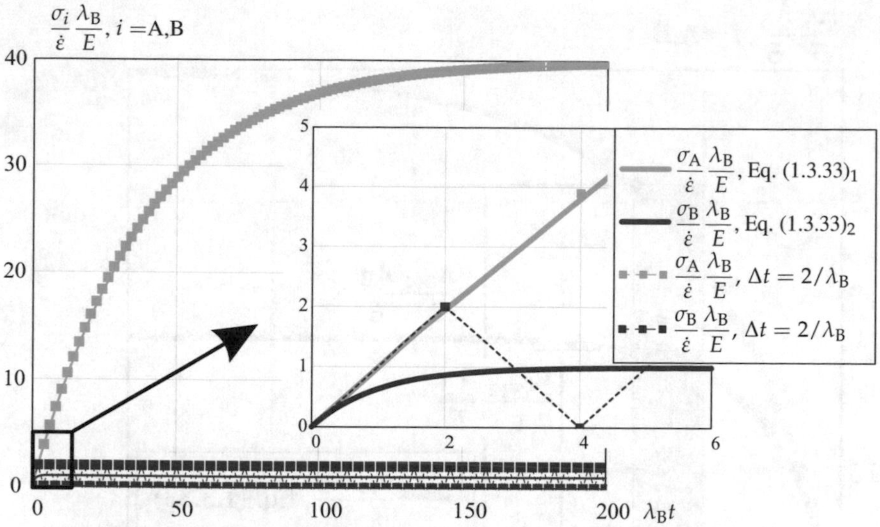

Fig. 1.6 Numerical solution by explicit Euler method for $\lambda_B = 40\lambda_A$

Otherwise the stress values would increase infinitely. Therefore a critical value of the time step size Δt_c exists such that the solutions are stable for $\Delta t < \Delta t_c$. From the inequalities (1.3.36) it follows

$$\Delta t_c = \min\left(\frac{2}{\lambda_A}, \frac{2}{\lambda_B}\right) \tag{1.3.37}$$

This is provided for all solutions presented in Fig. 1.5 with $\lambda_B = 2\lambda_A$ and the critical time step size $\Delta t_c = 2/\lambda_B$.

Although for the time step size $\Delta t = 1.5/\lambda_B < t_c$ the solutions for stresses in bars are stable, for σ_B the oscillatory solution is obtained, Fig. 1.5. From the series solutions (1.3.35) it is evident that in order to prevent oscillatory solutions, the following inequalities must be satisfied

$$1 - \lambda_A \Delta t > 0, \quad 1 - \lambda_B \Delta t > 0$$

From these inequalities we can compute the critical time step size. For the considered case $\lambda_B > \lambda_A$ and the critical time step size is $\Delta t_c = 1/\lambda_B$. In order to prevent non-physical oscillations one should select the time step size such that $\Delta t < \Delta t_{cO}$. An example for $\Delta t = 0.5/\lambda_B$ is shown in Fig. 1.5.

The analysis shows that the explicit Euler method is conditionally stable, i.e. the time step size has to be chosen sufficiently small to assure the solution stability. For many problems the stability would require extremely small time step sizes and rapid increase of computational time. As an example, let us assume that the bars have extremely different viscous properties. Figure 1.6 illustrates the numerical solution for $\lambda_B = 40\lambda_A$ with the time step size $\Delta t = 2/\lambda_B$. According to Eq. (1.3.37) this time

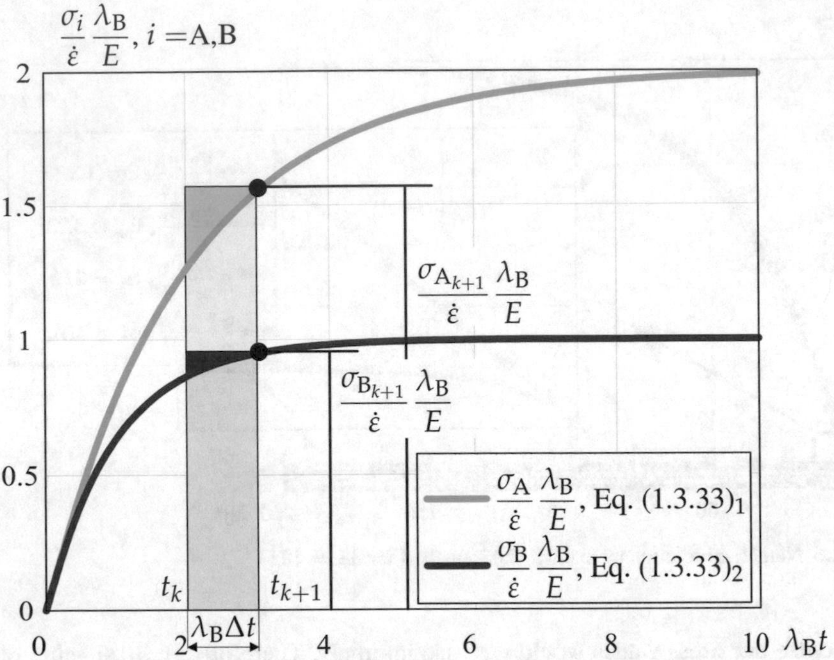

Fig. 1.7 Approximate evaluation of integrals based on solutions at time point t_{k+1} for $\lambda_B = 2\lambda_A$

step is critical for the bar B and leads to the oscillatory solution. However, for the bar A this time step size is too small. Indeed in order to compute the stress in the bar A within the whole transition range too many time steps are required. In this example the nearly steady stress state in the bar is only reached after 100 time steps. To avoid oscillations for the stress in the bar A the time step size should be decreased. This would reduce the efficiency of the explicit Euler method significantly. For mechanical systems with many degrees of freedom and extremely different inelastic properties the use of the explicit Euler method would lead to large number of time steps and high computational effort.

There are many problems in mechanics where explicit methods are not efficient or even not applicable. Examples include structural analysis problems of plasticity and visco-plasticity (Belytschko et al, 2014; Simo and Hughes, 2000), damage mechanics (Altenbach and Naumenko, 1997), rigid body dynamics (Altenbach et al, 2007) and many others. The corresponding differential equations are called stiff (Hairer and Wanner, 1996) and implicit methods are required to obtain stable numerical solutions. Below we discuss two examples of one-step implicit methods.

1.3.3.2 Implicit Euler Method

Figure 1.7 illustrates the approximations of the integrals in Eqs (1.3.31) by areas of rectangles based on solutions at time point t_{k+1}. The integrals can be expressed as follows

$$\int_{t_k}^{t_{k+1}} \sigma_{\mathrm{A}} \mathrm{d}t \approx \sigma_{\mathrm{A}_{k+1}} \Delta t_k, \qquad \int_{t_k}^{t_{k+1}} \sigma_{\mathrm{B}} \mathrm{d}t \approx \sigma_{\mathrm{B}_{k+1}} \Delta t_k \tag{1.3.38}$$

Now Equations (1.3.31) take the following form

$$\begin{aligned} \sigma_{\mathrm{A}_{k+1}} &= \sigma_{\mathrm{A}_k} - \lambda_{\mathrm{A}} \sigma_{\mathrm{A}_{k+1}} \Delta t_k + E \Delta \varepsilon_k, \\ \sigma_{\mathrm{B}_{k+1}} &= \sigma_{\mathrm{B}_k} - \lambda_{\mathrm{B}} \sigma_{\mathrm{B}_{k+1}} \Delta t_k + E \Delta \varepsilon_k \end{aligned} \tag{1.3.39}$$

In comparison to the explicit Euler method, the stresses at time step $k+1$ appear on the right-hand side of Eqs (1.3.39). Therefore Eqs (1.3.39) must be solved in order to obtain the results at the time step $k+1$. This leads to additional numerical operations. In the case of the considered two bar system and the linear viscous material behavior, the solutions are elementary and can be obtained in a closed analytical form as follows

$$\begin{aligned} \sigma_{\mathrm{A}_{k+1}} &= \frac{1}{1 + \lambda_{\mathrm{A}} \Delta t_k} \left(\sigma_{\mathrm{A}_k} + E \Delta \varepsilon_k \right), \\ \sigma_{\mathrm{B}_{k+1}} &= \frac{1}{1 + \lambda_{\mathrm{B}} \Delta t_k} \left(\sigma_{\mathrm{B}_k} + E \Delta \varepsilon_k \right) \end{aligned} \tag{1.3.40}$$

As in the previous example let us assume that the bars are subjected to a constant strain rate $\dot{\varepsilon}$ and the time step size is constant $\Delta t_k = \Delta t$ such that the strain increments are $\Delta \varepsilon_k = \dot{\varepsilon} \Delta t$. In this case the following series solutions can be derived from Eqs (1.3.40) by induction

$$\begin{aligned} \sigma_{\mathrm{A}_k} &= \frac{E \dot{\varepsilon} \Delta t}{1 + \lambda_{\mathrm{A}} \Delta t} \sum_{j=0}^{k-1} \left(\frac{1}{1 + \lambda_{\mathrm{A}} \Delta t} \right)^j, \\ \sigma_{\mathrm{B}_k} &= \frac{E \dot{\varepsilon} \Delta t}{1 + \lambda_{\mathrm{B}} \Delta t} \sum_{j=0}^{k-1} \left(\frac{1}{1 + \lambda_{\mathrm{B}} \Delta t} \right)^j \end{aligned} \tag{1.3.41}$$

Equations (1.3.41) are approximate solutions to the differential Eqs (1.3.15) for the stress values at the time step k obtained by the implicit Euler method. The convergence of the series (1.3.41) is provided for any value of the time step size. Therefore the implicit Euler method is unconditionally stable. Figure 1.8 illustrates the approximate solutions for the case $\lambda_{\mathrm{B}} = 2\lambda_{\mathrm{A}}$ and two different time steps. For the time increment $\Delta t = 1.5/\lambda_{\mathrm{B}}$ the explicit Euler method yields the oscillating solution for the stress in the bar B, Fig. 1.5, while the corresponding solution by the implicit Euler method is monotonic, Fig. 1.8.

Figure 1.9 illustrates the results for the bars with extremely different viscosities, $\lambda_{\mathrm{B}} = 40\lambda_{\mathrm{A}}$. The critical time step size for the explicit Euler method is $\Delta t = 2/\lambda_{\mathrm{B}}$. With this time step size the oscillating solution for the bar B was obtained, see Fig. 1.6. Applying the implicit Euler method the solutions remain monotonic, Fig. 1.9.

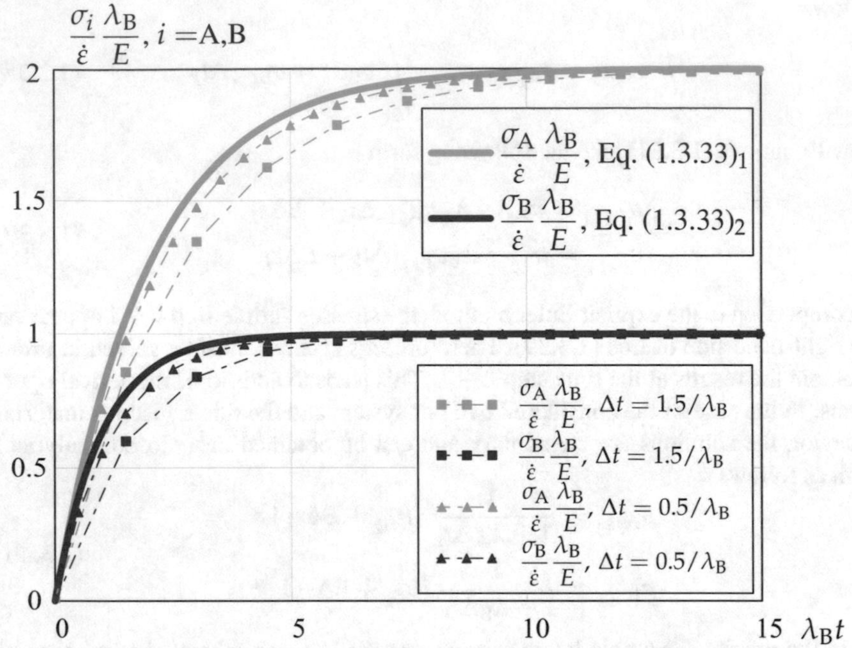

Fig. 1.8 Numerical solutions by implicit Euler method with different time steps for $\lambda_B = 2\lambda_A$

1.3.3.3 Trapezoidal Rule

Let us consider the following approximations to the integrals in Eqs (1.3.31)

$$\int_{t_k}^{t_{k+1}} \sigma_A \mathrm{d}t \approx \frac{1}{2}(\sigma_{A_k} + \sigma_{A_{k+1}})\Delta t_k, \quad \int_{t_k}^{t_{k+1}} \sigma_B \mathrm{d}t \approx \frac{1}{2}(\sigma_{B_k} + \sigma_{B_{k+1}})\Delta t_k \tag{1.3.42}$$

Figure 1.10 illustrates the approximations (1.3.42) of areas under the stress vs time curves by trapezoid areas based on solutions at time points t_k and t_{k+1}. With approximations (1.3.42), Eqs (1.3.31) take the following form

$$\begin{aligned} \sigma_{A_{k+1}} &= \sigma_{A_k} - \frac{\lambda_A}{2}(\sigma_{A_k} + \sigma_{A_{k+1}})\Delta t_k + E\Delta\varepsilon_k, \\ \sigma_{B_{k+1}} &= \sigma_{B_k} - \frac{\lambda_B}{2}(\sigma_{B_k} + \sigma_{B_{k+1}})\Delta t_k + E\Delta\varepsilon_k \end{aligned} \tag{1.3.43}$$

The trapezoidal rule leads to an implicit time integration method since the stress values at time step $k+1$ appear on the right hand side of Eqs (1.3.43). After solving the linear Eqs (1.3.43) we obtain

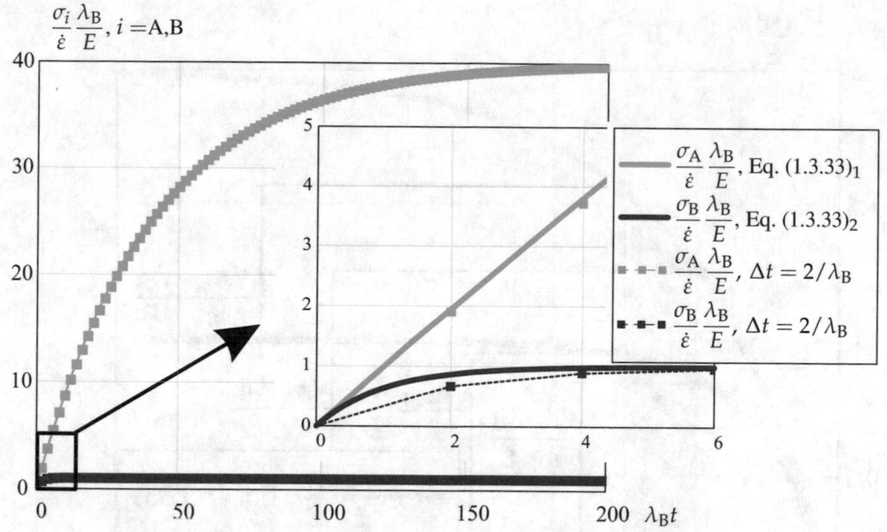

Fig. 1.9 Numerical solution by implicit Euler method for $\lambda_B = 40\lambda_A$

$$\sigma_{A_{k+1}} = \frac{1 - \frac{1}{2}\lambda_A \Delta t_k}{1 + \frac{1}{2}\lambda_A \Delta t_k}\sigma_{A_k} + \frac{1}{1 + \frac{1}{2}\lambda_A \Delta t_k} E\Delta\varepsilon_k,$$
$$\sigma_{B_{k+1}} = \frac{1 - \frac{1}{2}\lambda_B \Delta t_k}{1 + \frac{1}{2}\lambda_B \Delta t_k}\sigma_{B_k} + \frac{1}{1 + \frac{1}{2}\lambda_B \Delta t_k} E\Delta\varepsilon_k \qquad (1.3.44)$$

Assuming that the bars are subjected to a constant strain rate $\dot{\varepsilon}$ and the time step size is constant $\Delta t_k = \Delta t$ such that $\Delta\varepsilon_k = \dot{\varepsilon}\Delta t$, the following series solutions can be derived from Eqs (1.3.44)

$$\sigma_{A_k} = \frac{E\dot{\varepsilon}\Delta t}{1 + \frac{1}{2}\lambda_A \Delta t}\sum_{j=0}^{k-1}\left(\frac{1 - \frac{1}{2}\lambda_A \Delta t}{1 + \frac{1}{2}\lambda_A \Delta t}\right)^j,$$
$$\sigma_{B_k} = \frac{E\dot{\varepsilon}\Delta t}{1 + \frac{1}{2}\lambda_B \Delta t}\sum_{j=0}^{k-1}\left(\frac{1 - \frac{1}{2}\lambda_B \Delta t}{1 + \frac{1}{2}\lambda_B \Delta t}\right)^j \qquad (1.3.45)$$

The series (1.3.45) is convergent for any value of the time step size. Therefore the trapezoidal rule is unconditionally stable. Figure 1.11 illustrates the approximate solutions for the case $\lambda_B = 2\lambda_A$ and two different time steps. The solutions are more accurate than those obtained by the implicit Euler method. For example with

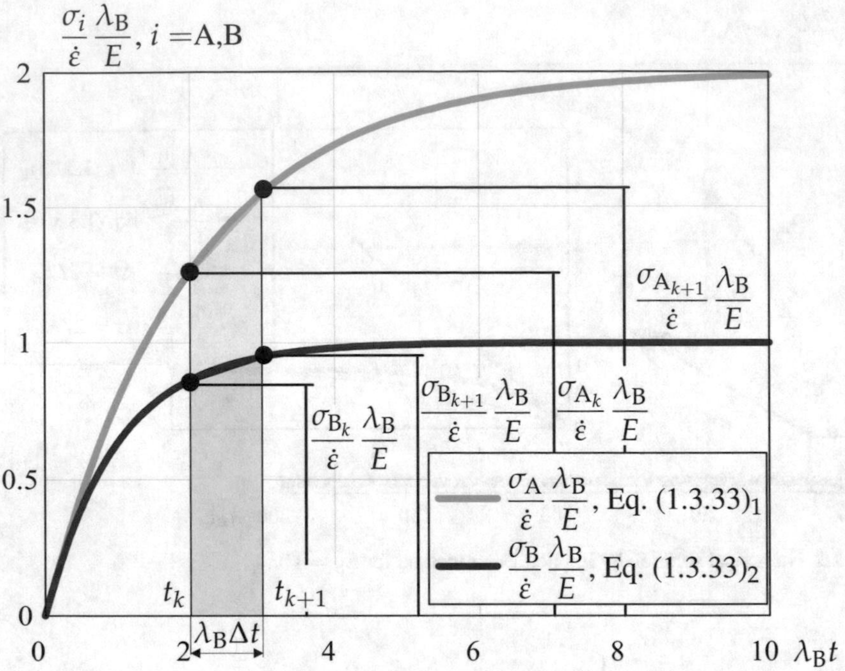

Fig. 1.10 Approximate evaluation of integrals based on solutions at time points t_k and t_{k+1} for $\lambda_B = 2\lambda_A$

the time step size $\Delta t = 1.5/\lambda_B$ the implicit Euler method leads to the error of nearly 20%, Fig. 1.8, while the trapezoidal rule provides the solution which is very close to the analytical one, Fig. 1.11. Below we analyze the accuracy of all considered methods.

1.3.3.4 Reviewing the Solutions

Let us compare the approximate analytical solutions for the two-bar system based on three introduced time-step methods. Assuming that the inequalities (1.3.36) are satisfied, the geometric series (1.3.35) can be evaluated providing the following approximate analytical solutions by the *explicit Euler method*

$$\sigma_{A_k} = \frac{E}{\lambda_A}\dot{\varepsilon}\left[1 - (1 - \lambda_A\Delta t)^k\right], \quad \sigma_{B_k} = \frac{E}{\lambda_B}\dot{\varepsilon}\left[1 - (1 - \lambda_B\Delta t)^k\right] \qquad (1.3.46)$$

The geometric series (1.3.41) leads to the following solutions by the *implicit Euler method*

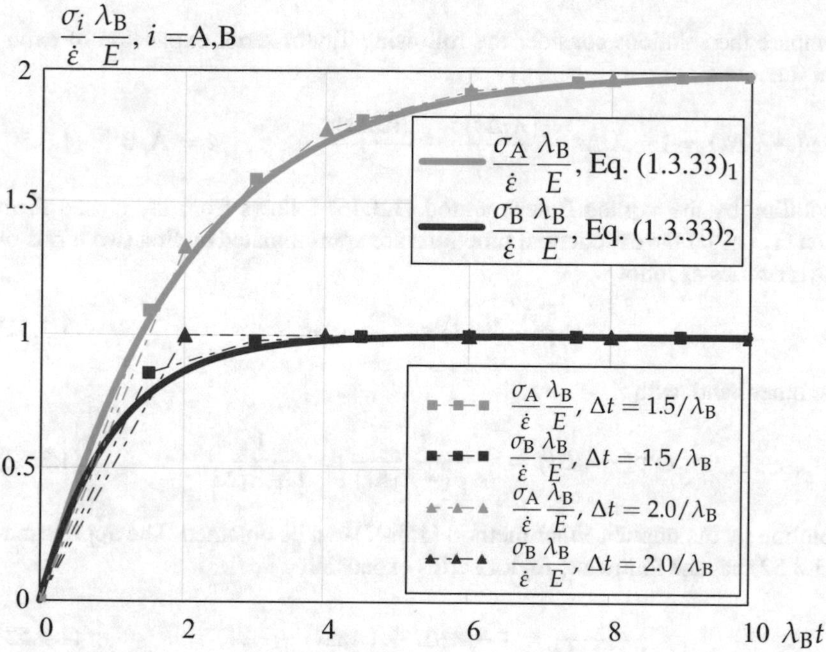

Fig. 1.11 Numerical solutions by trapezoidal rule with different time steps for $\lambda_B = 2\lambda_A$

$$\sigma_{A_k} = \frac{E}{\lambda_A}\dot\varepsilon\left[1-\left(\frac{1}{1+\lambda_A\Delta t}\right)^k\right], \qquad \sigma_{B_k} = \frac{E}{\lambda_B}\dot\varepsilon\left[1-\left(\frac{1}{1+\lambda_B\Delta t}\right)^k\right] \tag{1.3.47}$$

From the geometric series (1.3.45) the following solutions by the *trapezoidal rule* can be obtained

$$\sigma_{A_k} = \frac{E}{\lambda_A}\dot\varepsilon\left[1-\left(\frac{1-\frac{1}{2}\lambda_A\Delta t}{1+\frac{1}{2}\lambda_A\Delta t}\right)^k\right], \qquad \sigma_{B_k} = \frac{E}{\lambda_B}\dot\varepsilon\left[1-\left(\frac{1-\frac{1}{2}\lambda_B\Delta t}{1+\frac{1}{2}\lambda_B\Delta t}\right)^k\right] \tag{1.3.48}$$

On the other hand the *closed form solutions* (1.3.19) evaluated at the time step t_k are

$$\sigma_{A_k} = \sigma_A(t_k) = \frac{E}{\lambda_A}\dot\varepsilon\left[1-\exp(-\lambda_A k\Delta t)\right], \qquad \sigma_{B_k} = \sigma_B(t_k) = \frac{E}{\lambda_B}\dot\varepsilon\left[1-\exp(-\lambda_B k\Delta t)\right] \tag{1.3.49}$$

To compare the solutions consider the following Taylor series expansion of exponential functions

$$\exp(-\lambda_i \Delta t) = 1 - \lambda_i \Delta t + \frac{(\lambda_i \Delta t)^2}{2} - \frac{(\lambda_i \Delta t)^3}{6} + \ldots, \quad i = \mathrm{A}, \mathrm{B} \qquad (1.3.50)$$

The solution by the explicit Euler method (1.3.46) follows from the closed form solution (1.3.49) if the exponential functions are approximated by first two terms of the Taylor series as follows

$$\exp(-\lambda_i \Delta t) \approx 1 - \lambda_i \Delta t \qquad (1.3.51)$$

On the other hand, with

$$\exp(-\lambda_i \Delta t) = \frac{1}{\exp(-\lambda_i \Delta t)} \approx \frac{1}{1 + \lambda_i \Delta t} \qquad (1.3.52)$$

the solution by the implicit Euler method (1.3.47) can be obtained. The approximation (1.3.52) has the following Taylor series expansion

$$\frac{1}{1 + \lambda_i \Delta t} = 1 - \lambda_i \Delta t + (\lambda_i \Delta t)^2 + \ldots \qquad (1.3.53)$$

The exponential function (1.3.50) can only be approximated by (1.3.52) if the series in (1.3.53) are truncated after the second term, which is linear with respect to Δt. This can be accomplished for time step sizes satisfying the inequalities

$$(\lambda_i \Delta t)^2 \ll \lambda_i \Delta t < 1$$

With

$$\exp(-\lambda_i \Delta t) = \frac{\exp(-\lambda_i \Delta t/2)}{\exp(\lambda_i \Delta t/2)} \approx \frac{1 - \frac{1}{2}\lambda_i \Delta t}{1 + \frac{1}{2}\lambda_i \Delta t} \qquad (1.3.54)$$

the solutions by the trapezoidal rule (1.3.48) follow from the closed form solutions (1.3.49). Approximation (1.3.54) can be expanded as follows

$$\frac{1 - \frac{1}{2}\lambda_i \Delta t}{1 + \frac{1}{2}\lambda_i \Delta t} \approx 1 - \lambda_i \Delta t + \frac{(\lambda_i \Delta t)^2}{2} - \frac{(\lambda_i \Delta t)^3}{3} + \ldots \qquad (1.3.55)$$

It coincides with (1.3.50) if the series are truncated after the third term. Thus, the trapezoidal rule features a higher order accuracy if compared to the Euler methods. We refer to Hairer and Wanner (1996); Hairer et al (2008) for general analysis and classification of different time-step methods.

1.4 Non-linear Inelastic Behavior

The majority of engineering materials exhibit non-linear behavior in the inelastic range. Section 1.4 deals with idealized inelasticity by neglecting processes accompanying deformation such as hardening, softening and damage. Constitutive equations for inelastic response are introduced to capture stress range dependent creep, relaxation and tensile behavior. Examples for two bar system are introduced to illustrate stress redistributions and inelastic strain evolution for different force and displacement controlled loading paths. Finally basic features of numerical time step procedures for problems of non-linear inelasticity are discussed. Two examples are introduced including the explicit and implicit Euler methods. In addition to the numerical approaches presented in Sect. 1.3 local error estimations as well as stability analysis are presented. They are essential for time step controls and efficient numerical solutions of non-linear problems.

1.4.1 Constitutive Equations

First let us consider a single bar under different load and displacement controls. The constitutive Eqs (1.1.7) can be specified as follows

$$\dot{\varepsilon}^{\mathrm{pl}} = \dot{\varepsilon}_0 f\left(\frac{|\sigma|}{\sigma_0}\right) \operatorname{sgn}(\sigma) \tag{1.4.56}$$

with

$$\dot{\varepsilon}_0 = a \exp\left(-\frac{Q}{RT}\right) \tag{1.4.57}$$

In Eqs (1.4.56) and (1.4.57) $\dot{\varepsilon}_0$ and σ_0 are, in general functions of temperature. In Sect. 1.4.1 we assume isothermal conditions such that $\dot{\varepsilon}_0$ and σ_0 are constants. In Naumenko and Altenbach (2016, Sect. 5.4.4) various response functions of stress f were introduced to capture inelastic behavior in a wide stress range. For the analysis of bars let us consider the following examples

$$\begin{array}{ll} f(x) = x^n & \text{power law,} \\ f(x) = x^{n_1} + x^{n_2} & \text{double power law,} \\ f(x) = \sinh x & \text{Prandtl-Eyring law,} \\ f(x) = (\sinh x)^n & \text{Garofalo law,} \end{array} \tag{1.4.58}$$

where n, n_1 and n_2 are constants. Neglecting the thermal strain the total strain in the bar is

$$\varepsilon = \frac{\sigma}{E} + \varepsilon^{\mathrm{pl}} \tag{1.4.59}$$

Taking the time derivative of (1.4.59) and with (1.4.56) the following differential equation for the stress in the bar can be obtained

$$\dot{\sigma} + E\dot{\varepsilon}_0 f\left(\frac{|\sigma|}{\sigma_0}\right)\operatorname{sgn}(\sigma) = E\dot{\varepsilon} \tag{1.4.60}$$

For the analysis it is convenient to introduce the normalized stress Σ, the normalized strain rate $\dot{\epsilon}$ and the normalized time τ as follows

$$\Sigma = \frac{\sigma}{\sigma_0}, \quad \dot{\epsilon} = \frac{\dot{\varepsilon}}{\dot{\varepsilon}_0}, \quad \tau = t\frac{E\dot{\varepsilon}_0}{\sigma_0} \tag{1.4.61}$$

Then Eq. (1.4.60) takes the following form

$$\frac{\mathrm{d}\Sigma}{\mathrm{d}\tau} + f(|\Sigma|)\operatorname{sgn}(\Sigma) = \dot{\epsilon} \tag{1.4.62}$$

1.4.1.1 Stress Functions for Creep and Relaxation

Long term creep and relaxation processes are observed for loadings with moderate stress levels. To capture the dependence of creep rate on the stress level several response functions are proposed. Here we consider and compare the power law $(1.4.58)_1$, the double power law $(1.4.58)_2$ with $n_1 = 1$ and $n_2 = n$ and the hyperbolic sine law $(1.4.58)_3$ stress functions as they describe creep and relaxation processes. For $\Sigma = \text{const}$ Eq. (1.4.62) describes steady-state creep of the bar. Figure 1.12 shows the normalized strain rate as a function of the normalized stress. For moderate stress range, say $1 < \sigma/\sigma_0 < 2$, a typical power law creep regime is observed. For low stress levels $0 < \sigma/\sigma_0 < 1$ the stress exponent decreases towards the value of unity. This behavior is called stress range or stress regime dependent creep (Boyle, 2012; Chowdhury et al, 2018; Hosseini et al, 2013; Naumenko et al, 2009; Naumenko and Kostenko, 2009). With the double power law two stress ranges or stress regimes can be captured. For $\sigma/\sigma_0 \ll 1$ the double power law stress function provides the linear creep behavior while for $\sigma/\sigma_0 \gg 1$ the power law creep is described, Fig. 1.12. Examples of material parameters in the double power law are presented in Kloc and Sklenička (1997); Naumenko et al (2009, 2011a); Rieth (2007).

With the hyperbolic sine law the linear creep regime follows for $\sigma/\sigma_0 \ll 1$, while for $\sigma/\sigma_0 \gg 1$ the exponential creep with the strain rate

$$\dot{\varepsilon}^{\mathrm{pl}} = \frac{\dot{\varepsilon}_0}{2}\exp\left(\frac{|\sigma|}{\sigma_0}\right)$$

is described. Examples for material parameters in the hyperbolic sine law are presented in Dyson and McLean (2001); Kowalewski et al (1994); Perrin and Hayhurst (1994); Naumenko et al (2011b), among others. In order to compare different stress functions let us fit the hyperbolic sine law to the stress dependence as described by the double power law with $n = 5$ within the range $0 \leq \Sigma \leq 2$. This is accomplished by minimizing the following functional

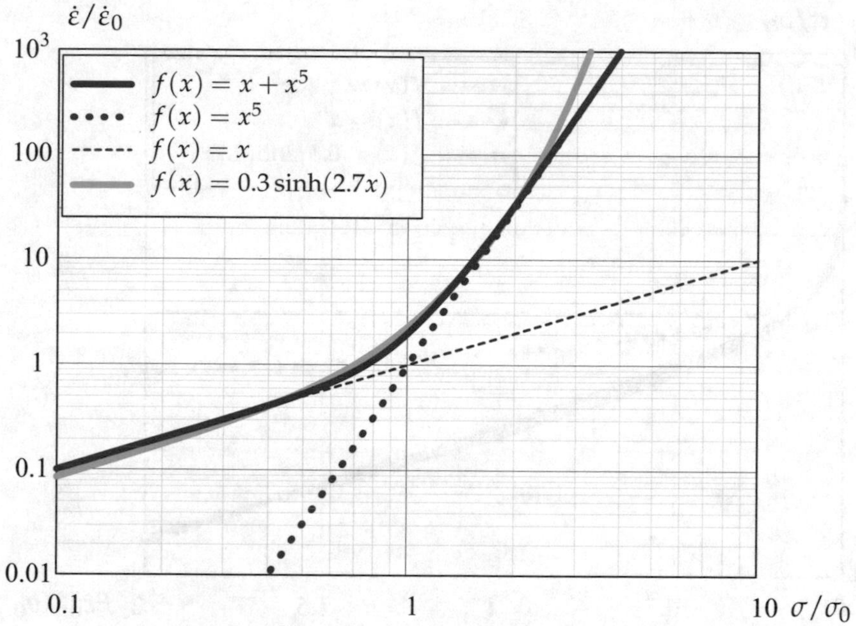

Fig. 1.12 Normalized steady-state creep rate vs normalized stress described with different stress functions

$$\Phi(a,b) = \int_{0.2}^{2} (a\sinh(bx) - x - x^5)^2 \mathrm{d}x$$

The result with $a = 0.3$ and $b = 2.7$ is shown in Fig. 1.12. It is obvious that both the double power law with $n = 5$ and the hyperbolic sine law lead to similar predictions. Many experimental data show that for higher stress levels both laws underestimate the inelastic strain rate (Frost and Ashby, 1982; Ilschner, 1973). Power law breakdown regime will be discussed in Subsubsect. 1.4.1.2.

For $\dot{\varepsilon} = 0$ Eq. (1.4.62) describes the stress relaxation. With the initial condition $\Sigma(0) = \Sigma_{\text{ref}} > 0$ the solution to Eq. (1.4.62) can be given as follows

$$g(\Sigma) = g(\Sigma_{\text{ref}}) - \tau, \quad g(x) = \int \frac{\mathrm{d}x}{f(x)} \tag{1.4.63}$$

For the *power law* stress function $f(x) = x^n$ Eq. (1.4.63) takes the form

$$\Sigma(\tau) = \begin{cases} [\Sigma_{\text{ref}}^{n-1} + (n-1)\tau]^{\frac{1}{1-n}} & \text{for } n > 1 \\ \Sigma_{\text{ref}} \exp(-\tau) & \text{for } n = 1 \end{cases} \tag{1.4.64}$$

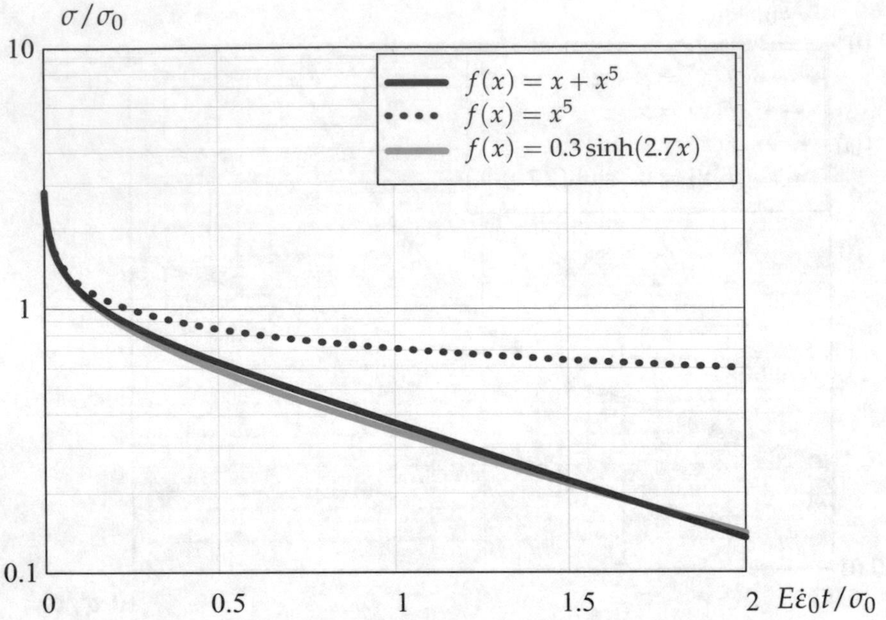

Fig. 1.13 Normalized stress vs normalized time. Stress relaxation with reference stress $\sigma/\sigma_0 = 3$ and different stress functions

For the *double power law* stress function $f(x) = x + x^n$, $n > 1$ Eq. (1.4.63) is specified as follows

$$\Sigma(\tau) = [(1 + \Sigma_{\text{ref}}^{n-1}) \exp((n-1)\tau) - 1]^{\frac{1}{1-n}} \tag{1.4.65}$$

For the *hyperbolic sine* stress function $f(x) = a\sinh(bx)$ the solution (1.4.63) reads

$$\Sigma(\tau) = \frac{2}{b}\operatorname{arctanh}\left(\tanh\left(\frac{b\Sigma_{\text{ref}}}{2}\right)\exp(-ab\tau)\right) \tag{1.4.66}$$

Figure 1.13 illustrates stress relaxation curves computed with Eqs (1.4.64) – (1.4.66) and the reference stress $\Sigma_{\text{ref}} = 3$. At the beginning of the relaxation process the stress decreases rapidly reaching the value of σ_0. Within this relaxation regime the stress relaxation curves obtained by the power law with $n = 5$, the double power law and the hyperbolic sine law are nearly the same. However, for stress levels below σ_0 different stress functions lead to different predictions. For this low stress regime the power law overestimates significantly the stress level. Figure 1.13 shows the stress relaxation curves for the reference stress $\Sigma_{\text{ref}} = 1$. The solutions based on the double power law (1.4.65) and the hyperbolic sine law (1.4.66) are nearly the same. They are close to the exponential stress vs time curve as described by the linear creep law. Examples illustrating the use of different stress functions in the analysis of stress relaxation are presented in Altenbach et al (2008); Kalyanasun-

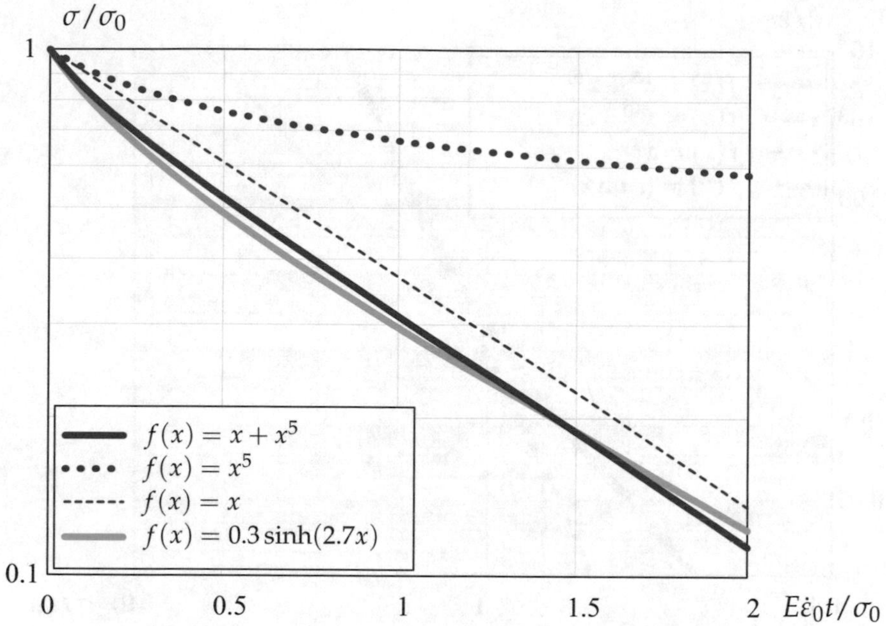

Fig. 1.14 Normalized stress vs normalized time. Stress relaxation with reference stress $\sigma/\sigma_0 = 1$ and different stress functions

daram and Holdsworth (2016); Kostenko and Naumenko (2017); Wang et al (2016) for different materials.

1.4.1.2 Power Law Breakdown and Monotonic Loading

For higher stress levels the power law breakdown regime is usually observed: the stress exponent increases as the stress increases. The power law underestimates the inelastic strain rate. Let us consider the power law $(1.4.58)_1$, the double power law $(1.4.58)_2$ and the Garofalo law $(1.4.58)_3$ as they describe the inelastic strain rate at constant stress and the stress response under monotonic loading with constant strain rate. Figure 1.15 shows plots of the normalized strain rate vs normalized stress curves with different stress functions. Note that the values of σ_0 and $\dot{\varepsilon}_0$ differ from those discussed in Subsubsect. 1.4.1.1. Here σ_0 designates the transition stress from power law to power law breakdown regime and $\dot{\varepsilon}_0$ is the corresponding strain rate.

For $\dot{\epsilon} = \text{const} > 0$ Eq. (1.4.62) describes the stress response under the monotonic tensile loading with constant strain rate. With the initial condition $\Sigma(0) = 0$ Eq. (1.4.62) can be solved numerically providing the stress-time or stress-strain curves for different given strain rates. Figure 1.16 illustrates an example for the normalized stress vs normalized time responses with $\dot{\epsilon} = 100$ and different stress functions. We observe that the double power law $(1.4.58)_2$ with $n_1 = 5$, $n_2 = 10$

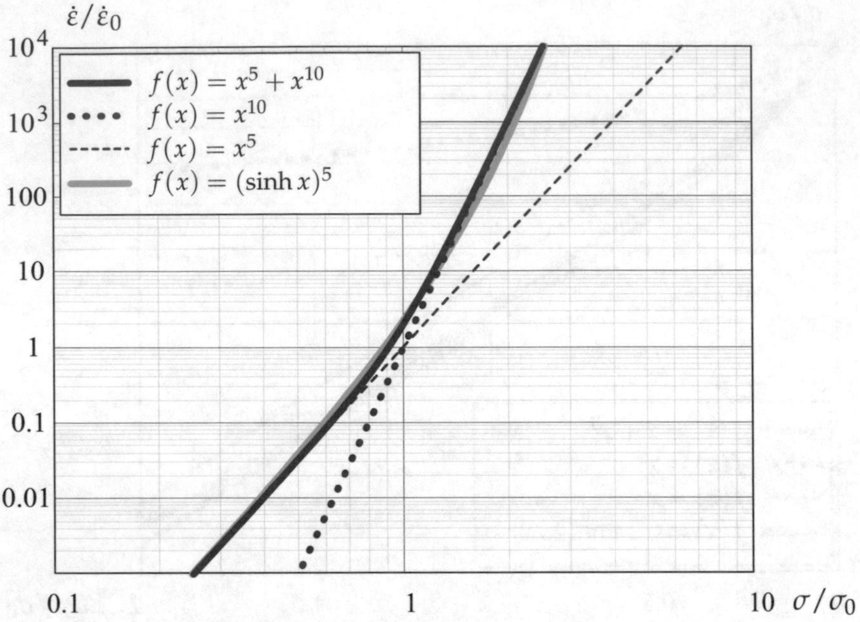

Fig. 1.15 Normalized steady-state inelastic strain rate vs normalized stress described with different stress functions

and the Garofalo law $(1.4.58)_3$ provide nearly the same stress responses for the same constant strain rate. The power law overestimates the steady-state stress value.

Setting the stress rate to zero, Eq. (1.4.62) gives the steady-state stress Σ_{ss} (yield stress) value as a function of the given strain rate as follows

$$\Sigma_{ss} = f^{-1}(\dot{\epsilon})$$

In essence, the strain rate dependence of the flow stress is illustrated in Fig. 1.15. It shows that the power law with $n = 5$ overestimates the stress value for $\sigma < \sigma_0$.

Experimental data illustrating the strain rate sensitivity of the flow stress under monotonic loading as well as the use of different stress functions to describe the stress-strain responses are presented in Chowdhury et al (2017); Eisenträger et al (2017, 2018); Kassner and Pérez-Prado (2004); Längler et al (2014); Miller (1987); Naumenko and Gariboldi (2014); Schmicker et al (2013) for various materials.

1.4.2 Governing Equations for Two-bar System

Let us apply the constitutive Eqs (1.4.56) and the response functions (1.4.58) for the analysis of the two-bar system. For bars A and B we have

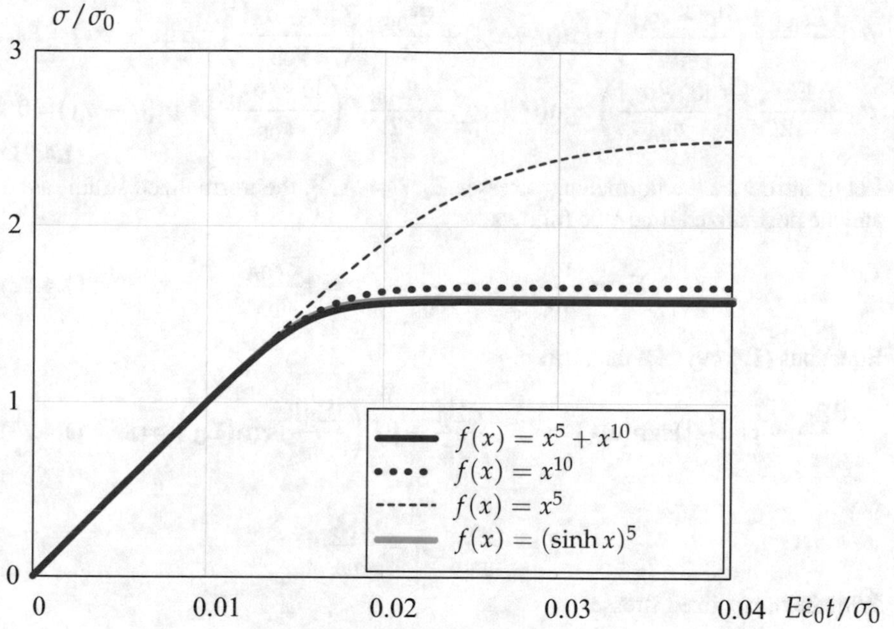

Fig. 1.16 Normalized stress vs normalized time. Stress response for monotonic loading with normalized strain rate $\dot{\varepsilon}/\dot{\varepsilon}_0 = 100$ and different stress functions

$$\dot{\varepsilon}_{\mathrm{A}}^{\mathrm{pl}} = \dot{\varepsilon}_{0\mathrm{A}}\, f\left(\frac{|\sigma_{\mathrm{A}}|}{\sigma_{0\mathrm{A}}}\right) \operatorname{sgn}(\sigma_{\mathrm{A}}), \quad \dot{\varepsilon}_{\mathrm{B}}^{\mathrm{pl}} = \dot{\varepsilon}_{0\mathrm{B}}\, f\left(\frac{|\sigma_{\mathrm{B}}|}{\sigma_{0\mathrm{B}}}\right) \operatorname{sgn}(\sigma_{\mathrm{B}}) \qquad (1.4.67)$$

with

$$\dot{\varepsilon}_{0\mathrm{A}} = a_{\mathrm{A}} \exp\left(-\frac{Q}{RT_{\mathrm{A}}}\right), \quad \dot{\varepsilon}_{0\mathrm{B}} = a_{\mathrm{B}} \exp\left(-\frac{Q}{RT_{\mathrm{B}}}\right) \qquad (1.4.68)$$

With Eqs (1.1.4), (1.1.5) and (1.4.67) the following differential equations for the stresses in the bars can be derived

$$\dot{\sigma}_{\mathrm{A}} + E\dot{\varepsilon}_{0\mathrm{A}}\, f\left(\frac{|\sigma_{\mathrm{A}}|}{\sigma_{0\mathrm{A}}}\right) \operatorname{sgn}(\sigma_{\mathrm{A}}) = E\dot{\varepsilon}, \quad \dot{\sigma}_{\mathrm{B}} + E\dot{\varepsilon}_{0\mathrm{B}}\, f\left(\frac{|\sigma_{\mathrm{B}}|}{\sigma_{0\mathrm{B}}}\right) \operatorname{sgn}(\sigma_{\mathrm{B}}) = E\dot{\varepsilon} \qquad (1.4.69)$$

With the new variables

$$\sigma_{\Delta} = \frac{\sigma_{\mathrm{A}} - \sigma_{\mathrm{B}}}{2}, \quad \sigma = \frac{\sigma_{\mathrm{A}} + \sigma_{\mathrm{B}}}{2} = \frac{F}{2A} \qquad (1.4.70)$$

Equations (1.4.69) can be rewritten as follows

$$
\begin{aligned}
&\dot{\sigma} + \frac{E\dot{\varepsilon}_{0A}}{2} f\left(\frac{|\sigma + \sigma_\Delta|}{\sigma_{0A}}\right) \operatorname{sgn}(\sigma + \sigma_\Delta) + \frac{E\dot{\varepsilon}_{0B}}{2} f\left(\frac{|\sigma - \sigma_\Delta|}{\sigma_{0B}}\right) \operatorname{sgn}(\sigma - \sigma_\Delta) = E\dot{\varepsilon}, \\
&\dot{\sigma}_\Delta + \frac{E\dot{\varepsilon}_{0A}}{2} f\left(\frac{|\sigma + \sigma_\Delta|}{\sigma_{0A}}\right) \operatorname{sgn}(\sigma + \sigma_\Delta) - \frac{E\dot{\varepsilon}_{0B}}{2} f\left(\frac{|\sigma - \sigma_\Delta|}{\sigma_{0B}}\right) \operatorname{sgn}(\sigma - \sigma_\Delta) = 0
\end{aligned}
\tag{1.4.71}
$$

Let us introduce the normalized stresses Σ_i, $i = \mathrm{A, B}$, the normalized strain rate $\dot{\epsilon}$ and the normalized time τ as follows

$$
\Sigma_i = \frac{\sigma_i}{\sigma_{0A}}, \quad \dot{\epsilon} = \frac{\dot{\varepsilon}}{\dot{\varepsilon}_{0A}}, \quad \tau = t\frac{E\dot{\varepsilon}_{0A}}{\sigma_{0A}} \tag{1.4.72}
$$

Equations (1.4.69) take the form

$$
\frac{\mathrm{d}\Sigma_A}{\mathrm{d}\tau} + f(|\Sigma_A|)\operatorname{sgn}(\Sigma_A) = \dot{\epsilon}, \quad \frac{\mathrm{d}\Sigma_B}{\mathrm{d}\tau} + \xi f\left(\frac{|\Sigma_B|}{\eta}\right)\operatorname{sgn}(\Sigma_B) = \dot{\epsilon}, \tag{1.4.73}
$$

where

$$
\xi = \frac{\dot{\varepsilon}_{0B}}{\dot{\varepsilon}_{0A}}, \quad \eta = \frac{\sigma_{0B}}{\sigma_{0A}}
$$

With the normalized stresses

$$
\Sigma = \frac{\sigma}{\sigma_{0A}}, \quad \Sigma_\Delta = \frac{\sigma_\Delta}{\sigma_{0A}} \tag{1.4.74}
$$

Eqs (1.4.71) read

$$
\begin{aligned}
&\frac{\mathrm{d}\Sigma}{\mathrm{d}\tau} + \frac{1}{2} f(|\Sigma + \Sigma_\Delta|)\operatorname{sgn}(\Sigma + \Sigma_\Delta) + \frac{1}{2}\xi f\left(\frac{|\Sigma - \Sigma_\Delta|}{\eta}\right) \operatorname{sgn}(\Sigma - \Sigma_\Delta) = \frac{\dot{\varepsilon}}{\dot{\varepsilon}_{0A}}, \\
&\frac{\mathrm{d}\Sigma_\Delta}{\mathrm{d}\tau} + \frac{1}{2} f(|\Sigma + \Sigma_\Delta|)\operatorname{sgn}(\Sigma + \Sigma_\Delta) - \frac{1}{2}\xi f\left(\frac{|\Sigma - \Sigma_\Delta|}{\eta}\right) \operatorname{sgn}(\Sigma - \Sigma_\Delta) = 0
\end{aligned}
\tag{1.4.75}
$$

1.4.3 Creep and Stress Redistribution

Let us assume that the bar system is uniformly heated and spontaneously loaded by the tensile force F. The force F is kept constant over the period of time. With $\sigma = \text{const}$, Eqs $(1.4.71)_1$ or Eqs $(1.4.75)_1$ describe the creep rate of the bar system while Eqs $(1.4.71)_1$ or Eqs $(1.4.75)_2$ provide the stress redistribution in the bars during creep. At the beginning of the creep process $\Sigma_\Delta = 0$. The initial (maximum) creep rate after the loading follows from Eq. $(1.4.71)_1$

$$
\dot{\varepsilon}_{\text{in}} = \frac{\dot{\varepsilon}_{0A}}{2} f\left(\frac{\sigma}{\sigma_{0A}}\right) + \frac{\dot{\varepsilon}_{0B}}{2} f\left(\frac{\sigma}{\sigma_{0B}}\right) \tag{1.4.76}
$$

We observe that the initial creep rate is the arithmetic mean of creep rates in the bars under the stress σ. Since the Young's moduli and cross section areas of the bars are assumed to be the same, the stresses have the same values after the loading. Therefore the initial strain rate (1.4.76) can be determined applying the iso-stress concept from the mechanics of composite materials.

To compute the minimum creep rate we have to evaluate steady state stress values in the bars after a certain period of time as a result of stress redistribution. By setting $\dot{\sigma}_\Delta = 0$, Eq. $(1.4.71)_2$ provides the following expression for the stress difference in the steady state

$$\dot{\varepsilon}_{0\text{A}} f\left(\frac{\sigma + \sigma_{\Delta_{\text{ss}}}}{\sigma_{0\text{A}}}\right) = \dot{\varepsilon}_{0\text{B}} f\left(\frac{\sigma - \sigma_{\Delta_{\text{ss}}}}{\sigma_{0\text{B}}}\right) \tag{1.4.77}$$

Equation (1.4.77) indicates that the steady creep state can be analyzed within the iso-strain rate concept from the mechanics of composite materials. With $\sigma_{\Delta_{\text{ss}}}$ evaluated from (1.4.77) the minimum creep rate follows from Eq. $(1.4.71)_1$

$$\dot{\varepsilon}_{\min} = \dot{\varepsilon}_{0\text{A}} f\left(\frac{\sigma + \sigma_{\Delta_{\text{ss}}}}{\sigma_{0\text{A}}}\right) = \dot{\varepsilon}_{0\text{B}} f\left(\frac{\sigma - \sigma_{\Delta_{\text{ss}}}}{\sigma_{0\text{B}}}\right) \tag{1.4.78}$$

For the power law stress function Eq. (1.4.77) can be solved in a closed analytical form as follows

$$\sigma_{\Delta_{\text{ss}}} = \frac{\alpha - 1}{\alpha + 1}\sigma, \quad \alpha = \frac{\zeta^{1/n}}{\eta} \tag{1.4.79}$$

For $\alpha = 1$ the stress difference is zero and the stresses in the bars remain constant over time. For $\alpha > 1$ stress redistribution occurs during the creep process. The maximum stress difference in the bars is a fraction of the applied stress. To analyze the stress redistribution Eq. $(1.4.71)_1$ is solved numerically with the initial condition $\sigma_\Delta = 0$. Figure 1.17 illustrates the stress difference as a function of the normalized time. For $\alpha > 0$ the creep rate of bar A is lower than that of bar B for the same stress level. Therefore, starting from the same stress values, the stress in the creep-hard bar A increases with time, while the stress in the creep-soft bar B relaxes. The stress difference increases towards the steady-state value $\sigma_{\Delta_{\text{ss}}}$. The duration of stress redistribution depends significantly on the value of the creep exponent. For the same value of α and the same applied stress σ, a larger value of the exponent n leads to a shorter time of stress redistribution, Fig. 1.17. Figure 1.18 shows the normalized inelastic strain rate as a function of the normalized inelastic strain of the bar system ε^{pl} defined as

$$\varepsilon^{\text{pl}} = \varepsilon - \frac{\sigma}{E}$$

Although primary creep in the bars is neglected, the bar system exhibits primary creep as a result of stress redistribution. The initial (maximum) creep rate continuously decreases towards the steady-state (minimum) value. For power law creep in the bars, the steady-state value of the stress difference $\sigma_{\Delta_{\text{ss}}}$ is a linear function of the applied stress. Therefore both the initial and the minimum creep rates are power law functions of the applied stress σ with the same stress exponent. These functions can be derived from Eqs (1.4.76), (1.4.78) and (1.4.79).

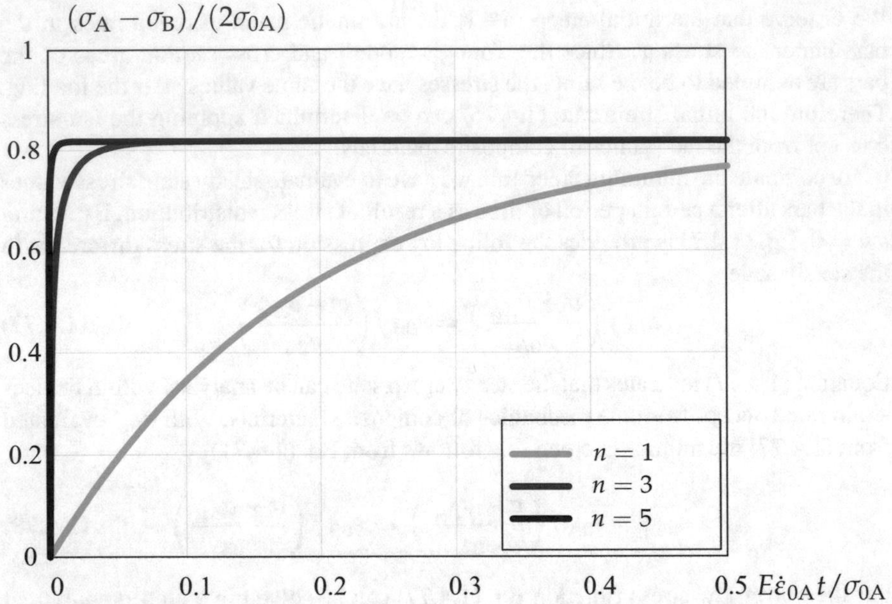

Fig. 1.17 Normalized stress difference vs normalized time for a two bar system with power law creep. Influence of the creep exponent for $\alpha = 10$

A more complex behavior of the two-bar system is observed if individual bars exhibit stress range dependent creep. As an example let us consider the hyperbolic sine creep laws for bars

$$\dot{\varepsilon}_i^{\rm pl} = \dot{\varepsilon}_{0i} \sinh\left(\frac{|\sigma_i|}{\sigma_{0i}}\right) \operatorname{sgn}(\sigma_i), \quad i = \mathrm{A}, \mathrm{B} \tag{1.4.80}$$

For the material parameters of the bars we assume the following relationships

$$\xi = \frac{\dot{\varepsilon}_{0\mathrm{B}}}{\dot{\varepsilon}_{0\mathrm{A}}} \geq 1, \quad \eta = \frac{\sigma_{0\mathrm{B}}}{\sigma_{0\mathrm{A}}} \leq 1$$

Figure 1.19 shows the steady-state value of the stress difference in the bars as a function of the applied stress. The graphs are obtained by the numerical solution of Eq. (1.4.77) for $\xi = 10^3$ and different values of η. Two stress ranges can be observed. For the low stress range, the asymptotic solution of Eq. (1.4.77) is

$$\sigma_{\Delta_{\rm ss}} = \frac{\sigma}{1 + \dfrac{\xi}{\eta}} \tag{1.4.81}$$

For the high stress range the asymptotic solution reads

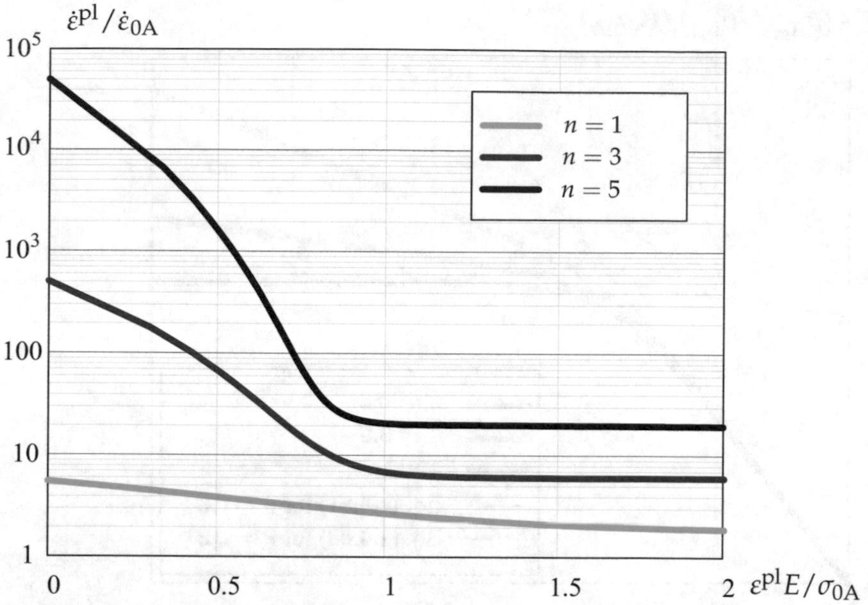

Fig. 1.18 Normalized inelastic strain rate vs normalized inelastic strain of a two bar system with power law creep. Influence of the creep exponent for $\alpha = 10$

$$\sigma_{\Delta_{\rm ss}} = \sigma_{0A} \frac{\ln \xi + \dfrac{\sigma}{\sigma_{0A}}\left(\dfrac{1}{\eta} - 1\right)}{1 + \dfrac{1}{\eta}} \tag{1.4.82}$$

Both the asymptotic solutions provide linear functions of the stress σ. The corresponding plots are illustrated in Fig. 1.19 for the case $\eta = 0.8$ by dashed lines. The slope of the curve within the second, high stress regime decreases with respect to the first, low stress regime. As Eq. (1.4.82) shows, the high stress regime slope depends on the value of η. For $\eta = 1$ the saturation stress does not depend on the applied stress in a high stress range.

The two slope $\sigma_{\Delta_{\rm ss}}$ vs σ behavior leads to the stress range dependent creep of the two bar system. Figure 1.20 illustrates the normalized initial (maximum) and steady state (minimum) strain rates as functions of the normalized stress. The graphs are plots of Eqs (1.4.76) and (1.4.78) for $\xi = 10^3$ and $\eta = 0.5$. One feature is that the difference between the initial and steady creep rates is stress range dependent and decreases with an increase in the stress level.

Figure 1.21 shows the normalized stress difference in bars as a function of the normalized time. The plots are presented for three stress levels. For $\sigma/\sigma_{0A} = 1$ and $\sigma/\sigma_{0A} = 2$ the duration of the stress redistribution is nearly the same. These stress levels are within the low stress regime, Fig. 1.19. The steady state stress difference in the bars can be determined from the linear stress dependence (1.4.81). The stress level $\sigma/\sigma_{0A} = 4$ is outside the low stress range. The slope of $\sigma_{\Delta_{\rm ss}}$ vs σ curve

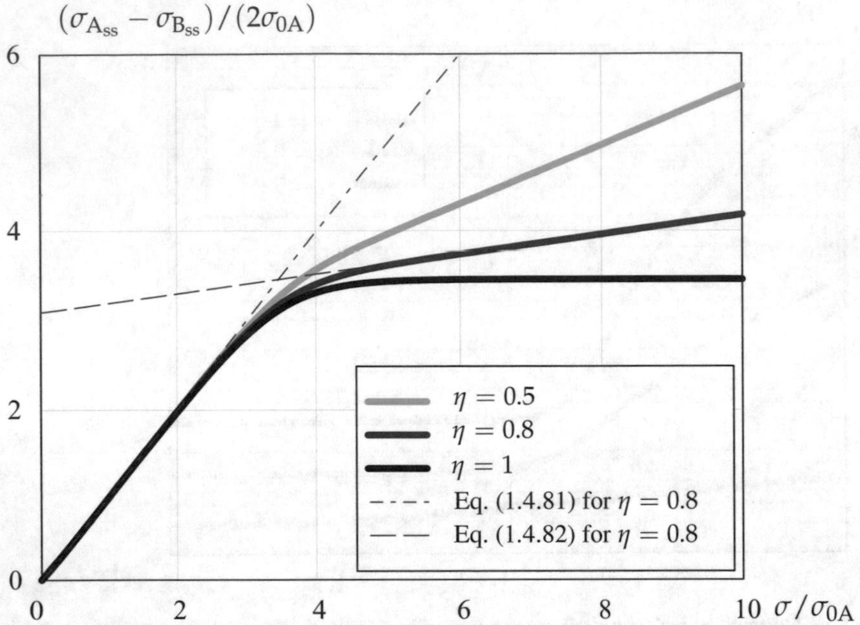

Fig. 1.19 Normalized stress difference in the steady creep state vs normalized stress for a two bar system with hyperbolic sine creep law. Influence of η stress level for $\xi = 10^3$

decreases, Fig. 1.19. As a result, the difference between the initial and minimum creep rates becomes lower, Fig. 1.20. The duration of stress redistribution decreases, Fig. 1.21.

Figure 1.22 illustrates the normalized inelastic strain rate as a function of the normalized time. The duration of the primary creep stage depends on the stress level and differs significantly for low stress and high stress regimes.

Let us note, that many high-temperature materials are composites with two or more constituents with different inelastic properties. The difference may arise due to different phases for example in composites (Roesler et al, 2007), different initial dislocation densities, for example in tempered martensitic steels (Blum, 2008), elongated grains in forged aluminium alloys (Gariboldi et al, 2016), or different grain sizes in different zones of welds (Naumenko and Altenbach, 2005). Such heterogeneous materials usually exhibit complex stress range dependent creep behavior. Composite models of inelastic deformation were developed to capture various creep phenomena (Meier and Blum, 1993). The considered two bar system gives a first insight to the stress range dependencies of primary and secondary creep in multi-phase materials.

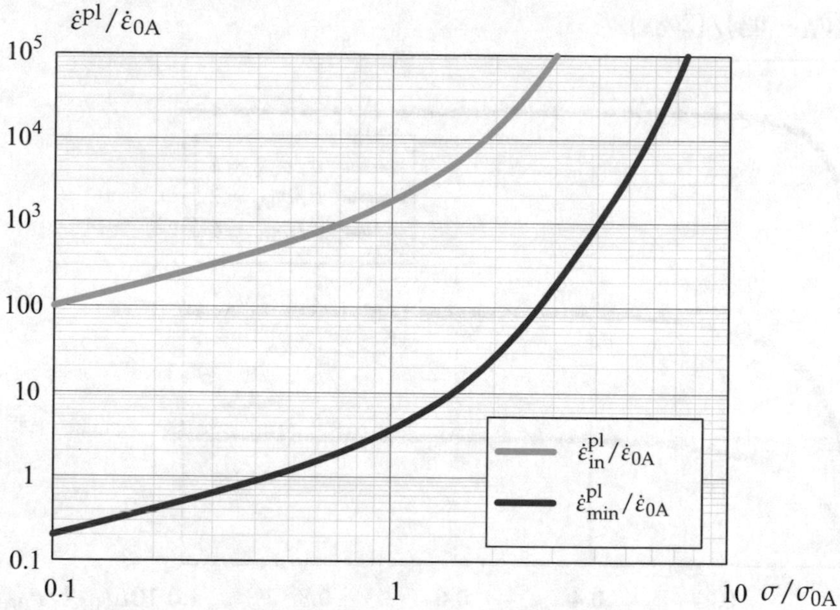

Fig. 1.20 Normalized inelastic strain rate vs normalized stress for a two bar system with hyperbolic sine creep law. Initial (maximum) and steady-state (minimum) strain rates for $\xi = 10^3$ and $\eta = 0.5$

1.4.4 Creep Followed by Recovery

Let us assume that the bar system was subjected to a constant tensile load F and constant temperature over a time interval $[0, t_{\rm L}]$. During this loading stage the behavior of the bar system is characterized by the creep strain accumulation and the stress redistribution, as discussed in Subsect. 1.4.3. Let the inelastic strain accumulated during the loading stage be $\varepsilon^{\rm pl}_{\rm L}$ and the stress difference in the bars $\sigma_{\Delta_{\rm L}} > 0$. Let us assume that at $t_{\rm L}$ the load F is rapidly removed. After the elastic springback the residual stresses in the bars are

$$\sigma_{\rm A_L} = -\sigma_{\rm B_L} = \sigma_{\Delta_{\rm L}}$$

When the bar system is kept at the same temperature without load, the residual stresses relax towards zero and the inelastic strain decreases towards a certain permanent value. To analyze this recovery process let us rewrite Eqs (1.4.75) with $\sigma = 0$ and $\sigma_{\Delta_{\rm L}} > 0$ as follows

$$\begin{aligned}\frac{d\varepsilon^{\rm pl}}{d\tau} &= \frac{1}{2}f(\Sigma_\Delta) - \frac{1}{2}\xi f\left(\frac{\Sigma_\Delta}{\eta}\right),\\ \frac{d\Sigma_\Delta}{d\tau} &= -\frac{1}{2}f(\Sigma_\Delta) - \frac{1}{2}\xi f\left(\frac{\Sigma_\Delta}{\eta}\right),\end{aligned} \tag{1.4.83}$$

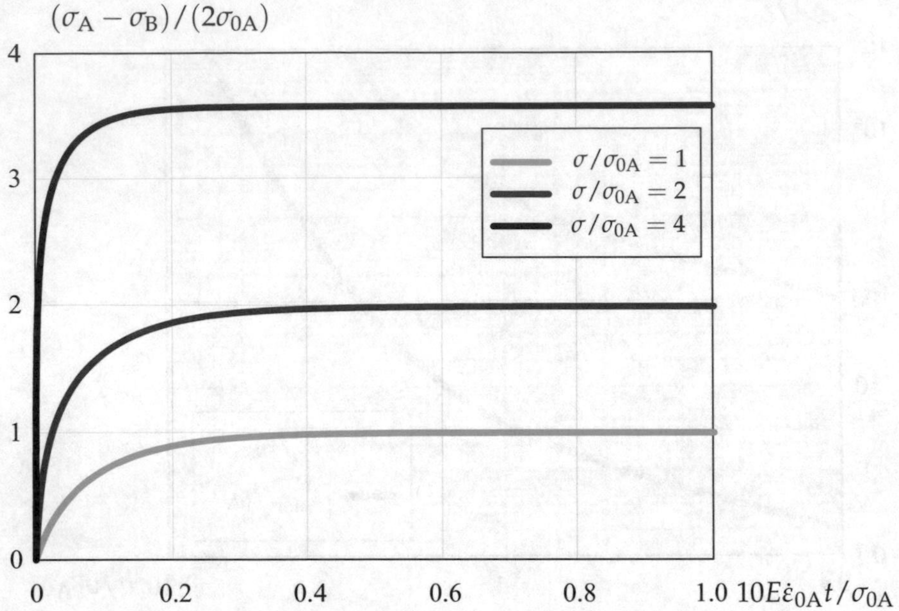

Fig. 1.21 Normalized stress difference vs normalized time for a two bar system with hyperbolic sine creep law. Influence of the stress level for $\xi = 10^3$ and $\eta = 0.5$

where

$$\epsilon^{pl} = \frac{E}{\sigma_{0A}} \varepsilon^{pl}$$

With the initial conditions $\epsilon^{pl}(t_L) = \varepsilon^{pl}_L$ and $\sigma_\Delta(t_L) = \sigma_{\Delta_L}$ the system (1.4.83) can be solved providing the evolution of the plastic strain and the stress difference after unloading. Figure 1.23 illustrates the numerical solution for the stress difference between the bars as a function of the normalized time applying the hyperbolic sine law (1.4.80) with $\xi = 10^3$ and $\eta = 0.5$. During the creep process over the time $\tau = 0.1$, the stresses reach the steady state. After removing the load, the stresses rapidly relax down to the zero values. Figure 1.24 shows the normalized inelastic strain vs normalized time. After unloading, creep recovery is observed: the inelastic strain decreases towards the permanent value. To evaluate the recoverable strain let us eliminate the time variable in Eqs (1.4.83) providing the following equation

$$\frac{\mathrm{d}\Sigma_\Delta}{\mathrm{d}\epsilon^{pl}} = -\frac{f(\Sigma_\Delta) + \xi f\left(\dfrac{\Sigma_\Delta}{\eta}\right)}{f(\Sigma_\Delta) - \xi f\left(\dfrac{\Sigma_\Delta}{\eta}\right)} \qquad (1.4.84)$$

By separation of variables we obtain

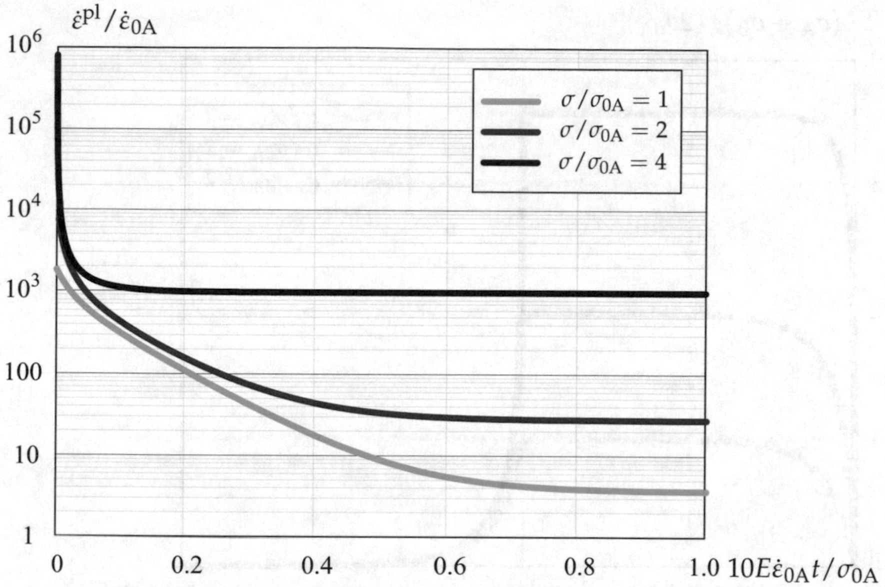

Fig. 1.22 Normalized inelastic strain rate vs normalized time for a two bar system with hyperbolic sine creep law. Influence of the stress level for $\xi = 10^3$ and $\eta = 0.5$

$$\epsilon_{\mathrm{L}}^{\mathrm{pl}} - \epsilon^{\mathrm{pl}} = \int\limits_{\Sigma_{\Delta_{\mathrm{L}}}}^{\Sigma_\Delta} \frac{f(x) - \xi f\left(\dfrac{x}{\eta}\right)}{f(x) + \xi f\left(\dfrac{x}{\eta}\right)} \mathrm{d}x \tag{1.4.85}$$

As $\Sigma_\Delta \to 0$ the plastic strain is $\epsilon^{\mathrm{pl}} \to \epsilon^{\mathrm{per}}$, where ϵ^{per} is the permanent strain. The recoverable inelastic strain is defined as $\epsilon^{\mathrm{rec}} = \epsilon_{\mathrm{L}}^{\mathrm{pl}} - \epsilon^{\mathrm{per}}$, Fig. 1.24. From Eq. (1.4.85) it can be evaluated as follows

$$\epsilon^{\mathrm{rec}} = g(\Sigma_{\Delta_{\mathrm{L}}}), \quad g(z) = \int\limits_0^z \frac{\xi f\left(\dfrac{x}{\eta}\right) - f(x)}{\xi f\left(\dfrac{x}{\eta}\right) + f(x)} \mathrm{d}x \tag{1.4.86}$$

For $\xi \ll 1$ and $\eta < 1$ the integral in (1.4.86) can be evaluated providing $g(z) = z$ with a good accuracy. Therefore the recoverable inelastic strain is proportional to the stress difference in the bars, reached during the creep loading.

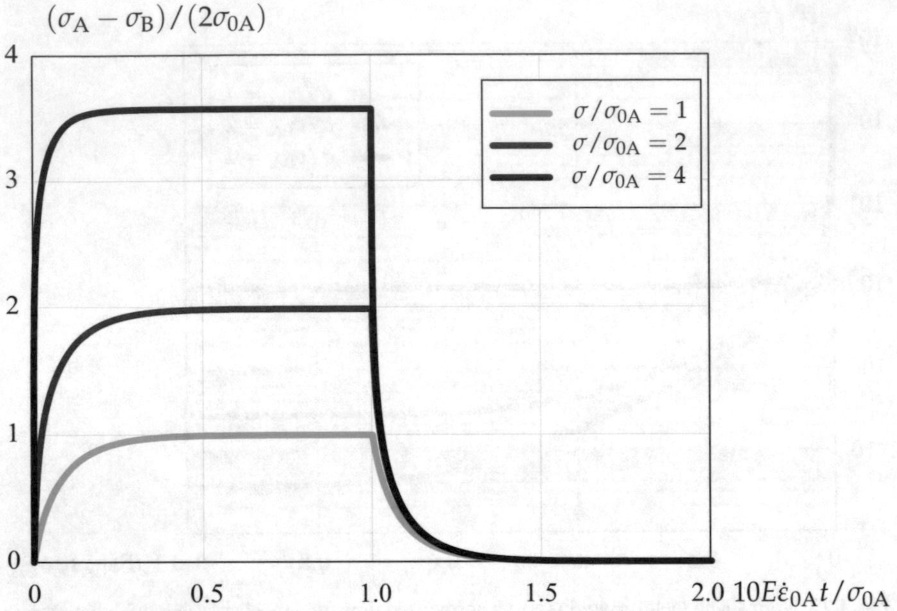

Fig. 1.23 Normalized stress difference vs normalized time for a two bar system applying hyperbolic sine creep law with $\xi = 10^3$ and $\eta = 0.5$

1.4.5 Stress Relaxation

Let us perform a relaxation test with the two-bar system. We assume that after a certain loading the strain is ϵ and the stresses in the bars are $\Sigma_{A_{ref}}$ and $\Sigma_{B_{ref}}$. When the strain is kept constant over time the stresses in bars relax. With the zero strain rate Eqs (1.4.73) take the following form

$$\frac{d\Sigma_A}{d\tau} + f(|\Sigma_A|)\text{sgn}(\Sigma_A) = 0, \quad \frac{d\Sigma_B}{d\tau} + \xi f\left(\frac{|\Sigma_B|}{\eta}\right)\text{sgn}(\Sigma_B) = 0 \quad (1.4.87)$$

For $\Sigma_{A_{ref}} > 0$ and $\Sigma_{B_{ref}} > 0$ the solutions to Eqs (1.4.87) can be formulated as follows

$$g(\Sigma_A) = g(\Sigma_{A_{ref}}) - \tau, \quad g\left(\frac{\Sigma_B}{\eta}\right) = g\left(\frac{\Sigma_{B_{ref}}}{\eta}\right) - \frac{\xi}{\eta}\tau, \quad (1.4.88)$$

where

$$g(x) = \int \frac{dx}{f(x)}$$

To keep the strain constant, the force applied to the bar system must be reduced according to the following law

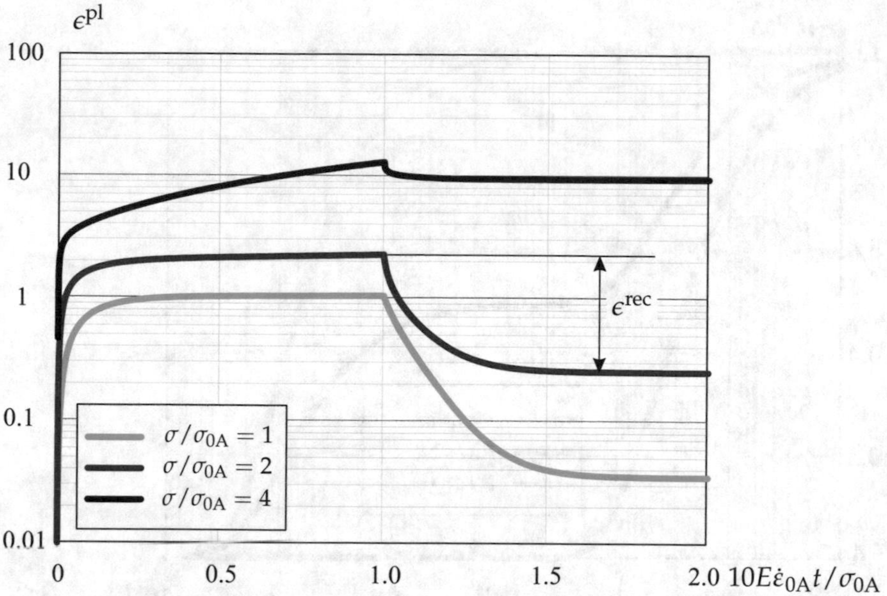

Fig. 1.24 Normalized inelastic strains vs normalized time for a two bar system applying hyperbolic sine creep law with $\xi = 10^3$ and $\eta = 0.5$

$$F(\tau) = 2A\sigma(\tau) = 2A\sigma_{0A}\Sigma(\tau), \tag{1.4.89}$$

where

$$\begin{aligned} 2\Sigma(\tau) &= \Sigma_A(\tau) + \Sigma_B(\tau) \\ &= g^{-1}\left[g(\Sigma_{A_{ref}}) - \tau\right] + \eta g^{-1}\left[g\left(\frac{\Sigma_{B_{ref}}}{\eta}\right) - \frac{\xi}{\eta}\tau\right] \end{aligned} \tag{1.4.90}$$

In the first example we assume that the bars exhibit the power law creep regime with $f(x) = x^n$ and $n > 1$. In this case

$$g(x) = \frac{1}{1-n}x^{1-n}, \quad g^{-1}(x) = [(1-n)x]^{\frac{1}{1-n}}$$

Equation (1.4.90) takes the following form

$$2\Sigma(\tau) = \left[\Sigma_{A_{ref}}^{1-n} + (n-1)\tau\right]^{\frac{1}{1-n}} + \left[\Sigma_{B_{ref}}^{1-n} + (n-1)\frac{\xi}{\eta^n}\tau\right]^{\frac{1}{1-n}} \tag{1.4.91}$$

In the second example let us consider the hyperbolic sine creep law. With

$$g(x) = \ln\left(\tanh\frac{x}{2}\right), \quad g^{-1}(x) = 2\operatorname{arctanh}(\exp x)$$

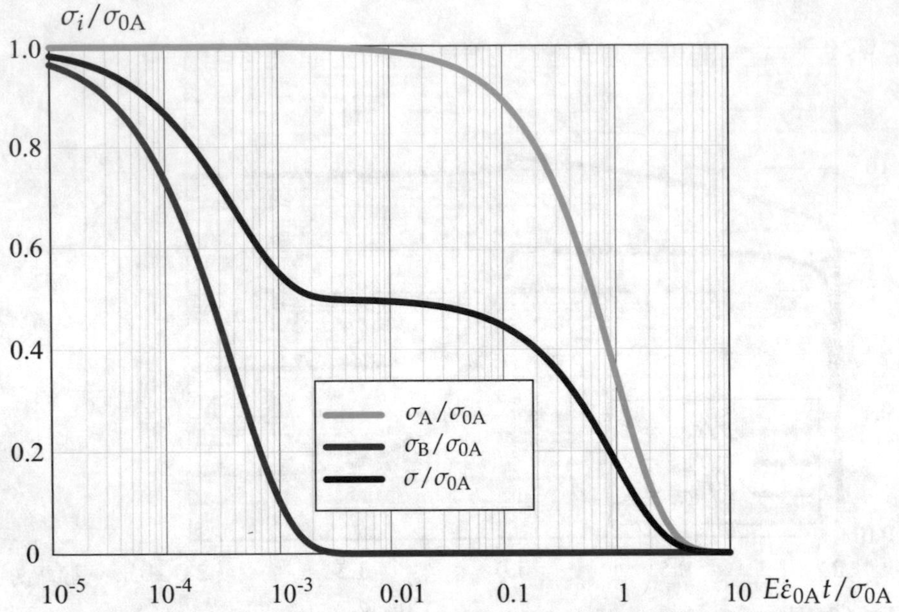

Fig. 1.25 Normalized stresses in the bars and normalized stress of two bar system vs normalized time applying hyperbolic sine creep law with $\xi = 10^3$, $\eta = 0.5$ and $\Sigma_{A_{ref}} = \Sigma_{B_{ref}} = 1$

Eq. (1.4.90) reads

$$2\Sigma(\tau) = 2\text{arctanh}\left[\tanh\left(\frac{\Sigma_{A_{ref}}}{2}\right)\exp(-\tau)\right] + 2\eta\text{arctanh}\left[\tanh\left(\frac{\Sigma_{B_{ref}}}{2\eta}\right)\exp\left(-\frac{\xi}{\eta}\tau\right)\right] \tag{1.4.92}$$

For loading within the elastic range the reference stresses are

$$\Sigma_{A_{ref}} = \Sigma_{B_{ref}} = \Sigma_{ref} = \epsilon, \tag{1.4.93}$$

where

$$\epsilon = \frac{E}{\sigma_{0A}}\varepsilon$$

Figure 1.25 shows normalized stresses in the bars and the normalized stress of the bar system as functions of normalized time according to Eq. (1.4.92). In the considered examples we set $\xi = 10^3$, $\eta = 0.5$ and $\Sigma_{A_{ref}} = \Sigma_{B_{ref}} = 1$. Due to extremely different inelastic properties of the bars, two relaxation stages of the stress Σ are observed. The first stage is characterized by the rapid stress relaxation in the inelastic-soft bar B with normalized relaxation time of approximately 2×10^{-3}. The second longer relaxation stage is due to inelastic-hard bar A with the characteristic time of approximately 10.

1.4.6 Displacement-controlled Monotonic and Cyclic Loadings

The displacement of the bar system δ is assumed to be a function of time. The force F required to move the bar system and the stresses in the bars have to be computed. For the monotonic loading the displacement is assumed to be a linear function of time

$$\delta(t) = v_\delta t,$$

where $v_\delta > 0$ is the velocity. From Eqs (1.1.2) the strains are computed as follows

$$\varepsilon_i = \varepsilon = \frac{v_\delta}{l} t, \quad i = \mathrm{A}, \mathrm{B}$$

The strain rates are

$$\dot{\varepsilon}_i = \dot{\varepsilon} = \frac{v_\delta}{l}$$

For linear viscous behavior of the bars the solutions for stresses can be given in the closed analytical form, see Subsect. 1.3.1. For non-linear inelastic bars Eqs (1.4.69) or (1.4.73) can be solved numerically providing stress responses. With the initial conditions $\Sigma_\mathrm{A} = \Sigma_\mathrm{B} = 0$, Eqs (1.4.73) give the initial slopes of stress-time curves

$$\frac{\mathrm{d}\Sigma_\mathrm{A}}{\mathrm{d}\tau} = \frac{\mathrm{d}\Sigma_\mathrm{B}}{\mathrm{d}\tau} = \dot{\varepsilon} \tag{1.4.94}$$

In the steady state flow stage the stress rates can be set to zero providing the stress values

$$\Sigma_{\mathrm{A_{ss}}} = f^{-1}(\dot{\varepsilon}), \quad \Sigma_{\mathrm{B_{ss}}} = \eta f^{-1}\left(\frac{\dot{\varepsilon}}{\xi}\right) \tag{1.4.95}$$

The force applied to the bar system in the steady state can be computed as folllows

$$F_\mathrm{ss} = 2A\sigma_\mathrm{ss} = 2A\sigma_{0\mathrm{A}}\Sigma_\mathrm{ss}, \tag{1.4.96}$$

where

$$2\Sigma_\mathrm{ss} = \Sigma_{\mathrm{A_{ss}}} + \Sigma_{\mathrm{B_{ss}}} = f^{-1}(\dot{\varepsilon}) + \eta f^{-1}\left(\frac{\dot{\varepsilon}}{\xi}\right) \tag{1.4.97}$$

As an example let us consider the Garofalo laws for the bars

$$\dot{\varepsilon}_i^\mathrm{pl} = \dot{\varepsilon}_{0i}\left[\sinh\left(\frac{|\sigma_i|}{\sigma_{0i}}\right)\right]^n \mathrm{sgn}(\sigma_i), \quad i = \mathrm{A}, \mathrm{B} \tag{1.4.98}$$

For the material parameters of the bars we assume the following relationships

$$\xi = \frac{\dot{\varepsilon}_{0\mathrm{B}}}{\dot{\varepsilon}_{0\mathrm{A}}} \geq 1, \quad \eta = \frac{\sigma_{0\mathrm{B}}}{\sigma_{0\mathrm{A}}} \leq 1$$

Figure 1.26 illustrates normalized inelastic strain rates as functions of normalized stresses in bars for $\xi = 10$, $\eta = 0.5$ and $n = 5$. Solid lines are plots of Garo-

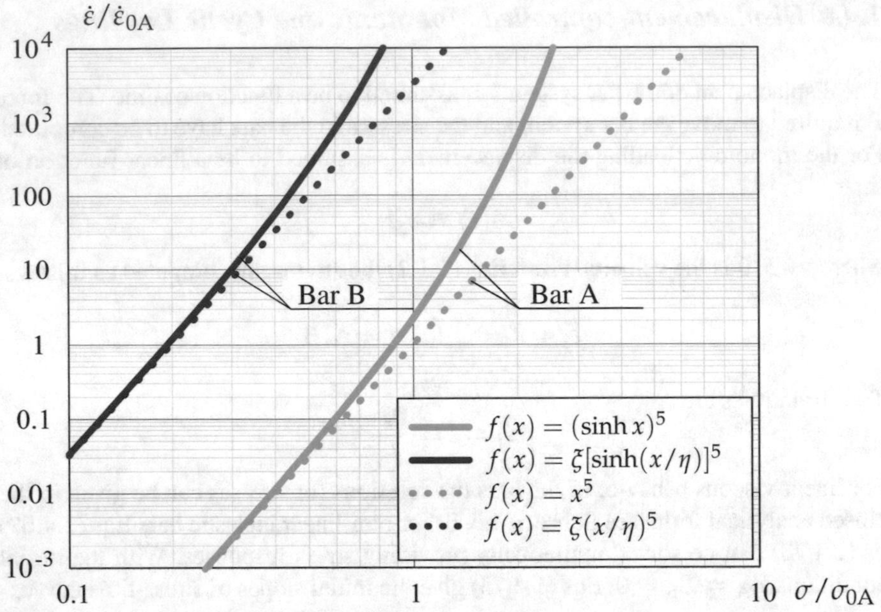

Fig. 1.26 Normalized inelastic strain rate vs normalized stress in bars applying Garofalo's and power stress functions with $\xi = 10$, $\eta = 0.5$ and $n = 5$

falo stress functions. For the comparison power law stress functions are plotted by broken lines. Figure 1.27 illustrates the numerical solutions of Eqs (1.4.73) for two strain rates. The steady-state stresses in the bars and the bar system can be computed by Eqs (1.4.95) and (1.4.96) with

$$f^{-1}(x) = \operatorname{arcsinh}\left(x^{\frac{1}{n}}\right) \tag{1.4.99}$$

For the strain rate $\dot{\epsilon} = 10$ the steady-state flow of the bar A is in the power law breakdown regime while the bar B in the power law regime, Fig. 1.26. For the higher strain rate $\dot{\epsilon} = 10$ both the bars flow in the power law breakdown range. The stress-strain curve of the bar system exhibits three stages. The first one is linear-elastic with $\Sigma_i = \Sigma = \epsilon$. As the stress attains the flow stress of the inelastic-soft bar B, the slope of the stress-strain curve decreases towards one half of the initial one. The stress-strain response in this stage is determined by $\Sigma = \epsilon/2 + \Sigma_{B_{ss}}/2$. The final stage is the steady-state flow stage of both the bars A and B. The behavior of the bar system depends significantly on the strain rate, Fig. 1.27. The decrease of the strain rate leads to a decrease of the steady-state flow stresses in the bars and in the bar system.

Let as analyze the cyclic response of the bar system applying the triangular strain profile, as shown in Fig. 1.28. The absolute value of the strain rate is $|\dot{\epsilon}| = 10$ and the stress amplitude is $\epsilon_a = 2$. By reaching the maximum strain value in the loading stage both the bars are in the steady flow regime, Fig. 1.28. By unloading

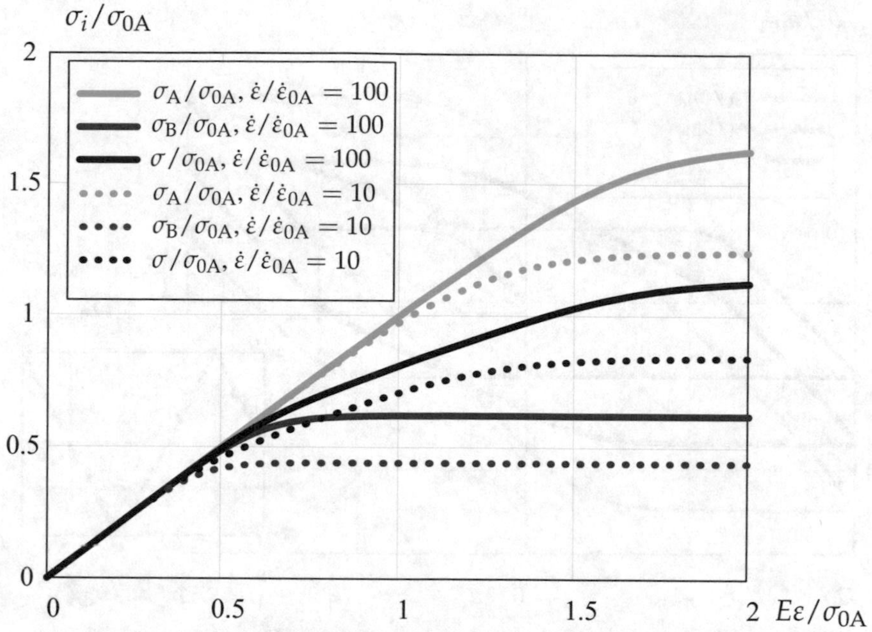

Fig. 1.27 Normalized stresses in the bars and normalized stress of two bar system vs normalized strain for two strain rates applying Garofalo's law with $\xi = 10$, $\eta = 0.5$ and $n = 5$

the stresses first decrease nearly linearly with the initial elastic slope. After reaching the compressive flow stress, the bar system starts to yield in the compressive range. As Fig. 1.28 shows, the compressive flow stress value $\Sigma_{f_{compr}}$ is lower than the initial flow stress $\Sigma_{f_{tens}}$ during loading. This Bauschinger effect of the bar system can be explained by residual stresses in the bars accumulated during the loading stage. From Fig. 1.28 it is obvious that $\Sigma_{f_{compr}} = \Sigma_{f_{tens}} - \Sigma_{res}$ with Σ_{res} being the residual stress in the bars at $\Sigma = 0$. Note that for the bar system unloaded to $\Sigma = 0$ the inelastic-hard bar A is under tension with $\Sigma_A = \Sigma_{res}$, while the inelastic-soft bar B is under compression with $\Sigma_B = -\Sigma_{res}$. On the other hand $\Sigma_{res} = \Sigma_{A_{ss}} - \Sigma_{ss}$ and with Eq. (1.4.97) we can compute the residual stress as follows

$$\Sigma_{res} = \frac{1}{2}\left[f^{-1}(\dot{\varepsilon}) - \eta f^{-1}\left(\frac{\dot{\varepsilon}}{\xi}\right)\right] \tag{1.4.100}$$

It is obvious that the Bauschinger effect of the bar system depends on the rate of loading. With Eqs (1.4.99) and (1.4.100) we obtain

$$\Sigma_{res} = \frac{1}{2}\left\{\operatorname{arcsinh}\left(\dot{\varepsilon}^{\frac{1}{n}}\right) - \eta \operatorname{arcsinh}\left[\left(\frac{\dot{\varepsilon}}{\xi}\right)^{\frac{1}{n}}\right]\right\} \tag{1.4.101}$$

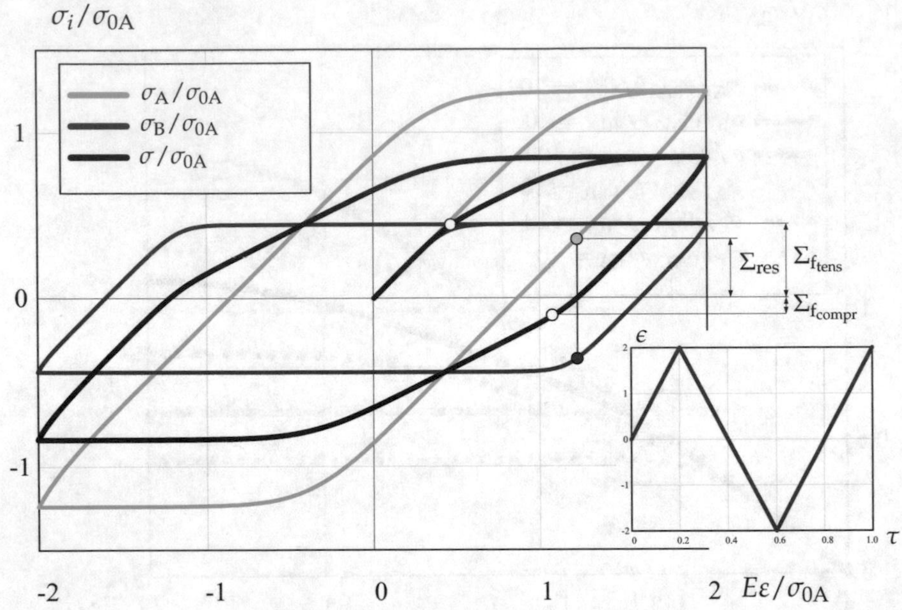

Fig. 1.28 Normalized stresses in the bars and normalized stress of two bar system vs normalized strain for triangular strain profile applying Garofalo's law with $\xi = 10$, $\eta = 0.5$ and $n = 5$

1.4.7 Time-step Methods

Time-step methods were introduced in Subect. 1.3.3 to solve linear ordinary differential equations. Subsection 1.4.7 gives basic features of time-step methods for the use in problems with non-linear inelastic material behavior. As an example let us consider the ordinary differential equation (1.4.60) describing stress response in a bar for the given strain rate

$$\frac{\mathrm{d}\Sigma}{\mathrm{d}\tau} + \Psi(\Sigma) = \dot{\epsilon}, \quad \Psi(\Sigma) = f(|\Sigma|)\operatorname{sgn}\Sigma \tag{1.4.102}$$

with the initial condition $\Sigma(0) = 0$. Any numerical solution procedure can be constructed by integration of Eq. (1.4.102) in a time interval $[t_k, t_{k+1}]$. Following the approaches presented in Subsect. 1.3.3, we obtain

$$\Delta\Sigma_k = -\int\limits_{\tau_k}^{\tau_{k+1}} \Psi(\Sigma)\mathrm{d}\tau + \Delta\epsilon_k, \quad \Delta\Sigma_k = \Sigma_{k+1} - \Sigma_k, \tag{1.4.103}$$

for the given strain increment $\Delta\epsilon_k = \epsilon_{k+1} - \epsilon_k$. The integral on the right-hand side of Eq. (1.4.103) is evaluated approximately leading to different time-step methods depending on the type of applied quadrature rule.

1.4.7.1 Explicit Euler Method

The integral in Eq. (1.4.103) is approximated by the following rectangular rule

$$\int_{\tau_k}^{\tau_{k+1}} \Psi(\Sigma)\mathrm{d}\tau \approx \Psi(\Sigma_k)\Delta\tau_k, \quad \Delta\tau_k = \tau_{k+1} - \tau_k \tag{1.4.104}$$

Equation (1.4.103) takes the following form

$$\Sigma_{k+1} = \Sigma_k - \Psi(\Sigma_k)\Delta\tau_k + \Delta\epsilon_k \tag{1.4.105}$$

Starting with the initial data $\Sigma_0 = 0$, Eq. (1.4.105) can be used to compute approximate solutions for the stress Σ_k in discrete time points. The accuracy of the solution within the time interval $[\tau_k, \tau_{k+1}]$ is related to the accuracy of the rectangular rule (1.4.104). The local error can be estimated applying the following inequality

$$\left| \int_{\tau_k}^{\tau_{k+1}} \Psi(\Sigma)\mathrm{d}\tau - \Psi(\Sigma_k)\Delta\tau_k \right| \leq R, \quad R = \frac{\Delta\tau_k^2}{2} \sup_{\Sigma_k \leq \Sigma \leq \Sigma_{k+1}} \left| \frac{\mathrm{d}\Psi}{\mathrm{d}\tau} \right| \tag{1.4.106}$$

Specifying the error tolerance by Errtol/2, the right hand side of the inequality (1.4.106) can be used to estimate the time step size

$$\sup_{\Sigma_k \leq \Sigma \leq \Sigma_{k+1}} \left| \frac{\mathrm{d}\Psi}{\mathrm{d}\tau} \right| \Delta\tau_k^2 \leq \mathrm{Errtol} \tag{1.4.107}$$

The time derivative of Ψ can be computed approximately providing the following expression

$$\sup_{\Sigma_k \leq \Sigma \leq \Sigma_{k+1}} \left| \frac{\mathrm{d}\Psi}{\mathrm{d}\tau} \right| \Delta\tau_k^2 \approx |\Psi(\Sigma_{k+1}) - \Psi(\Sigma_k)|\Delta\tau_k \leq \mathrm{Errtol} \tag{1.4.108}$$

Alternatively the inequality (1.4.108) can be derived as follows. Assume that the integral in (1.4.104) is evaluated by the trapezoidal rule having the higher order accuracy (see Subsubsect. 1.3.3.3)

$$\int_{\tau_k}^{\tau_{k+1}} \Psi(\Sigma)\mathrm{d}\tau \approx \frac{\Psi(\Sigma_k) + \Psi(\Sigma_{k+1})}{2}\Delta\tau_k \tag{1.4.109}$$

Subtracting (1.4.108) from (1.4.109) provides the approximate value of the local error

$$R \approx \frac{|\Psi(\Sigma_{k+1} - \Psi(\Sigma_k)|}{2}\Delta\tau_k \tag{1.4.110}$$

With the error tolerance Errtol/2 we obtain

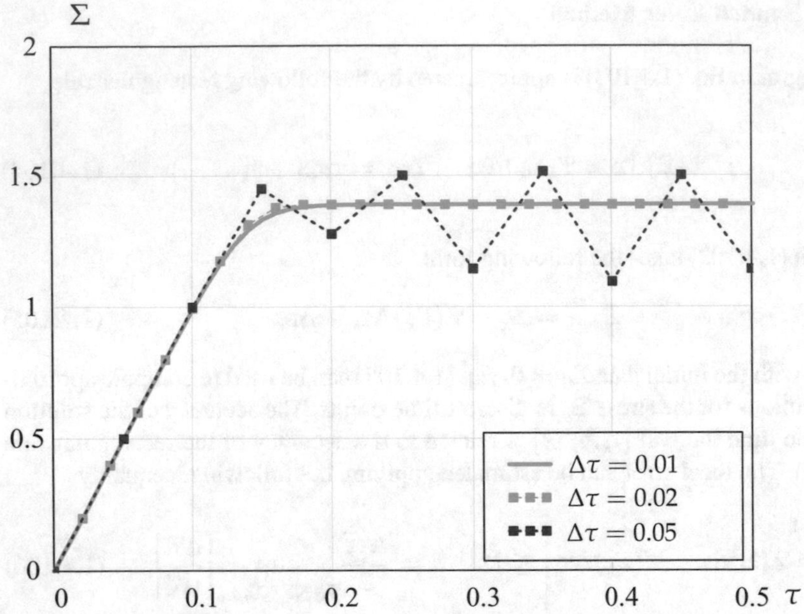

Fig. 1.29 Numerical solutions of Eq. (1.4.102) by explicit Euler method with different time step sizes for $\dot{\varepsilon} = 10$ and power law stress function with $n = 7$

$$|\Psi(\Sigma_{k+1}) - \Psi(\Sigma_k)|\Delta\tau_k \leq \text{Errtol} \qquad (1.4.111)$$

Remembering that $\Psi(\Sigma)$ is the inelastic strain rate the inequality (1.4.111) can also be formulated as follows

$$\left|\dot{\varepsilon}^{\text{pl}}_{k+1} - \dot{\varepsilon}^{\text{pl}}_{k}\right| \Delta t_k \leq \text{Errtol} \qquad (1.4.112)$$

Figure 1.29 illustrates the numerical solution of Eq. (1.4.102) with $f(x) = x^n$, $n = 7$, $\dot{\varepsilon} = 10$ and three constant time step sizes. Figure 1.30 shows the local error R computed with Eq. (1.4.110). The maximum error is in the transition from elastic stage to steady-state inelastic behavior. With a decrease of the time step size the local error decreases. For the time step size $\Delta\tau = 0.05$ the non-stable oscillatory solution is obtained, Fig. 1.29. Approaches to evaluate the critical time step size for linear ordinary differential equations are discussed in Subsubsect. 1.3.3.1. For non-linear equations one might expect that the critical time step size depends on properties of function $f(x)$, for example its first derivative, the rate of loading and the current values of inelastic strain and/or time. To estimate the critical time step size let us linearize Eq. (1.4.102) about Σ_k with the following approximation for the function Ψ

$$\Psi(\Sigma) \approx \Psi(\Sigma_k) + \Psi'(\Sigma_k)(\Sigma - \Sigma_k), \quad \Psi' = \frac{\mathrm{d}\Psi}{\mathrm{d}\Sigma} \qquad (1.4.113)$$

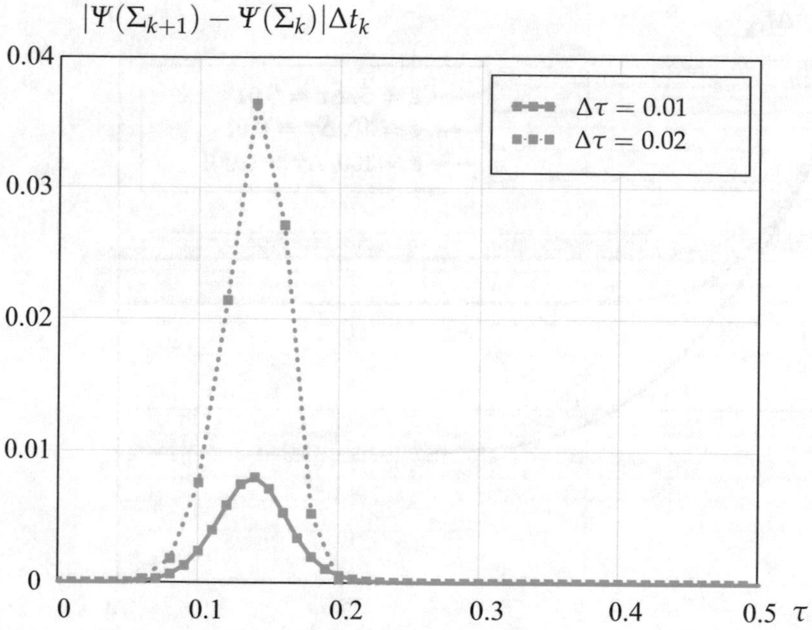

Fig. 1.30 Local error of explicit Euler by solving Eq. (1.4.102) with different time step sizes for $\dot{\varepsilon} = 10$ and power law stress function with $n = 7$

Equation (1.4.102) takes the following form

$$\frac{\mathrm{d}\Sigma}{\mathrm{d}\tau} + \lambda_k \Sigma = \dot{\varepsilon} - \Psi(\Sigma_k) + \lambda_k \Sigma_k, \tag{1.4.114}$$

where $\lambda_k = \Psi'(\Sigma_k)$. Applying the approaches discussed in Subsubsect. 1.3.3.1 for linear ordinary differential equations, the critical time step size can be obtained as follows

$$\Delta\tau_{\mathrm{c}_k} = \frac{2}{\lambda_k} \tag{1.4.115}$$

In computations with the explicit Euler method the current time step size must be selected such that

$$\Delta\tau_k < \Delta\tau_{\mathrm{c}_k} \tag{1.4.116}$$

To prevent oscillations one should use the following value $\Delta\tau_{\mathrm{cO}_k}$ as the critical time step, see Subsubsect. 1.3.3.1

$$\Delta\tau_{\mathrm{cO}_k} = \frac{1}{\lambda_k} \tag{1.4.117}$$

Figure 1.31 illustrates critical time step sizes estimated by Eq. (1.4.115) during the computations by the explicit Euler method. In the calculations the time step sizes were assumed constant and less than the minimum critical values. For the known steady-state stress value the minimum critical time step size can be computed from

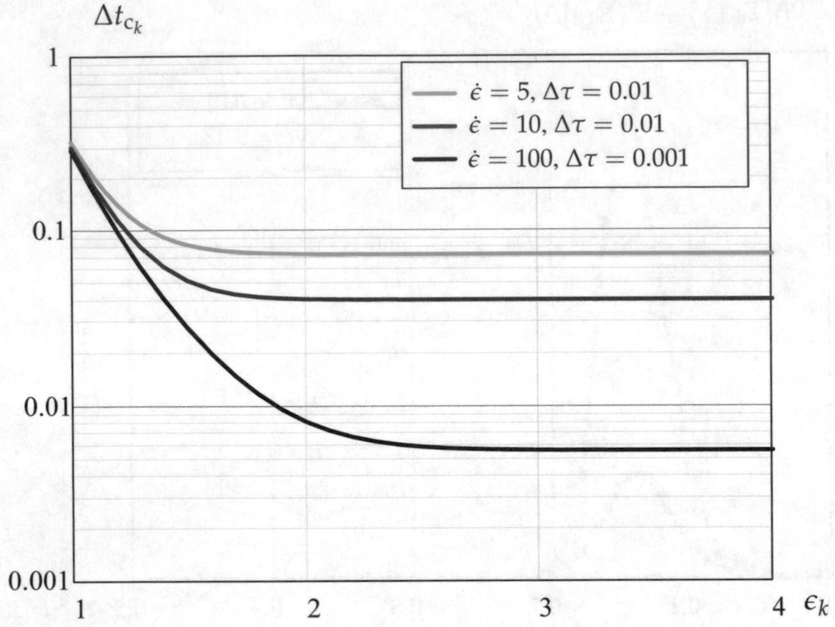

Fig. 1.31 Critical time step size estimated by Eq. (1.4.115) vs normalized strain for different strain rates

Eq. (1.4.115) as follows

$$\Delta\tau_{c_{\min}} = \frac{2}{\lambda_{\max}}, \quad \lambda_{\max} = \Psi'(\Sigma_{ss}), \quad \Sigma_{ss} = \Psi^{-1}(\dot{\epsilon}) \tag{1.4.118}$$

For the power law function $f(x)$ and $\Sigma > 0$ we can compute

$$\lambda_{\max} = n\Sigma_{ss}^{n-1}, \quad \Sigma_{ss} = \dot{\epsilon}^{\frac{1}{n}}$$

Thus the minimum time step size is

$$\tau_{c_{\min}} = \frac{2}{n}\dot{\epsilon}^{\frac{1-n}{n}} \tag{1.4.119}$$

Contrary to the linear ordinary differential equations discussed in Subsubsect. 1.3.3.1 the critical time step size is a function of the strain rate and the strain, Fig. 1.31. For the normalized strain rates of 5, 10 and 100 the minimum critical time step sizes computed from Eq (1.4.119) are 0.072, 0.04 and $5.52 \cdot 10^{-3}$ respectively. These values coincide with the values obtained by numerical analysis, Fig. 1.31. It is obvious that an increase of the strain rate leads to a decrease of the critical time step size. Furthermore, Eq. (1.4.119) shows that $\tau_{c_{\min}}$ depends on the exponent n. For the linear viscous flow $n = 1$ and $\tau_{c_{\min}} = 2$. This result is already discussed in Subsubsect. 1.3.3.1. On the other hand, for the case of rigid plasticity $n \to \infty$

and $\tau_{c_{\min}} \rightarrow 0$. Therefore, the explicit Euler method cannot be used for the structural analysis with rate-independent plasticity or visco-plasticity under high strain rate loadings. For creep, relaxation, monotonic and cyclic loadings with moderate and low strain rates the explicit Euler method is found to be efficient and applied in Altenbach and Naumenko (1997); Boyle and Spence (1983); Lin et al (1998); Zienkiewicz and Cormeau (1974), among many others. Let us note that the numerical stability procedure presented here for a single equation can be generalized to multi-axial loading situations and systems of ordinary differential equations providing theoretical estimates for critical time step sizes (Cormeau, 1975). Explicit Euler method with the time step size control based on inequalities (1.4.112) and (1.4.115) is available in commercial finite element codes, e.g. Abaqus User's Guide (2017) and can be used for inelastic structural analysis.

1.4.7.2 Implicit Euler Method

To approximate the integral in Eq. (1.4.103) the following rectangular rule is applied

$$\int_{\tau_k}^{\tau_{k+1}} \Psi(\Sigma)\mathrm{d}\tau \approx \Psi(\Sigma_{k+1})\Delta\tau_k, \quad \Delta\tau_k = \tau_{k+1} - \tau_k \tag{1.4.120}$$

Equation (1.4.103) takes the following form

$$\Sigma_{k+1} = \Sigma_k - \Psi(\Sigma_{k+1})\Delta\tau_k + \Delta\epsilon_k \tag{1.4.121}$$

The implicit (backward) Euler method is discussed in Subsubsect. 1.3.3.2 in application to linear ordinary differential equations. It is shown that an advantage of the method is the unconditional stability. Furthermore in Subsubsect. 1.3.3.4 we demonstrated that both explicit and implicit Euler methods provide first order accuracy. The local error can be estimated by the following inequality

$$\left| \int_{\tau_k}^{\tau_{k+1}} \Psi(\Sigma)\mathrm{d}\tau - \Psi(\Sigma_k)\Delta\tau_k \right| \leq R, \quad R = \frac{\Delta\tau_k^2}{2} \sup_{\Sigma_{k+1} \leq \Sigma \leq \Sigma_{k+1}} \left| \frac{\mathrm{d}\Psi}{\mathrm{d}\tau} \right| \tag{1.4.122}$$

With the given error tolerance $\mathrm{Errtol}/2$ and inequality (1.4.122) the following inequality can be derived (see Subsubsect. 1.4.7.2 for details)

$$|\Psi(\Sigma_{k+1}) - \Psi(\Sigma_k)|\Delta\tau_k \leq \mathrm{Errtol} \tag{1.4.123}$$

Similar to the explicit Euler method the inequality (1.4.123) can be used for the time step control. However, to compute Σ_{k+1} we have to solve non-linear Eq. (1.4.121). With $\Sigma_{k+1} = y$, Eq. (1.4.121) reads

$$y = \Sigma_k - \Psi(y)\Delta\tau_k + \Delta\epsilon_k \tag{1.4.124}$$

Equation (1.4.124) can be solved in a closed analytical form for the case of linear viscous creep, Subsubsect. 1.3.3.2, and rate-independent plasticity (Simo and Hughes, 2000). In the general case, non-linear Eq. (1.4.124) must be solved numerically applying known iteration methods. The simplest approach is the fixed-point iteration with the following scheme

$$y^{m+1} = \Sigma_k - \Psi(y^m)\Delta\tau_k + \Delta\epsilon_k, \tag{1.4.125}$$

where m is the iteration number. As the initial guess one may take the stress in the previous time step, i.e. $y^0 = \Sigma_k$. For small time step sizes, few iterations are required to obtain Σ_{k+1} with desired accuracy. Another popular approach is to apply the Newton-Raphson iteration scheme. To this end the function $\Psi(y)$ is linearized about the value y^m as follows

$$\Psi(y) \approx \Psi(y^m) + \Psi'(y^m)(y - y^m), \quad \Psi' = \frac{\mathrm{d}\Psi}{\mathrm{d}y} \tag{1.4.126}$$

Equation (1.4.124) takes the form

$$y = \Sigma_k - \Psi'(y^m)(y - y^m)\Delta\tau_k - \Psi(y^m)\Delta\tau_k + \Delta\epsilon_k \tag{1.4.127}$$

Now for the $m+1$-th iteration we have

$$y^{m+1}[1 + \Psi'(y^m)\Delta\tau_k] = \Sigma_k + \Psi'(y^m)y^m\Delta\tau_k - \Psi(y^m)\Delta\tau_k + \Delta\epsilon_k \tag{1.4.128}$$

Starting with the initial guess $y^0 = \Sigma_k$, Eq. (1.4.128) can be used for the iteration process. Finally we set $\Sigma_{k+1} = y^l$, where l is the number of iterations. Usually the root of the non-linear equation (1.4.124) is close to the initial guess y_k because the time step size $\Delta\tau_k$ is sufficiently small to guarantee the accuracy of the Euler method. Therefore with an appropriate time step control only one iteration can be enough. In this case Eq. (1.4.128) can be simplified as follows

$$\Sigma_{k+1}[1 + \Psi'(\Sigma_k)\Delta\tau_k] = \Sigma_k[1 + \Psi'(\Sigma_k)\Delta\tau_k] - \Psi(\Sigma_k)\Delta\tau_k + \Delta\epsilon_k \tag{1.4.129}$$

The solution provides Σ_{k+1}

$$\Sigma_{k+1} = \Sigma_k + [1 + \Psi'(\Sigma_k)\Delta\tau_k]^{-1}[-\Psi(\Sigma_k)\Delta\tau_k + \Delta\epsilon_k] \tag{1.4.130}$$

With

$$\Delta\epsilon_k^{\mathrm{pl}} = \Psi(\Sigma_k)\Delta\tau_k, \quad C_k = [1 + \Psi'(\Sigma_k)\Delta\tau_k]^{-1} \tag{1.4.131}$$

we can rewrite Eq. (1.4.130) as follows

$$\Delta\Sigma_k = C_k(\Delta\epsilon_k - \Delta\epsilon_k^{\mathrm{pl}}) \tag{1.4.132}$$

Equation (1.4.132) relates the stress increment to the elastic strain increment. Therefore C_k plays the role of the tangent stiffness.

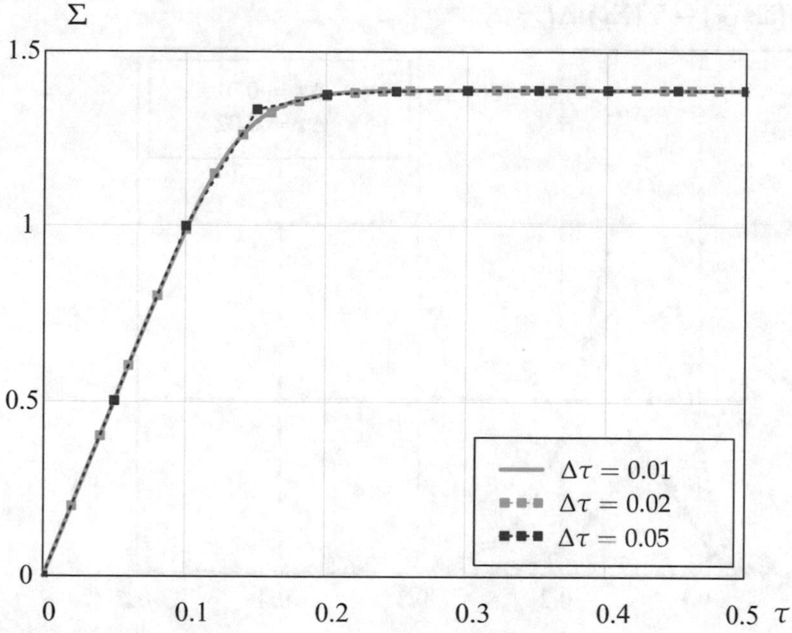

Fig. 1.32 Numerical solutions of Eq. (1.4.102) by implicit Euler method with different time step sizes for $\dot{c} = 10$ and power law stress function with $n = 7$

Figure 1.32 shows numerical solutions of Eq. (1.4.102) with constant time step sizes for $\dot{c} = 10$ and power law stress function with $n = 7$. The numerical scheme is according to Eqs (1.4.131) and (1.4.132). It is based on the implicit Euler method and the Newton-Raphson scheme with only one iteration. Figure 1.33 shows the numerical accuracy estimated by the left-hand side of inequality (1.4.123). By comparison to the results presented in Fig. 1.30 we may conclude that both explicit and implicit Euler methods provide the same order of accuracy for the time step sizes $\Delta t = 0.01$ and $\Delta t = 0.02$. For the time step size $\Delta t = 0.05$ non-stable oscillatory solution was obtained by the explicit Euler method, Fig. 1.29 while the solution based on the implicit Euler method is monotonic. Therefore one may use larger time step sizes which is an advantage of implicit methods for problems with many degrees of freedom. Furthermore, stability controls are not required.

The presented approach can be generalized to the multi-axial stress states and applied to the structural analysis. The corresponding formulations will be discussed in Subsect. 2.2.1. Implicit one-step methods together with Newton-Raphson iteration method with one iteration were originally proposed in Peirce et al (1984) for numerical solutions of visco-plasticity problems.

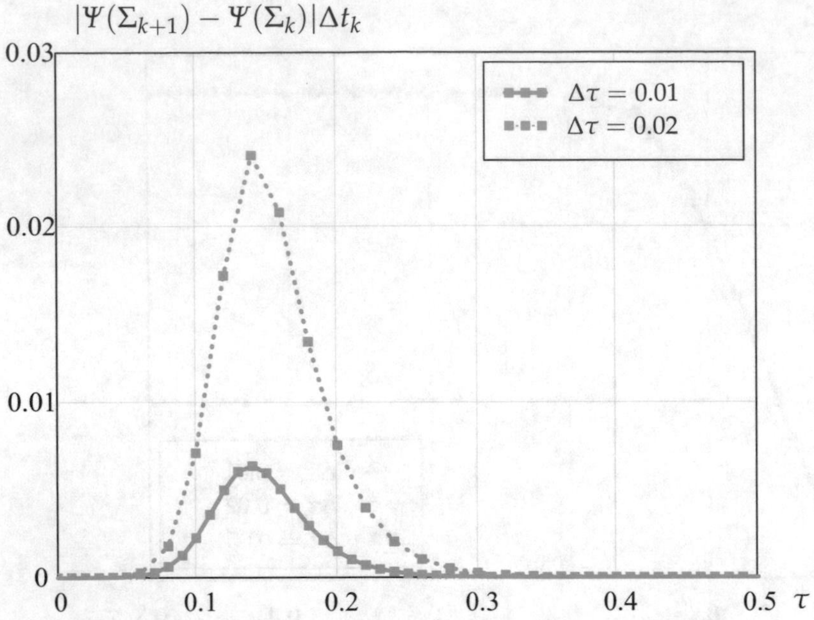

Fig. 1.33 Local error of implicit Euler by solving Eq. (1.4.102) with different time step sizes for $\dot{\epsilon} = 10$ and power law stress function with $n = 7$

References

Abaqus User's Guide (2017) Abaqus Analysis User's Guide. Volume III: Materials

Altenbach H, Naumenko K (1997) Creep bending of thin-walled shells and plates by consideration of finite deflections. Computational Mechanics 19:490 – 495

Altenbach H, Naumenko K, Pylypenko S, Renner B (2007) Influence of rotary inertia on the fiber dynamics in homogeneous creeping flows. ZAMM-Journal of Applied Mathematics and Mechanics/Zeitschrift für Angewandte Mathematik und Mechanik 87(2):81 – 93

Altenbach H, Naumenko K, Gorash Y (2008) Creep analysis for a wide stress range based on stress relaxation experiments. International Journal of Modern Physics B 22:5413 – 5418

Belytschko T, Liu WK, Moran B, Elkhodary K (2014) Nonlinear Finite Elements for Continua and Structures. Wiley

Blum W (2008) Mechanisms of creep deformation in steel. In: Abe F, Kern TU, Viswanathan R (eds) Creep-Resistant Steels, Woodhead Publishing, Cambridge, pp 365 – 402

Boyle JT (2012) The creep behavior of simple structures with a stress range-dependent constitutive model. Archive of Applied Mechanics 82(4):495 – 514

Boyle JT, Spence J (1983) Stress Analysis for Creep. Butterworth, London

Butcher JC (2016) Numerical Methods for Ordinary Differential Equations. John Wiley & Sons

Chowdhury H, Naumenko K, Altenbach H, Krueger M (2017) Rate dependent tension-compression-asymmetry of Ti-61.8 at% Al alloy with long period superstructures at 1050°C. Materials Science and Engineering: A 700:503–511

Chowdhury H, Naumenko K, Altenbach H (2018) Aspects of power law flow rules in crystal plasticity with glide-climb driven hardening and recovery. International Journal of Mechanical Sciences 146-147:486 – 496

Cormeau I (1975) Numerical stability in quasi-static elasto/visco-plasticity. International Journal for Numerical Methods in Engineering 9(1):109–127
Dyson BF, McLean M (2001) Micromechanism-quantification for creep constitutive equations. In: Murakami S, Ohno N (eds) IUTAM Symposium on Creep in Structures, Kluwer, Dordrecht, pp 3 – 16
Eisenträger J, Naumenko K, Altenbach H, Gariboldi E (2017) Analysis of temperature and strain rate dependencies of softening regime for tempered martensitic steel. The Journal of Strain Analysis for Engineering Design 52(4):226–238
Eisenträger J, Naumenko K, Altenbach H (2018) Calibration of a phase mixture model for hardening and softening regimes in tempered martensitic steel over wide stress and temperature ranges. The Journal of Strain Analysis for Engineering Design 53(3):156–177
Frost HJ, Ashby MF (1982) Deformation-Mechanism Maps. Pergamon, Oxford
Gariboldi E, Naumenko K, Ozhoga-Maslovskaja O, Zappa E (2016) Analysis of anisotropic damage in forged Al-Cu-Mg-Si alloy based on creep tests, micrographs of fractured specimen and digital image correlations. Materials Science and Engineering: A 652:175 – 185
Hairer E, Wanner G (1996) Solving Ordinary Differential Equations. Stiff and Differential-Algebraic Problems II, vol. 14 of Springer Series in Computational Mathematics. Springer
Hairer E, Nørsett S, Wanner G (2008) Solving Ordinary Differential Equations I: Nonstiff Problems. Springer Series in Computational Mathematics, Springer
Hosseini E, Holdsworth SR, Mazza E (2013) Stress regime-dependent creep constitutive model considerations in finite element continuum damage mechanics. International Journal of Damage Mechanics 22(8):1186 – 1205
Ilschner B (1973) Hochtemperatur-Plastizität. Springer, Berlin et al.
Kalyanasundaram V, Holdsworth SR (2016) Prediction of forward creep behaviour from stress relaxation data for a 10% Cr steel at 600 C. Transactions of the Indian Institute of Metals 69(2):573–578
Kassner ME, Pérez-Prado MT (2004) Fundamentals of Creep in Metals and Alloys. Elsevier, Amsterdam
Kloc L, Sklenička V (1997) Transition from power-law to viscous creep beahviour of P 91 type heat-resistant steel. Materials Science and Engineering A234-A236:962 – 965
Kostenko Y, Naumenko K (2017) Prediction of stress relaxation in power plant components based on a constitutive model. In: ASME Turbo Expo 2017: Turbomachinery Technical Conference and Exposition, American Society of Mechanical Engineers, p V008T29A017
Kowalewski ZL, Hayhurst DR, Dyson BF (1994) Mechanisms-based creep constitutive equations for an aluminium alloy. The Journal of Strain Analysis for Engineering Design 29(4):309 – 316
Längler F, Naumenko K, Altenbach H, Ievdokymov M (2014) A constitutive model for inelastic behavior of casting materials under thermo-mechanical loading. The Journal of Strain Analysis for Engineering Design 49:421 – 428
Lin J, Dunne F, Hayhurst D (1998) Approximate method for the analysis of components undergoing ratchetting and failure. The Journal of Strain Analysis for Engineering Design 33(1):55–65
Meier M, Blum W (1993) Modelling high temperature creep of academic and industrial materials using the composite model. Materials Science and Engineering: A 164(1-2):290–294
Miller AK (ed) (1987) Unified Constitutive Equations for Creep and Plasticity. Elsevier, London, New York
Naumenko K, Altenbach H (2005) A phenomenological model for anisotropic creep in a multi-pass weld metal. Archive of Applied Mechanics 74:808 – 819
Naumenko K, Altenbach H (2016) Modeling High Temperature Materials Behavior for Structural Analysis: Part I: Continuum Mechanics Foundations and Constitutive Models, Advanced Structured Materials, vol 28. Springer
Naumenko K, Gariboldi E (2014) A phase mixture model for anisotropic creep of forged Al-Cu-Mg-Si alloy. Materials Science and Engineering: A 618:368 – 376
Naumenko K, Kostenko Y (2009) Structural analysis of a power plant component using a stress-range-dependent creep-damage constitutive model. Materials Science and Engineering A510-A511:169 – 174

Naumenko K, Altenbach H, Gorash Y (2009) Creep analysis with a stress range dependent constitutive model. Archive of Applied Mechanics 79:619 – 630

Naumenko K, Altenbach H, Kutschke A (2011a) A combined model for hardening, softening and damage processes in advanced heat resistant steels at elevated temperature. International Journal of Damage Mechanics 20:578 – 597

Naumenko K, Kutschke A, Kostenko Y, Rudolf T (2011b) Multi-axial thermo-mechanical analysis of power plant components from 9-12%Cr steels at high temperature. Engineering Fracture Mechanics 78:1657 – 1668

Peirce D, Shih CF, Needleman A (1984) A tangent modulus method for rate dependent solids. Computers & Structures 18:875 – 887

Perrin IJ, Hayhurst DR (1994) Creep constitutive equations for a 0.5Cr-0.5Mo-0.25V ferritic steel in the temperature range 600-675°C. The Journal of Strain Analysis for Engineering Design 31(4):299 – 314

Rieth M (2007) A comprising steady-state creep model for the austenitic AISI 316 L(N) steel. Journal of Nuclear Materials 360-370(2):915 – 919

Roesler J, Harders H, Baeker M (2007) Mechanical Behaviour of Engineering Materials: Metals, Ceramics, Polymers, and Composites. Springer

Schmicker D, Naumenko K, Strackeljan J (2013) A robust simulation of Direct Drive Friction Welding with a modified Carreau fluid constitutive model. Computer Methods in Applied Mechanics and Engineering 265:186 – 194

Simo J, Hughes T (2000) Computational Inelasticity. Interdisciplinary Applied Mathematics, Springer

Skelton RP (2003) Creep-fatigue interactions (crack initiation). In: Karihaloo I, Milne R, Ritchie B (eds) Comprehensive Structural Integrity, Pergamon, Oxford, pp 25 – 112

Wang Y, Spindler M, Truman C, Smith D (2016) Critical analysis of the prediction of stress relaxation from forward creep of type 316H austenitic stainless steel. Materials & Design 95:656 – 668

Zienkiewicz O, Cormeau I (1974) Visco-plasticity-plasticity and creep in elastic solids-a unified numerical solution approach. International Journal for Numerical Methods in Engineering 8(4):821–845

Chapter 2
Initial-boundary Value Problems and Solution Procedures

The objective of Chapt. 2 is to introduce the governing mechanical equations to describe inelastic behavior in three-dimensional solids and to discuss numerical solution procedures. The set of equations includes material independent equations, constitutive and evolution equations, as well as the initial and boundary conditions. The formulated initial-boundary value problem (IBVP) can be solved by numerical methods. Explicit and implicit time integration methods were introduced in Chapt. 1 for bars. In Chapt. 2 they are generalized to analyze three-dimensional solids. Applying time-step procedures, linearized boundary value problems should be solved within time and/or iteration steps. The attention will be given to the variational formulations and the use of direct variational methods.

2.1 Governing Equations for Structural Analysis

2.1.1 Preliminary Remarks and Assumptions

Let us consider a solid occupying the volume V with the surface A. We assume that the solid is fixed on the surface part A_u and loaded by surface forces on the part A_f, as illustrated in Fig. 2.1. Let $\boldsymbol{\mathcal{R}}$ be the position vector for a point P in the reference state of the solid, Fig. 2.1, and $\boldsymbol{r}$ be the position vector of this point (designated by P') in the actual configuration. The displacement vector $\boldsymbol{u}$ connects the points P and P', Fig. 2.1. The position vector $\boldsymbol{\mathcal{R}}$ can be parameterized with the Cartesian coordinate system including the orthonormal basis $\boldsymbol{i}, \boldsymbol{j}, \boldsymbol{k}$ and the coordinates X, Y, Z, i.e.

$$\boldsymbol{\mathcal{R}}(X,Y,Z) = X\boldsymbol{i} + Y\boldsymbol{j} + Z\boldsymbol{k}$$

For specific solids it may be more convenient to use curvilinear coordinates, for example cylindrical, spherical, skew etc. Specifying the curvilinear coordinates by $X^1 = q^1, X^2 = q^2, X^3 = q^3$, see Naumenko and Altenbach (2016, Sect. 4.1), the position vector is parameterized as follows

K. Naumenko and H. Altenbach, *Modeling High Temperature Materials Behavior for Structural Analysis*, Advanced Structured Materials 112,
https://doi.org/10.1007/978-3-030-20381-8_2

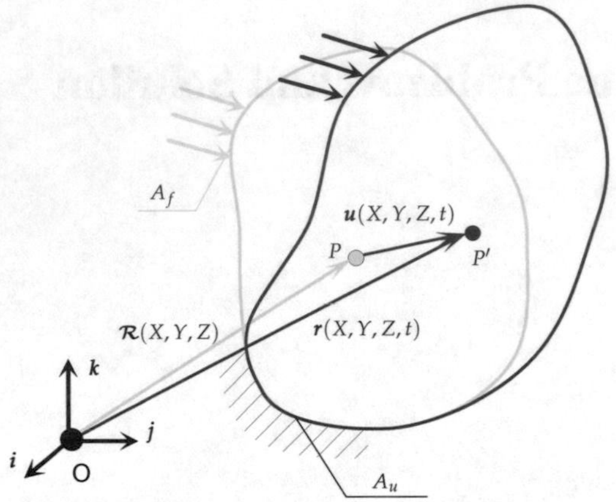

Fig. 2.1 Three-dimensional solid with specified external loads and constraints

$$\mathcal{R}(X^1, X^2, X^3) = X(X^1, X^2, X^3)\boldsymbol{i} + Y(X^1, X^2, X^3)\boldsymbol{j} + Z(X^1, X^2, X^3)\boldsymbol{k} \quad (2.1.1)$$

The motion of the continuum is defined by the following mapping

$$\boldsymbol{r} = \boldsymbol{\Phi}(\mathcal{R}, t) \quad (2.1.2)$$

The displacement vector $\boldsymbol{u}$ is defined as follows (Fig. 2.1)

$$\boldsymbol{u} = \boldsymbol{r} - \mathcal{R} \quad (2.1.3)$$

The vector $\boldsymbol{r}$ can be specified with the basis $\boldsymbol{i}, \boldsymbol{j}, \boldsymbol{k}$

$$\boldsymbol{r}(X^1, X^2, X^3, t) = x(X^1, X^2, X^3, t)\boldsymbol{i} + y(X^1, X^2, X^3, t)\boldsymbol{j} + z(X^1, X^2, X^3, t)\boldsymbol{k},$$

where x, y, z are the actual Cartesian coordinates. The basic problem of continuum mechanics is the identification of the function $\boldsymbol{\Phi}$ for all vectors $\mathcal{R}$ within the body, for the given time interval $t_0 \leq t \leq t_n$ as well as for given external loads.

The governing equations of continuum mechanics are discussed in e.g. Altenbach (2018); Backhaus (1983); Bertram (2012); Betten (2001); Eringen (1999); Haupt (2002); Lai et al (1993); Lurie (1990, 2005); Maugin (2013). Constitutive equations describing inelastic material processes have been introduced in Naumenko and Altenbach (2016, Chapt. 5).

To give a short summary of the governing equations and to illustrate the solution procedures, a linearized theory in the sense of infinitesimal strains and displacements will be introduced. Constitutive equations for inelastic material behavior usually the inelastic strain tensor $\boldsymbol{\varepsilon}^{\mathrm{pl}}(X^i, t)$ and a set of internal state variables, for example hardening, softening and damage variables $\zeta_k(X^i, t), k = 1, \ldots, n$. They are introduced to capture the current state of the material microstructure and to reflect the history of the inelastic process.

2.1.2 Summary of Governing Equations

By analogy to the uni-axial case, Sect. 1.1, the governing equations can be summarized as follows

- kinematical equations
  - strain-displacement relation

$$\boldsymbol{\varepsilon} = \frac{1}{2}\left[\nabla \boldsymbol{u} + (\nabla \boldsymbol{u})^{\mathrm{T}}\right], \quad q^i \in V, \tag{2.1.4}$$

  where $\boldsymbol{\varepsilon}$ is the tensor of infinitesimal strains.
  - compatibility condition

$$\nabla \times (\nabla \times \boldsymbol{\varepsilon})^{\mathrm{T}} = \boldsymbol{0}, \quad q^i \in V, \tag{2.1.5}$$

- equilibrium conditions

$$\nabla \cdot \boldsymbol{\sigma} + \rho \bar{\boldsymbol{f}} = \boldsymbol{0}, \quad \boldsymbol{\sigma} = \boldsymbol{\sigma}^{\mathrm{T}}, \quad q^i \in V, \tag{2.1.6}$$

  where ρ is the material density and $\bar{\boldsymbol{f}}$ is the density of volumetric forces
- boundary conditions

$$\begin{aligned} \boldsymbol{u} &= \bar{\boldsymbol{u}}, & q^i &\in A_u, \\ \boldsymbol{\sigma} \cdot \boldsymbol{\nu} &= \bar{\boldsymbol{p}}, & q^i &\in A_p, \end{aligned} \tag{2.1.7}$$

where $\bar{\boldsymbol{u}}$ is the given displacement vector, $\bar{\boldsymbol{p}}$ is the vector of given surface forces and $\boldsymbol{\nu}$ is the outward unit normal to A_p. The vectors $\bar{\boldsymbol{f}}$, $\bar{\boldsymbol{p}}$ and $\bar{\boldsymbol{u}}$ can, in general, be functions of coordinates and time.

With the assumption of infinitesimal strains, the linear strain tensor can be additively decomposed into elastic, thermal and creep parts

$$\boldsymbol{\varepsilon} = \boldsymbol{\varepsilon}^{\mathrm{el}} + \boldsymbol{\varepsilon}^{\mathrm{th}} + \boldsymbol{\varepsilon}^{\mathrm{pl}} \tag{2.1.8}$$

The constitutive equation for the stress tensor can be assumed in the form of the generalized Hooke's law as follows

$$\boldsymbol{\sigma} = {}^{(4)}\boldsymbol{C} \cdot\cdot (\boldsymbol{\varepsilon} - \boldsymbol{\varepsilon}^{\mathrm{th}} - \boldsymbol{\varepsilon}^{\mathrm{pl}}) \tag{2.1.9}$$

In the case of isotropic elasticity the tensor ${}^{(4)}\boldsymbol{C}$ takes the form

$${}^{(4)}\boldsymbol{C} = \lambda \boldsymbol{I} \otimes \boldsymbol{I} + \mu(\boldsymbol{e}_k \otimes \boldsymbol{I} \otimes \boldsymbol{e}^k + \boldsymbol{e}_i \otimes \boldsymbol{e}_k \otimes \boldsymbol{e}^i \otimes \boldsymbol{e}^k), \tag{2.1.10}$$

where λ and μ are the Lamé's parameters

$$\mu = G = \frac{E}{2(1+\nu)}, \quad \lambda = \frac{\nu E}{(1+\nu)(1-2\nu)}$$

E is the Young's modulus, G is the shear modulus and ν is the Poisson's ratio.

When an isotropic solid is heated up or cooled down from the reference temperature T_0 to T, the thermal part of the strain tensor is

$$\boldsymbol{\varepsilon}^{\text{th}} = \alpha_T \Theta \boldsymbol{I}, \quad \Theta \equiv T - T_0, \tag{2.1.11}$$

where α_T is the coefficient of thermal expansion. The temperature difference Θ can be a function of coordinates and time, too.

The constitutive equation for the inelastic strain rate and evolution equations for internal state variables are discussed in Naumenko and Altenbach (2016, Chapt. 5). The rate equations can be given in the following form

$$\begin{aligned} \dot{\boldsymbol{\varepsilon}}^{\text{pl}} &= \boldsymbol{g}(\boldsymbol{\sigma}, \boldsymbol{\varepsilon}^{\text{pl}}, \boldsymbol{\varsigma}_k, \dot{\boldsymbol{\varsigma}}_k, T), \quad k = 1, \ldots, n, \\ \dot{\boldsymbol{\varsigma}}_k &= \boldsymbol{h}_k(\boldsymbol{\sigma}, \boldsymbol{\varepsilon}^{\text{pl}}, \dot{\boldsymbol{\varepsilon}}^{\text{pl}}, \boldsymbol{\varsigma}_l, T), \quad l = 1, \ldots, n \end{aligned} \tag{2.1.12}$$

The variables $\boldsymbol{\varsigma}_k$ include hardening, softening, damage or other internal state variables and can be scalars and/or tensors. For the inelastic strain as well as for the set of hardening and damage variables the initial conditions must be specified

$$\boldsymbol{\varepsilon}^{\text{pl}}\Big|_{t=0} = \boldsymbol{\varepsilon}^{\text{pl}^0}, \quad \boldsymbol{\varsigma}_k\Big|_{t=0} = \boldsymbol{\varsigma}_k^0 \tag{2.1.13}$$

where $\boldsymbol{\varepsilon}^{\text{pl}^0}$ and $\boldsymbol{\varsigma}_k^0$ are initial values of the inelastic strain tensor and internal state variables.

Assuming isotropic material behavior, Eqs (2.1.9) and (2.1.11) yield

$$\boldsymbol{\sigma} = \lambda(\operatorname{tr}\boldsymbol{\varepsilon})\boldsymbol{I} + 2\mu(\boldsymbol{\varepsilon} - \boldsymbol{\varepsilon}^{\text{pl}}) - (3\lambda + 2\mu)\alpha_{\text{th}}\Theta\boldsymbol{I}, \tag{2.1.14}$$

With the decomposition of the stress tensor into the spherical and deviatoric parts

$$\boldsymbol{\sigma} = \sigma_{\text{m}}\boldsymbol{I} + \boldsymbol{s}, \quad \sigma_{\text{m}} = \frac{1}{3}\operatorname{tr}\boldsymbol{\sigma},$$

where σ_{m} is the mean stress and $\boldsymbol{s}$ is the stress deviator, Eq. (2.1.14) yields

$$\sigma_{\text{m}} = K\left(\varepsilon_V - \varepsilon_V^{\text{pl}} - 3\alpha_{\text{th}}\Theta\right), \quad \boldsymbol{s} = 2\mu(\boldsymbol{\epsilon} - \boldsymbol{\epsilon}^{\text{pl}}), \tag{2.1.15}$$

where

$$K = \frac{3\lambda + 2\mu}{3} = \frac{E}{3(1 - 2\nu)}$$

is the bulk modulus, $\varepsilon_V = \operatorname{tr}\boldsymbol{\varepsilon}$, $\varepsilon_V^{\text{pl}} = \operatorname{tr}\boldsymbol{\varepsilon}^{\text{pl}}$ are the volumetric strains and

$$\boldsymbol{\epsilon} = \operatorname{dev}\boldsymbol{\varepsilon}, \quad \boldsymbol{\epsilon}^{\text{pl}} = \operatorname{dev}\boldsymbol{\varepsilon}^{\text{pl}} \tag{2.1.16}$$

are strain deviators.

By analogy to the uni-axial case of the bars discussed in Chapt. 1, let us derive the governing equations for the stress redistribution. For the sake of brevity let us assume isothermal loading. The time derivative of Eqs (2.1.15) yields

$$\dot{\sigma}_{\mathrm{m}} = K\left(\dot{\varepsilon}_V - \dot{\varepsilon}_V^{\mathrm{pl}}\right), \quad \dot{\boldsymbol{s}} = 2\mu(\dot{\boldsymbol{\epsilon}} - \dot{\boldsymbol{\epsilon}}^{\mathrm{pl}}), \tag{2.1.17}$$

With the constitutive model for inelastic flow (2.1.12) we obtain

$$\begin{aligned} \dot{\sigma}_{\mathrm{m}} + K \operatorname{tr} \boldsymbol{g}(\sigma_{\mathrm{m}}, \boldsymbol{s}, \boldsymbol{\varepsilon}^{\mathrm{pl}}, \varsigma_k, \dot{\varsigma}_k, T) &= K\dot{\varepsilon}_V, \\ \dot{\boldsymbol{s}} + 2\mu \operatorname{dev} \boldsymbol{g}(\sigma_{\mathrm{m}}, \boldsymbol{s}, \boldsymbol{\varepsilon}^{\mathrm{pl}}, \varsigma_k, \dot{\varsigma}_k, T) &= 2\mu\dot{\boldsymbol{\epsilon}} \end{aligned} \tag{2.1.18}$$

Equations (2.1.18) must be supplemented by evolution equations for internal state variables. The stress tensor must satisfy the equilibrium conditions (2.1.6) while the strain tensor, the compatibility conditions (2.1.5). For many materials the inelastic flow does not produce considerable change in volume. In this case $\operatorname{tr} \boldsymbol{g}(\sigma_{\mathrm{m}}, \boldsymbol{s}, \boldsymbol{\varepsilon}^{\mathrm{pl}}, \varsigma_k, \dot{\varsigma}_k, T) = 0$ and Eq. $(2.1.18)_1$ reduces to the linear-elastic relation between the mean stress and the volumetric strain.

As an example let us consider the Frederick-Armstrong type model (Frederick and Armstrong, 2007) for inelastic flow which is widely used in the inelastic structural analysis. This model includes a constitutive equation for the inelastic strain rate tensor and a kinetic equation for the backstress tensor to capture kinematic hardening. The constitutive equation and the non-linear kinematic hardening rule can be derived within the framework of continuum thermodynamics of dissipative processes (Chaboche, 2008; Naumenko and Altenbach, 2016). Alternatively, a phase mixture (composite or fraction) model of inelastic deformation can be applied to introduce the backstress tensors and to formulate the non linear kinematic hardening rule. Indeed, many high-temperature materials can be considered as composites with constituents having different inelastic properties. Differences in inelastic strain rates can be significant, for example due to different initial dislocation densities in subgrain interiors and subgrain boundaries (Blum, 2008), elongated grains in forged aluminium alloys (Gariboldi et al, 2016) or different grain sizes in different zones of welds (Naumenko and Altenbach, 2005), or lamellar morphology of titanium aluminide based alloys (Chowdhury et al, 2017). The basic idea is to idealize the heterogeneous inelastic deformation at the microscale by considering a mixture with two or more constituents with different inelastic properties (Besseling, 1958; Besseling and van der Giessen, 1994). Assuming the total deformation of constituents to be the same, redistribution of stresses would take place, leading to the decrease of the overall inelastic strain rate. Composite models with only two constituents (inelastic-hard and inelastic-soft constituents) and with constant volume fraction can be reduced to the Frederick-Armstrong type model with one backstress tensor. If the volume fraction of the hard constituent is assumed to decrease during inelastic deformation, softening processes can be taken into account (Naumenko and Altenbach, 2016; Naumenko et al, 2011b). The material parameters in Frederick-Armstrong type constitutive models are available for many materials, for example, tempered martensite steel (Zhang and Xuan, 2017), cast iron (Längler et al, 2014) and aluminum alloys (Gariboldi et al, 2016; Naumenko and Gariboldi,

2014). The constitutive equation for the inelastic strain rate tensor can be formulated as follows (Naumenko and Altenbach, 2016, Chapt. 5)

$$\dot{\boldsymbol{\varepsilon}}^{\text{pl}} = \frac{3}{2}\frac{\dot{\varepsilon}_{\text{vM}}}{\overline{\sigma}_{\text{vM}}}\overline{\boldsymbol{s}}, \tag{2.1.19}$$

where $\overline{\boldsymbol{s}} = \boldsymbol{s} - \boldsymbol{\beta}$ is the active stress deviator, $\boldsymbol{\beta}$ is the backstress deviator and

$$\overline{\sigma}_{\text{vM}} = \sqrt{\frac{3}{2}\text{tr}\left(\overline{\boldsymbol{s}}^2\right)} \tag{2.1.20}$$

is the von Mises equivalent stress related to the active stress deviator. Equation (2.1.19) is applicable for a class of isotropic materials that show inelastic incompressibility. The von Mises equivalent inelastic strain rate

$$\dot{\varepsilon}_{\text{vM}} = \sqrt{\frac{2}{3}\text{tr}\left(\dot{\boldsymbol{\varepsilon}}^{\text{pl}^2}\right)} \tag{2.1.21}$$

is related to the active von Mises equivalent stress, the absolute temperature and a set of internal state variables that reflect the actual state of the microstructure. Examples include isotropic hardening, softening, ageing and damage variables (Chowdhury et al, 2018; Naumenko et al, 2011a; Lemaitre and Chaboche, 1990). Below we limit our analysis to kinematic hardening. In this case the von Mises equivalent inelastic strain rate is defined as follows

$$\dot{\varepsilon}_{\text{vM}} = R(T)f\left(\overline{\sigma}_{\text{vM}}\right), \tag{2.1.22}$$

where $R(T)$ is a function of the absolute temperature and $f\left(\overline{\sigma}_{\text{vM}}\right)$ is a function of the active stress. These functions can be identified from inelastic material response under monotonic loading.

For isothermal loading conditions, the evolution equation for the backstress deviator can be formulated as follows

$$\dot{\boldsymbol{\beta}} = \frac{2}{3}C_h(T)\left[\dot{\boldsymbol{\varepsilon}}^{\text{pl}} - \frac{3}{2}\dot{\varepsilon}_{\text{vM}}\frac{\boldsymbol{\beta}}{h\left(\sigma_{\text{vM}}\right)}\right], \tag{2.1.23}$$

where $C_h(T)$ is a function of temperature and $h\left(\sigma_{\text{vM}}\right)$ is a function of the von Mises equivalent stress

$$\sigma_{\text{vM}} = \sqrt{\frac{3}{2}\text{tr}\ \left(\boldsymbol{s}^2\right)} \tag{2.1.24}$$

Equations (2.1.18) and (2.1.23) can be put in the following form

$$\sigma_{\mathrm{m}} = K\varepsilon_V,$$

$$\dot{\boldsymbol{s}} = 2\mu\dot{\boldsymbol{e}} - 2\mu\frac{3}{2}R(T)f(\overline{\sigma}_{\mathrm{vM}})\frac{\overline{\boldsymbol{s}}}{\overline{\sigma}_{\mathrm{vM}}}, \tag{2.1.25}$$

$$\dot{\boldsymbol{\beta}} = C_h(T)R(T)f(\overline{\sigma}_{\mathrm{vM}})\left[\frac{\overline{\boldsymbol{s}}}{\overline{\sigma}_{\mathrm{vM}}} - \frac{\boldsymbol{\beta}}{h(\overline{\sigma}_{\mathrm{vM}})}\right]$$

2.1.3 Steady-state Creep and Elastic Analogy

When external loads and the temperature are stationary and the material exhibits stable inelastic flow,[1] a steady creep state for a solid exists for which stresses do not change in time. Let us demonstrate the existence of a steady creep state for the case of Frederick-Armstrong constitutive model. Setting the rate of the backstress deviator in Eq. (2.1.25)$_3$ to zero yields

$$\frac{\overline{\boldsymbol{s}}_{\mathrm{ss}}}{\overline{\sigma}_{\mathrm{vM_{ss}}}} = \frac{\boldsymbol{\beta}_{\mathrm{ss}}}{h_{\mathrm{ss}}}, \tag{2.1.26}$$

where

$$\overline{\boldsymbol{s}}_{\mathrm{ss}} = \boldsymbol{s}_{\mathrm{ss}} - \boldsymbol{\beta}_{\mathrm{ss}}, \quad h_{\mathrm{ss}} = h(\overline{\sigma}_{\mathrm{vM_{ss}}}),$$

with $\boldsymbol{s}_{\mathrm{ss}}$ and $\boldsymbol{\beta}_{\mathrm{ss}}$ being steady-state stress and backstress deviators, respectively. From Eq. (2.1.26) it follows

$$\frac{3}{2}\frac{\overline{\boldsymbol{s}}^2_{\mathrm{ss}}}{\overline{\sigma}^2_{\mathrm{vM_{ss}}}} = \frac{3}{2}\frac{\boldsymbol{\beta}^2_{\mathrm{ss}}}{h^2_{\mathrm{ss}}} \tag{2.1.27}$$

Taking the trace of Eq. (2.1.27) we obtain

$$h_{\mathrm{ss}} = \beta_{\mathrm{vM_{ss}}}, \quad \beta_{\mathrm{vM_{ss}}} = \sqrt{\frac{3}{2}\mathrm{tr}\left(\boldsymbol{\beta}^2_{\mathrm{ss}}\right)} \tag{2.1.28}$$

On the other hand Eq. (2.1.26) can be rewritten as follows

$$\frac{\boldsymbol{s}_{\mathrm{ss}} - \boldsymbol{\beta}_{\mathrm{ss}}}{\overline{\sigma}_{\mathrm{vM_{ss}}}} = \frac{\boldsymbol{\beta}_{\mathrm{ss}}}{h_{\mathrm{ss}}} \quad \Rightarrow \quad \frac{\boldsymbol{s}_{\mathrm{ss}}}{\overline{\sigma}_{\mathrm{vM_{ss}}}} = \boldsymbol{\beta}_{\mathrm{ss}}\left(\frac{1}{h_{\mathrm{ss}}} + \frac{1}{\overline{\sigma}_{\mathrm{vM_{ss}}}}\right) \tag{2.1.29}$$

With Eq. (2.1.28) we obtain

$$\frac{3}{2}\frac{\boldsymbol{s}^2_{\mathrm{ss}}}{\overline{\sigma}^2_{\mathrm{vM_{ss}}}} = \frac{3}{2}\boldsymbol{\beta}^2_{\mathrm{ss}}\left(\frac{1}{\beta_{\mathrm{vM_{ss}}}} + \frac{1}{\overline{\sigma}_{\mathrm{vM_{ss}}}}\right)^2 \tag{2.1.30}$$

[1] This can only be assumed by neglecting softening and damage processes as well as within the geometrically linear analysis.

Taking the trace of Eq. (2.1.30) we obtain

$$\frac{\sigma^2_{\mathrm{vM_{ss}}}}{\overline{\sigma}^2_{\mathrm{vM_{ss}}}} = \left(\frac{\beta_{\mathrm{vM_{ss}}} + \overline{\sigma}_{\mathrm{vM_{ss}}}}{\overline{\sigma}_{\mathrm{vM_{ss}}}}\right)^2 \quad \Rightarrow \quad \overline{\sigma}_{\mathrm{vM_{ss}}} = \sigma_{\mathrm{vM_{ss}}} - \beta_{\mathrm{vM_{ss}}} \tag{2.1.31}$$

With Eqs (2.1.29) and (2.1.31) we obtain the steady state values of the backstress and the active stress deviators as follows

$$\boldsymbol{\beta}_{\mathrm{ss}} = h_{\mathrm{ss}} \frac{\boldsymbol{s}_{\mathrm{ss}}}{\sigma_{\mathrm{vM_{ss}}}}, \quad \bar{\boldsymbol{s}}_{\mathrm{ss}} = (h_{\mathrm{ss}} - \sigma_{\mathrm{vM_{ss}}}) \frac{\boldsymbol{s}_{\mathrm{ss}}}{\sigma_{\mathrm{vM_{ss}}}} \tag{2.1.32}$$

To analyze the stress state let us set the rate of the stress deviator in Eq. $(2.1.25)_2$ to zero. As a result we obtain the following constitutive equation

$$\dot{\boldsymbol{\varepsilon}}_{\mathrm{ss}} = \frac{3}{2} R(T) f\left(\overline{\sigma}_{\mathrm{vM_{ss}}}\right) \frac{\boldsymbol{s}_{\mathrm{ss}}}{\sigma_{\mathrm{vM_{ss}}}}, \tag{2.1.33}$$

In what follows let us omit the index ss remembering that all quantities are defined in the steady-state. With the abbreviation

$$\tilde{f}(x) = f[h(x) - x]$$

Eq. (2.1.31) takes the following form

$$\dot{\boldsymbol{\varepsilon}} = \frac{3}{2} R(T) \tilde{f}\left(\sigma_{\mathrm{vM}}\right) \frac{\boldsymbol{s}}{\sigma_{\mathrm{vM}}} \tag{2.1.34}$$

Equation (2.1.34) is known as the von Mises-Odqvist flow rule for steady-state creep. It relates the total strain rate with the stress deviator and is non-linear except the case of viscous fluid with $\tilde{f}(x) = x$. Let $\boldsymbol{v} = \dot{\boldsymbol{u}}$ be the velocity vector. Since $\dot{\varepsilon}_V = 0$ it follows from Eq. (2.1.4)

$$\nabla \cdot \boldsymbol{v} = 0, \quad \dot{\boldsymbol{\varepsilon}} = \frac{1}{2}\left(\nabla \boldsymbol{v} + (\nabla \boldsymbol{v})^{\mathrm{T}}\right) \tag{2.1.35}$$

The mean stress in the steady state is not defined by the constitutive equation. It should be determined from the equilibrium condition (2.1.6). Indeed we can write

$$\nabla \sigma_{\mathrm{m}} + \nabla \cdot \boldsymbol{s} + \rho \bar{\boldsymbol{f}} = \boldsymbol{0} \tag{2.1.36}$$

Equations (2.1.34) – (2.1.36) can be considered as equations of non-linear incompressible elasticity if the velocity vector and the strain rate tensor are formally replaced by the displacement vector and strain tensor, respectively. This non-linear elastic analogy provides a way to find efficient solutions to steady-state flow problems by the use of solutions of the non-linear theory of elasticity (Antman, 1995; Lurie, 2005, 1990). Many closed-form solutions to steady-state creep problems are presented in Boyle and Spence (1983); Hyde et al (2013); Hult (1966); Odqvist and Hult (1962); Odqvist (1974).

2.1.4 Matrix Representation

To formulate initial-boundary value problems and numerical solution procedures let us rewrite Eqs (2.1.4) – (2.1.13) in the matrix notation. For the sake of brevity we introduce the Cartesian coordinates x_1, x_2, x_3. The Cartesian components of vectors and tensors can be collected into the following vectors and matrices:

Stress vector $\boldsymbol{\sigma}$	$\boldsymbol{\sigma}^{\mathrm{T}}$	$= [\sigma_{11}\ \sigma_{22}\ \sigma_{33}\ \sigma_{12}\ \sigma_{23}\ \sigma_{31}]$
Strain vector $\boldsymbol{\varepsilon}$	$\boldsymbol{\varepsilon}^{\mathrm{T}}$	$= [\varepsilon_{11}\ \varepsilon_{22}\ \varepsilon_{33}\ \gamma_{12}\ \gamma_{23}\ \gamma_{31}]$
Displacement vector $\boldsymbol{u}$	$\boldsymbol{u}^{\mathrm{T}}$	$= [u_1\ u_2\ u_3]$

Vector of creep strains $\boldsymbol{\varepsilon}^{\mathrm{pl}}$	$\boldsymbol{\varepsilon}^{\mathrm{pl}^{\mathrm{T}}}$	$= [\varepsilon_{11}^{\mathrm{pl}}\ \varepsilon_{22}^{\mathrm{pl}}\ \varepsilon_{33}^{\mathrm{pl}}\ \gamma_{12}^{\mathrm{pl}}\ \gamma_{23}^{\mathrm{pl}}\ \gamma_{31}^{\mathrm{pl}}]$
Vector of internal variables $\boldsymbol{\xi}$	$\boldsymbol{\xi}^{\mathrm{T}}$	$= [\varsigma_1\ \varsigma_2\ \dots\varsigma_n]$
Vector of thermal strains $\boldsymbol{\varepsilon}^{\mathrm{th}}$	$\boldsymbol{\varepsilon}^{\mathrm{th}^{\mathrm{T}}}$	$= [\alpha_T\Delta T\ \alpha_T\Delta T\ \alpha_T\Delta T\ 0\ 0\ 0]$

Vector of body forces $\bar{\boldsymbol{f}}$	$\bar{\boldsymbol{f}}^{\mathrm{T}}$	$= [\bar{f}_1\ \bar{f}_2\ \bar{f}_3]$
Vector of surface forces $\bar{\boldsymbol{p}}$	$\bar{\boldsymbol{p}}^{\mathrm{T}}$	$= [\bar{p}_1\ \bar{p}_2\ \bar{p}_3]$

Stress vector $\boldsymbol{\sigma}_{\boldsymbol{\nu}}$ on $\boldsymbol{\nu}\mathrm{d}A$	$\boldsymbol{\sigma}_{\boldsymbol{\nu}}^{\mathrm{T}}$	$= [\sigma_{\nu_1}\ \sigma_{\nu_2}\ \sigma_{\nu_3}]$
Normal vector $\boldsymbol{\nu}$	$\boldsymbol{\nu}^{\mathrm{T}}$	$= [\nu_1\ \nu_2\ \nu_3]$,
	ν_i	$= \cos(\boldsymbol{\nu}, x_i)$

$$\text{Transformation matrix} \quad \overset{u}{\boldsymbol{T}} \qquad \overset{u}{\boldsymbol{T}} = \begin{bmatrix} 1 & 0 & 0 \\ 0 & 1 & 0 \\ 0 & 0 & 1 \end{bmatrix}$$

$$\text{Transformation matrix} \quad \overset{\sigma}{\boldsymbol{T}} \qquad \overset{\sigma}{\boldsymbol{T}} = \begin{bmatrix} \nu_1 & 0 & 0 & \nu_2 & 0 & \nu_3 \\ 0 & \nu_2 & 0 & \nu_1 & \nu_3 & 0 \\ 0 & 0 & \nu_3 & 0 & \nu_2 & \nu_1 \end{bmatrix}$$

Differential matrix $\boldsymbol{D}$

$$\boldsymbol{D} = \begin{bmatrix} \partial_1 & 0 & 0 & \partial_2 & 0 & \partial_3 \\ 0 & \partial_2 & 0 & \partial_1 & \partial_3 & 0 \\ 0 & 0 & \partial_3 & 0 & \partial_2 & \partial_1 \end{bmatrix}$$

Differential matrix $\boldsymbol{D}_1$

$$\boldsymbol{D}_1 = \begin{bmatrix} 0 & \partial_3^2 & \partial_2^2 & 0 & -\partial_2\partial_3 & 0 \\ \partial_3^2 & 0 & \partial_1^2 & 0 & 0 & -\partial_1\partial_3 \\ \partial_2^2 & \partial_1^2 & 0 & -\partial_1\partial_2 & 0 & 0 \\ 0 & 0 & -\partial_1\partial_2 & -\frac{1}{2}\partial_3^2 & \frac{1}{2}\partial_1\partial_3 & \frac{1}{2}\partial_2\partial_3 \\ -\partial_2\partial_3 & 0 & 0 & \frac{1}{2}\partial_1\partial_3 & -\frac{1}{2}\partial_1^2 & \frac{1}{2}\partial_1\partial_2 \\ 0 & -\partial_1\partial_3 & 0 & \frac{1}{2}\partial_2\partial_3 & \frac{1}{2}\partial_1\partial_2 & -\frac{1}{2}\partial_2^2 \end{bmatrix}$$

with

$$\partial_i = \frac{\partial(\ldots)}{\partial x_i}, \quad \partial_i^2 = \frac{\partial^2(\ldots)}{\partial x_i^2}$$

Elasticity matrix (stiffness matrix) $\boldsymbol{E}$

$$\boldsymbol{E} = \begin{bmatrix} 2\mu+\lambda & \lambda & \lambda & 0 & 0 & 0 \\ & 2\mu+\lambda & \lambda & 0 & 0 & 0 \\ & & 2\mu+\lambda & 0 & 0 & 0 \\ & & & \mu & 0 & 0 \\ & & & & \mu & 0 \\ \text{SYM} & & & & & \mu \end{bmatrix}$$

Reciprocal elasticity matrix (compliance matrix) $\boldsymbol{E}^{-1}$

$$\boldsymbol{E}^{-1} = \frac{1}{E} \begin{bmatrix} 1 & -\nu & -\nu & 0 & 0 & 0 \\ & 1 & -\nu & 0 & 0 & 0 \\ & & 1 & 0 & 0 & 0 \\ & & & 2(1+\nu) & 0 & 0 \\ & & & & 2(1+\nu) & 0 \\ \text{SYM} & & & & & 2(1+\nu) \end{bmatrix}$$

With the introduced notations and $\boldsymbol{x}^{\mathrm{T}} = [x_1\ x_2\ x_3]$ we can rewrite the governing equations (2.1.4) – (2.1.13) as follows

Kinematical equations:
Strain-displacement relation

$$\boldsymbol{\varepsilon} = \boldsymbol{D}^{\mathrm{T}}\boldsymbol{u}, \quad \boldsymbol{x} \in V \tag{2.1.37}$$

Compatibility condition

$$\boldsymbol{D}_1\boldsymbol{\varepsilon} = \boldsymbol{0}, \quad \boldsymbol{x} \in V \tag{2.1.38}$$

Prescribed boundary displacements $\bar{\boldsymbol{u}}$ on A_u

$$\overset{u}{\boldsymbol{T}}\boldsymbol{u} = \bar{\boldsymbol{u}}, \quad \boldsymbol{x} \in A_u \tag{2.1.39}$$

Equilibrium conditions:

$$\boldsymbol{D}\boldsymbol{\sigma} + \bar{\boldsymbol{f}} = \boldsymbol{0}, \quad \boldsymbol{x} \in V \tag{2.1.40}$$

Prescribed surface forces $\bar{\boldsymbol{p}}$ on A_p

$$\overset{\sigma}{\boldsymbol{T}}\boldsymbol{\sigma} = \boldsymbol{\sigma}_v = \bar{\boldsymbol{p}}, \quad \boldsymbol{x} \in A_p \tag{2.1.41}$$

Constitutive and evolution equations:

$$\boldsymbol{\sigma} = \boldsymbol{E}(\boldsymbol{\varepsilon} - \boldsymbol{\varepsilon}^{\mathrm{th}} - \boldsymbol{\varepsilon}^{\mathrm{pl}}), \quad \boldsymbol{x} \in V \tag{2.1.42}$$

$$\begin{aligned} \dot{\boldsymbol{\varepsilon}}^{\mathrm{pl}} &= \boldsymbol{g}(\boldsymbol{\sigma}, \boldsymbol{\varepsilon}^{\mathrm{pl}}\boldsymbol{\xi}, \dot{\boldsymbol{\xi}}, T) \\ \dot{\boldsymbol{\xi}} &= \boldsymbol{h}(\boldsymbol{\sigma}, \boldsymbol{\varepsilon}^{\mathrm{pl}}, \dot{\boldsymbol{\varepsilon}}^{\mathrm{pl}}, \boldsymbol{\xi}, T) \end{aligned} \tag{2.1.43}$$

Initial conditions

$$\boldsymbol{\varepsilon}^{\mathrm{pl}}(\boldsymbol{x}, 0) = \boldsymbol{\varepsilon}^{\mathrm{pl}^0}, \quad \boldsymbol{\xi}(\boldsymbol{x}, 0) = \boldsymbol{\xi}^0 \tag{2.1.44}$$

The function $\boldsymbol{g}$ can be formulated for the given constitutive law of inelastic deformation, see Naumenko and Altenbach (2016, Chapt. 5). The vector $\boldsymbol{\xi}$ and and the function $\boldsymbol{h}$ can be defined for the selected internal state variables and the corresponding evolution equations.

2.2 Numerical Solution Techniques

Let us assume that the inelastic strain vector and the vector of internal state variables are known functions of the coordinates for a fixed time. With the strain-displacement relations (2.1.37), the constitutive equations (2.1.42) can be written as follows

$$\boldsymbol{\sigma} = \boldsymbol{E}\left(\boldsymbol{D}^{\mathrm{T}}\boldsymbol{u} - \boldsymbol{\varepsilon}^{\mathrm{th}} - \boldsymbol{\varepsilon}^{\mathrm{pl}}\right) \tag{2.2.45}$$

Taking into account the equilibrium conditions (2.1.40) and the static boundary conditions (2.1.41) we obtain

$$\begin{aligned} \boldsymbol{DED}^{\mathrm{T}}\boldsymbol{u} &= -\bar{\boldsymbol{f}} + \boldsymbol{DE}\boldsymbol{\varepsilon}^{\mathrm{th}} + \boldsymbol{DE}\boldsymbol{\varepsilon}^{\mathrm{pl}}, \boldsymbol{x} \in V, \\ \overset{\sigma}{\boldsymbol{T}}\,\boldsymbol{ED}^{\mathrm{T}}\boldsymbol{u} &= \quad \bar{\boldsymbol{p}} + \overset{\sigma}{\boldsymbol{T}}\,\boldsymbol{E}\boldsymbol{\varepsilon}^{\mathrm{th}} + \overset{\sigma}{\boldsymbol{T}}\,\boldsymbol{E}\boldsymbol{\varepsilon}^{\mathrm{pl}}, \boldsymbol{x} \in A_p \end{aligned} \tag{2.2.46}$$

The partial differential equations $(2.2.46)_1$, the kinematic boundary conditions (2.1.39) and the static boundary conditions $(2.2.46)_2$ represent the BVP with respect to the unknown displacement vector $\boldsymbol{u}$ for the given inelastic strain field. Introducing the fictitious force vectors corresponding to the given thermal and inelastic strains at fixed time we can write Eqs (2.2.46) as follows

$$\begin{aligned} \boldsymbol{DED}^{\mathrm{T}}\boldsymbol{u} &= -\bar{\boldsymbol{f}} + \boldsymbol{f}^{\mathrm{th}} + \boldsymbol{f}^{\mathrm{pl}}, \boldsymbol{f}^{\mathrm{th}} = \boldsymbol{DE}\boldsymbol{\varepsilon}^{\mathrm{th}}, \boldsymbol{f}^{\mathrm{pl}} = \boldsymbol{DE}\boldsymbol{\varepsilon}^{\mathrm{pl}}, \\ \overset{\sigma}{\boldsymbol{T}}\,\boldsymbol{ED}^{\mathrm{T}}\boldsymbol{u} &= \quad \bar{\boldsymbol{p}} + \boldsymbol{p}^{\mathrm{th}} + \boldsymbol{p}^{\mathrm{pl}}, \boldsymbol{p}^{\mathrm{th}} = \overset{\sigma}{\boldsymbol{T}}\,\boldsymbol{E}\boldsymbol{\varepsilon}^{\mathrm{th}}, \boldsymbol{p}^{\mathrm{pl}} = \overset{\sigma}{\boldsymbol{T}}\,\boldsymbol{E}\boldsymbol{\varepsilon}^{\mathrm{pl}} \end{aligned} \tag{2.2.47}$$

These equations are the equilibrium conditions expressed in terms of three unknown components of the displacement vector. After the solution of Eqs (2.2.47), one can obtain the six components of the stress vector from Eq. (2.2.45). Inserting the stress vector into the inelastic constitutive equations (2.1.43), one can calculate the rates of inelastic strains and of the internal variables. Based on the equations introduced, the IBVP of the type $\dot{\boldsymbol{Y}} = \boldsymbol{G}(\boldsymbol{Y})$ can be formulated, where $\boldsymbol{Y}$ includes the vectors of inelastic strains and internal variables. The operator $\boldsymbol{G}$ involves the solution of the linearized boundary value problem for the fixed inelastic strains and internal variables. The initial conditions are given by Eqs (2.1.44).

An alternative formulation can be derived based on the compatibility condition (2.1.38). The time derivative of constitutive equations (2.1.42) provides

$$\dot{\boldsymbol{\sigma}} = \boldsymbol{E}(\dot{\boldsymbol{\varepsilon}} - \dot{\boldsymbol{\varepsilon}}^{\mathrm{th}} - \dot{\boldsymbol{\varepsilon}}^{\mathrm{pl}}) = \boldsymbol{E}\left[\dot{\boldsymbol{\varepsilon}} - \dot{\boldsymbol{\varepsilon}}^{\mathrm{th}} - \boldsymbol{g}(\boldsymbol{\sigma}, \boldsymbol{\xi}, T)\right]$$

Reordering this equation, the total strain rate vector can be computed from

$$\dot{\boldsymbol{\varepsilon}} = \boldsymbol{E}^{-1}\dot{\boldsymbol{\sigma}} + \dot{\boldsymbol{\varepsilon}}^{\mathrm{th}} + \boldsymbol{g}(\boldsymbol{\sigma}, \boldsymbol{\xi}, T) \tag{2.2.48}$$

For isothermal processes the rate of thermal strains vanishes, i.e. $\dot{\boldsymbol{\varepsilon}}^{\mathrm{th}} = \boldsymbol{0}$. The compatibility condition (2.1.38) can be rewritten in terms of the strain rate vector

$$\boldsymbol{D}_1\dot{\boldsymbol{\varepsilon}} = \boldsymbol{0} \tag{2.2.49}$$

After inserting (2.2.48) into (2.2.49) we obtain

$$\boldsymbol{D}_1\boldsymbol{E}^{-1}\dot{\boldsymbol{\sigma}} + \boldsymbol{D}_1\boldsymbol{g}(\boldsymbol{\sigma}, \boldsymbol{\xi}, T) = \boldsymbol{0} \tag{2.2.50}$$

The six equations (2.2.50) describe the stress redistribution during the creep process. The initial conditions are the solutions of the linear elastic problem for the stresses

$$\boldsymbol{D}_1\boldsymbol{E}^{-1}\boldsymbol{\sigma}(\boldsymbol{x},0) = \boldsymbol{0},$$

as well as $\boldsymbol{\xi}(\boldsymbol{x},0) = \boldsymbol{\xi}_0$. The IBVP can be formulated again as $\dot{\boldsymbol{Y}} = \boldsymbol{G}(\boldsymbol{Y})$, where $\boldsymbol{Y}$ includes now the stress vector and the vector of internal state variables. The stress redistribution equation (2.2.50) can also be formulated in terms of stress functions. A variety of stress functions can be found in such a way that the equilibrium conditions (2.1.40) are identically satisfied. In general, a tensor stress function is required to satisfy the equilibrium conditions (Lurie, 2005). For special problems of structural mechanics, scalar-valued stress functions lead to efficient solutions. Examples include the Airy stress function for plane stress/strain problems, the Prandtl stress function for torsion problems, three Maxwell's stress functions for three-dimensional problems in Cartesian coordinates (Hahn, 1985; Pilkey and Wunderlich, 1994). Introducing the vector of stress functions $\boldsymbol{\psi}$, such that $\boldsymbol{\sigma} = \boldsymbol{D}_1\boldsymbol{\psi}$, the equilibrium conditions $\boldsymbol{D\sigma} = \boldsymbol{DD}_1\boldsymbol{\psi} = \boldsymbol{0}$ are identically satisfied in the absence of body forces. With the stress functions $\boldsymbol{\psi}$, we can write Eq. (2.2.50) as follows

$$\boldsymbol{D}_1\boldsymbol{E}^{-1}\boldsymbol{D}_1\dot{\boldsymbol{\psi}} + \boldsymbol{D}_1\boldsymbol{g}(\boldsymbol{D}_1\boldsymbol{\psi},\boldsymbol{\xi},T) = \boldsymbol{0}$$

Because there exist identities between the six compatibility conditions (only three of them are independent), see e.g. Hahn (1985), it is possible to transform the six equations (2.2.50) into three independent equations. After inserting into (2.2.50) one can obtain three equations for three unknown stress functions.

In addition to the displacement formulation (2.2.47) and the stress formulation (2.2.50), it is possible to express the governing equations in terms of displacements and stresses. Such mixed formulations can be useful for solving inelastic problems of beams, plates and shells.

2.2.1 Time-step Methods

The governing equations include first-order time derivatives and the prescribed initial conditions. The unknown displacements in Eqs (2.2.47) or the unknown stresses in Eq. (2.2.50) are functions of coordinates and time. The exact integration of these equations with respect to the time variable is possible only for one-dimensional problems, e.g. for rods and elementary inelastic constitutive laws, Chapt. 1. In the general case of the structural analysis, numerical time integration methods must be applied for solving non-linear IBVP. The commonly used solution technique in mechanics and thermodynamics are the time-step methods. Examples for time-step methods are discussed in Subsect. 1.3.3 for elastic-linear viscous bars and in Subsect. 1.4.7 for bars with non-linear inelastic material behavior. A variety of time integration algorithms can be found in textbooks on numerical methods, e.g. Curnier

(1994); Engeln-Müllges and Reutter (1991); Hairer et al (1987); Schwetlick and Kretzschmar (1991).

Here we discuss examples of time integration procedures widely used in inelastic analysis. Following the approaches presented in Subsect. 1.3.3, let us integrate the rate equations (2.1.42) within a time interval $[t_n, t_{n+1}]$ as follows

$$\Delta\boldsymbol{\varepsilon}_n^{\mathrm{pl}} = \int\limits_{t_n}^{t_{n+1}} \boldsymbol{g}(\boldsymbol{\sigma}, \boldsymbol{\varepsilon}^{\mathrm{pl}} \boldsymbol{\zeta}, \dot{\boldsymbol{\zeta}}, T)\mathrm{d}t, \quad \Delta\boldsymbol{\zeta}_n = \int\limits_{t_n}^{t_{n+1}} \boldsymbol{h}(\boldsymbol{\sigma}, \boldsymbol{\varepsilon}^{\mathrm{pl}}, \dot{\boldsymbol{\varepsilon}}^{\mathrm{pl}}, \boldsymbol{\zeta}, T)\mathrm{d}t, \tag{2.2.51}$$

where n is the step number and the increments are defined as follows

$$\Delta\boldsymbol{\varepsilon}_n^{\mathrm{pl}} = \boldsymbol{\varepsilon}_{n+1}^{\mathrm{pl}} - \boldsymbol{\varepsilon}_n^{\mathrm{pl}}, \quad \Delta\boldsymbol{\zeta}_n = \boldsymbol{\zeta}_{n+1} - \boldsymbol{\zeta}_n, \quad \Delta t_n = t_{n+1} - t_n \tag{2.2.52}$$

The integral on the right-hand side of Eqs (2.2.51) can be evaluated approximately by a variety of quadrature rules leading to different time-step methods. Let us discuss examples of time integration procedures widely used in inelastic analysis.

2.2.1.1 Explicit Methods

With a left rectangular rule, see Subsec 1.3.3 and 1.4.7, the approximate integration of Eqs (2.2.52) results in the explicit Euler method

$$\Delta\boldsymbol{\varepsilon}_n^{\mathrm{pl}} = \Delta t_n \boldsymbol{g}(\boldsymbol{\sigma}_n, \boldsymbol{\zeta}_n, T_n), \quad \Delta\boldsymbol{\zeta}_n = \Delta t_n \boldsymbol{h}(\boldsymbol{\sigma}_n, \boldsymbol{\zeta}_n, T_n) \tag{2.2.53}$$

As shown in Subsect. 1.4.7, the explicit Euler method is efficient for problems of creep, stress redistribution and stress relaxation. Let us illustrate the use of Eqs (2.2.53) in IBVP of creep for structures subjected to stationary external loads and temperature. To this end we apply the displacement formulation of the governing equations and neglect the thermal strains for the sake of brevity. As initial condition the solution of the elasticity problem

$$\boldsymbol{D}\boldsymbol{E}\boldsymbol{D}^{\mathrm{T}}\boldsymbol{u}_0 = -\bar{\boldsymbol{f}}, \quad \boldsymbol{\sigma}_0 = \boldsymbol{E}\boldsymbol{D}^{\mathrm{T}}\boldsymbol{u}_0 \tag{2.2.54}$$

with $\boldsymbol{u}_0 = \boldsymbol{u}(\boldsymbol{x}, 0)$ and $\boldsymbol{\sigma}_0 = \boldsymbol{\sigma}(\boldsymbol{x}, 0)$. can be applied. Applying the explicit Euler method and using Eqs (2.2.45) and (2.2.47), the displacements and stresses at t_{n+1} can be updated for the known values at t_n. The step-by-step solution is continued until the final time value t_N is reached. In problems where creep damage analysis and structural life estimations are required one should check, whether the critical damage state is reached at the current time step. Let $0 \leq \omega \leq \omega_* < 1$ be a scalar damage variable and ω_* is the corresponding critical value. Then the following time integration scheme can be applied:

```
set n = 0, ε_0^pl = 0, ξ_0 = 0
solve BVP DED^T u_0 = −f̄, calculate σ_0 = ED^T u_0
1: calculate
```

$$\Delta \boldsymbol{\varepsilon}^{\text{pl}}_n = \Delta t_n \boldsymbol{g}(\boldsymbol{\sigma}_n, \boldsymbol{\xi}_n, T_n), \qquad \Delta \boldsymbol{\xi}_n = \Delta t_n \boldsymbol{h}(\boldsymbol{\sigma}_n, \boldsymbol{\xi}_n, T_n)$$

$$\boldsymbol{\varepsilon}^{\text{pl}}_{n+1} = \boldsymbol{\varepsilon}^{\text{pl}}_n + \Delta \boldsymbol{\varepsilon}^{\text{pl}}_n, \qquad \boldsymbol{\xi}_{n+1} = \boldsymbol{\xi}_n + \Delta \boldsymbol{\xi}_n,$$

```
solve BVP
```

$$\begin{aligned} \boldsymbol{DED}^{\text{T}}\boldsymbol{u}_{n+1} &= -\bar{\boldsymbol{f}} + \boldsymbol{DE}\boldsymbol{\varepsilon}^{\text{pl}}_{n+1}, \\ \boldsymbol{\sigma}_{n+1} &= \boldsymbol{E}(\boldsymbol{D}^{\text{T}}\boldsymbol{u}_{n+1} - \boldsymbol{\varepsilon}^{\text{pl}}_{n+1}) \end{aligned} \tag{2.2.55}$$

```
if t_{n+1} < t_N and ω_{n+1} < ω_*, then set n := n+1 go to 1
else finish
```

The explicit Euler method is widely used in inelastic analysis because of simplicity, e.g. Altenbach et al (2008); Boyle and Spence (1983); Schmicker et al (2013), and is available in commercial finite element codes, e.g. Abaqus User's Guide (2017). The accuracy of the method depends on the time step size. Furthermore, this method is conditionally stable that means that the stability on the time step size. Therefore stable results can be obtained only for $\Delta t \leq \Delta t_c$. Approaches to estimate the critical time step size and to control time step sizes based on the given error tolerance are discussed in Subsect. 1.4.7 for one-dimensional cases. A critical time step size can be mathematically derived for each explicit time integration method based on the Courant-Friedrichs-Lewy condition (Courant et al, 1928). Several empirical formulae were proposed to determine the critical time step size. For example, in Zienkiewicz and Taylor (1991) it is recommended to estimate the critical time step size from the condition that the increment of the creep strain does not exceed one half of the elastic strain, i.e.

$$\Delta t_n \boldsymbol{g}(\boldsymbol{\sigma}_n, \boldsymbol{\xi}_n, T_n) \leq \frac{1}{2}\boldsymbol{E}^{-1}\boldsymbol{\sigma}_n$$

The explicit Euler method can be recommended for creep analysis of structures subjected to constant or monotonic loading conditions. In the case of loading jumps or cyclic loading changes, very small time steps are necessary in order to provide a stable solution.

One way to improve the accuracy of time-dependent solutions is the use of multistep methods of the Runge-Kutta type, see e.g. Boyle and Spence (1983); Engeln-Müllges and Reutter (1991); Hairer et al (1987). These explicit methods are conditionally stable as well. However, they provide higher-order accuracy if compared with the one-step forward difference method. For creep-damage related analysis, the so-called embedded methods (Hairer et al, 1987), which include the time step size control, can be recommended. For example, in Altenbach et al (1996, 1997a) the embedded fourth-order Kutta-Merson method has been applied to creep problems of shells of revolution.

2.2.1.2 Implicit Methods

For many problems in mechanics explicit methods are not efficient or even not applicable. Examples include structural analysis of rate-independent plasticity (Belytschko et al, 2014; Simo and Hughes, 2000), damage mechanics (Altenbach and Naumenko, 1997), non-linear rigid body dynamics (Altenbach et al, 2007) and many others. The governing differential equations are stiff (Hairer and Wanner, 1996) and implicit methods are required to obtain stable numerical solutions.

Applying the generalized trapezoidal rule (Curnier, 1994) to Eqs (2.2.52) we obtain

$$\begin{aligned}\boldsymbol{\varepsilon}^{\mathrm{pl}}_{n+1} &= \boldsymbol{\varepsilon}^{\mathrm{pl}}_n + \Delta t\left[(1-\theta)\dot{\boldsymbol{\varepsilon}}^{\mathrm{pl}}_n + \theta\dot{\boldsymbol{\varepsilon}}^{\mathrm{pl}}_{n+1}\right],\\ \boldsymbol{\xi}_{n+1} &= \boldsymbol{\xi}_n + \Delta t\left[(1-\theta)\dot{\boldsymbol{\xi}}_n + \theta\dot{\boldsymbol{\xi}}_{n+1}\right],\end{aligned} \tag{2.2.56}$$

where θ ($0 \leq \theta \leq 1$) is the parameter controlling the stability. With Eqs (2.2.56) different time-step methods methods can be obtained. Setting $\theta = 0$ the explicit Euler method (2.2.53) follows. For $\theta > 0$ we obtain a variety of implicit methods: for $\theta = 1/2$ - the trapezoidal rule (Crank-Nicolson method), and for $\theta = 1$ the backward difference method (implicit Euler method). The advantage of the implicit methods is their unconditional stability, see Subsec 1.3.3 and 1.4.7. The price for this unconditional stability is the necessity to solve non-linear equations at each time step. Equations (2.2.56) can be rewritten as follows

$$\begin{aligned}\boldsymbol{\varepsilon}^{\mathrm{pl}}_{n+1} &= \boldsymbol{\varepsilon}^{\mathrm{pl}}_n + \Delta\boldsymbol{\varepsilon}^{\mathrm{pl}}_n,\\ \boldsymbol{\xi}_{n+1} &= \boldsymbol{\xi}_n + \Delta\boldsymbol{\xi}_n,\\ \Delta\boldsymbol{\varepsilon}^{\mathrm{pl}}_n &= \Delta t_n\left[(1-\theta)\boldsymbol{g}(\boldsymbol{\sigma}_n, \boldsymbol{\xi}_n, T_n) + \theta\boldsymbol{g}(\boldsymbol{\sigma}_{n+1}, \boldsymbol{\xi}_{n+1}, T_{n+1})\right],\\ \Delta\boldsymbol{\xi}_n &= \Delta t_n\left[(1-\theta)\boldsymbol{h}(\boldsymbol{\sigma}_n, \boldsymbol{\xi}_n, T_n) + \theta\boldsymbol{h}(\boldsymbol{\sigma}_{n+1}, \boldsymbol{\xi}_{n+1}, T_{n+1})\right]\end{aligned} \tag{2.2.57}$$

Equations (2.2.57) are non-linear with respect to $\boldsymbol{\xi}_{n+1}$ for $\theta > 0$. Note that for a material model with strain hardening, the vector $\boldsymbol{\xi}_n$ includes the equivalent creep strain. In this case Eqs (2.2.57) are non-linear with respect to $\boldsymbol{\varepsilon}^{\mathrm{pl}}_{n+1}$. These equations can be solved using known iteration methods. Applying the fixed point iteration method, the following scheme at the time step $[t_n, t_{n+1}]$ can be obtained:

$$
\begin{array}{l}
\texttt{set } i=0,\ \boldsymbol{\varepsilon}^{cr^0}_{n+1}=\boldsymbol{\varepsilon}^{\mathrm{pl}}_n,\quad \boldsymbol{\xi}^i_n=\boldsymbol{\xi}_n,\quad \boldsymbol{\sigma}^0_{n+1}=\boldsymbol{\sigma}_n\\
\texttt{1: calculate}\\
\Delta\boldsymbol{\varepsilon}^{cr^i}_n=\Delta t_n\left[(1-\theta)\boldsymbol{g}(\boldsymbol{\sigma}_n,\boldsymbol{\xi}_n,T_n)+\theta\boldsymbol{g}(\boldsymbol{\sigma}^i_{n+1},\boldsymbol{\xi}^i_{n+1},T_{n+1})\right],\\
\Delta\boldsymbol{\xi}^i_n=\Delta t_n\left[(1-\theta)\boldsymbol{h}(\boldsymbol{\sigma}_n,\boldsymbol{\xi}_n,T_n)+\theta\boldsymbol{h}(\boldsymbol{\sigma}^i_{n+1},\boldsymbol{\xi}^i_{n+1},T_{n+1})\right],\\
\boldsymbol{\varepsilon}^{cr^{i+1}}_{n+1}=\boldsymbol{\varepsilon}^{\mathrm{pl}}_n+\Delta\boldsymbol{\varepsilon}^{cr^i}_n,\qquad \boldsymbol{\xi}^{i+1}_{n+1}=\boldsymbol{\xi}_n+\Delta\boldsymbol{\xi}^i_n,\\
\texttt{if } |\boldsymbol{\varepsilon}^{cr^{i+1}}_{n+1}-\boldsymbol{\varepsilon}^{cr^i}_{n+1}|>\epsilon \texttt{ and } |\boldsymbol{\xi}^{i+1}_{n+1}-\boldsymbol{\xi}^i_{n+1}|>\epsilon\\
\texttt{then solve BVP}\\
\qquad \boldsymbol{DED}^{\mathrm{T}}\boldsymbol{u}^{i+1}_{n+1}=-\bar{\boldsymbol{f}}_{n+1}+\boldsymbol{DE}\boldsymbol{\varepsilon}^{cr^{i+1}}_{n+1},\\
\qquad\qquad \boldsymbol{\sigma}^{i+1}_{n+1}=\boldsymbol{E}\left(\boldsymbol{D}^{\mathrm{T}}\boldsymbol{u}^{i+1}_{n+1}-\boldsymbol{\varepsilon}^{cr^{i+1}}_{n+1}\right)\\
\texttt{set } i:=i+1 \texttt{ go to 1}\\
\texttt{else}\\
\texttt{set } \boldsymbol{\varepsilon}^{\mathrm{pl}}_{n+1}=\boldsymbol{\varepsilon}^{cr^{i+1}}_{n+1},\ \boldsymbol{\xi}_{n+1}=\boldsymbol{\xi}^{i+1}_{n+1}
\end{array}
\qquad (2.2.58)
$$

The accuracy and the efficiency of the implicit method in connection with the introduced iteration scheme is now additionally dependent on the tolerance ϵ and the convergence rate of the fixed point iterations. The initial guess in the above introduced scheme is the forward difference predictor. Since the convergence rate of the fixed point iterations is highly dependent on the "quality" of the initial guess, the efficiency of this scheme is determined again by the time step size. If the desired accuracy ϵ is not reached within $3-4$ iterations, the time step size should be decreased and the calculations repeated starting from step 1. The slow convergence of the fixed point iterations is the drawback of the proposed algorithm. However, in the case of creep-damage studies, this algorithm is more efficient in comparison with the explicit forward method. Examples are discussed in Altenbach and Naumenko (1997). To provide higher convergence rates the implicit time integration scheme can be combined the Newton-Raphson iteration method or its modifications (Belytschko et al, 2014; Zienkiewicz and Taylor, 1991; Wriggers, 2008).

In many cases the roots of the non-linear equations (2.2.57) are close to the initial guesses because the time step size Δt_n must be sufficiently small to guarantee the accuracy of the generalized trapezoidal rule. An example is presented in Subsect. 1.4.7 illustrating that with an appropriate time step control only one Newton-Raphson iteration can be enough. This leads to a semi-implicit scheme with a modified tangent stiffness, see e.g. Bassani and Hawk (1990); Peirce et al (1984). To derive the method Eqs (2.2.57) are linearized with respect to $\dot{\boldsymbol{\xi}}_{n+1}$. For the sake of brevity let us assume that the functions $\boldsymbol{g}$ and $\boldsymbol{h}$ are independent from the temperature T. Then we can write

$$\begin{aligned}\dot{\boldsymbol{\varepsilon}}^{\rm pl}_{n+1} &\cong \boldsymbol{g}(\boldsymbol{\sigma}_n,\boldsymbol{\xi}_n)+\boldsymbol{g}_{,\sigma}(\boldsymbol{\sigma}_n,\boldsymbol{\xi}_n)\Delta\boldsymbol{\sigma}_n+\boldsymbol{g}_{,\xi}(\boldsymbol{\sigma}_n,\boldsymbol{\xi}_n)\Delta\boldsymbol{\xi}_n,\\ \dot{\boldsymbol{\xi}}_{n+1} &\cong \boldsymbol{h}(\boldsymbol{\sigma}_n,\boldsymbol{\xi}_n)+\boldsymbol{h}_{,\sigma}(\boldsymbol{\sigma}_n,\boldsymbol{\xi}_n)\Delta\boldsymbol{\sigma}_n+\boldsymbol{h}_{,\xi}(\boldsymbol{\sigma}_n,\boldsymbol{\xi}_n)\Delta\boldsymbol{\xi}_n\end{aligned} \qquad (2.2.59)$$

with

$$\boldsymbol{g}_{,\sigma}=\frac{\partial\boldsymbol{g}}{\partial\boldsymbol{\sigma}},\quad \boldsymbol{g}_{,\xi}=\frac{\partial\boldsymbol{g}}{\partial\boldsymbol{\xi}},\quad \boldsymbol{h}_{,\sigma}=\frac{\partial\boldsymbol{h}}{\partial\boldsymbol{\sigma}},\quad \boldsymbol{h}_{,\xi}=\frac{\partial\boldsymbol{h}}{\partial\boldsymbol{\xi}}$$

From (2.2.57) we obtain

$$\begin{aligned}\Delta\boldsymbol{\varepsilon}^{\rm pl}_n &= \Delta t_n\left(\boldsymbol{g}_n+\theta\boldsymbol{g}_{n,\sigma}\Delta\boldsymbol{\sigma}_n+\theta\boldsymbol{g}_{n,\xi}\Delta\boldsymbol{\xi}_n\right),\\ \Delta\boldsymbol{\xi}_n &= \Delta t_n\left(\boldsymbol{h}_n+\theta\boldsymbol{h}_{n,\sigma}\Delta\boldsymbol{\sigma}_n+\theta\boldsymbol{h}_{n,\xi}\Delta\boldsymbol{\xi}_n\right),\end{aligned} \qquad (2.2.60)$$

where

$$\begin{aligned}\boldsymbol{g}_n &\equiv \boldsymbol{g}(\boldsymbol{\sigma}_n,\boldsymbol{\xi}_n), & \boldsymbol{h}_n &\equiv \boldsymbol{h}(\boldsymbol{\sigma}_n,\boldsymbol{\xi}_n),\\ \boldsymbol{g}_{n,\sigma} &\equiv \frac{\partial\boldsymbol{g}}{\partial\boldsymbol{\sigma}}(\boldsymbol{\sigma}_n,\boldsymbol{\xi}_n), & \boldsymbol{g}_{n,\xi} &\equiv \frac{\partial\boldsymbol{g}}{\partial\boldsymbol{\xi}}(\boldsymbol{\sigma}_n,\boldsymbol{\xi}_n),\\ \boldsymbol{h}_{n,\sigma} &\equiv \frac{\partial\boldsymbol{h}}{\partial\boldsymbol{\sigma}}(\boldsymbol{\sigma}_n,\boldsymbol{\xi}_n), & \boldsymbol{h}_{n,\xi} &\equiv \frac{\partial\boldsymbol{h}}{\partial\boldsymbol{\xi}}(\boldsymbol{\sigma}_n,\boldsymbol{\xi}_n)\end{aligned}$$

The second equation (2.2.60) can be rewritten as

$$\Delta\boldsymbol{\xi}_n=\Delta t_n\left[\boldsymbol{I}-\Delta t_n\theta\boldsymbol{h}_{n,\xi}\right]^{-1}\left[\boldsymbol{h}_n+\theta\boldsymbol{h}_{n,\sigma}\Delta\boldsymbol{\sigma}_n\right] \qquad (2.2.61)$$

Inserting this equation into the first equation (2.2.60) we obtain

$$\begin{aligned}\Delta\boldsymbol{\varepsilon}^{\rm pl}_n =& \Delta t_n(\boldsymbol{g}_n+\theta\boldsymbol{g}_{n,\sigma}\Delta t\boldsymbol{\sigma}_n)\\ &+\Delta t_n^2\boldsymbol{h}_{n,\xi}\left[\boldsymbol{I}-\Delta t_n\theta\boldsymbol{h}_{n,\xi}\right]^{-1}\left[\boldsymbol{h}_n+\theta\boldsymbol{h}_{n,\sigma}\Delta\boldsymbol{\sigma}_n\right]\end{aligned} \qquad (2.2.62)$$

Neglecting the last term in the right-hand side of Eq. (2.2.62), Eq. $(2.2.46)_1$ takes the form

$$\boldsymbol{DED}^{\rm T}\Delta\boldsymbol{u}_n=\boldsymbol{DE}\Delta\boldsymbol{\varepsilon}^{\rm pl}_n\cong\Delta t_n\boldsymbol{DE}\left[\boldsymbol{g}_n+\theta\boldsymbol{g}_{n,\sigma}\Delta\boldsymbol{\sigma}_n\right] \qquad (2.2.63)$$

Here $\bar{\boldsymbol{f}}=$ const and $\boldsymbol{\varepsilon}^{\rm th}=$ const are assumed. From (2.2.45) the increment of the stress vector can be computed as follows

$$\Delta\boldsymbol{\sigma}_n=\boldsymbol{E}\left[\boldsymbol{D}^{\rm T}\Delta\boldsymbol{u}_n-\Delta t_n\boldsymbol{g}_n-\Delta t_n\theta\boldsymbol{g}_{n,\sigma}\Delta\boldsymbol{\sigma}_n\right],$$

or

$$\Delta\boldsymbol{\sigma}_n=\left[\boldsymbol{I}+\Delta t_n\theta\boldsymbol{E}\boldsymbol{g}_{n,\sigma}\right]^{-1}\boldsymbol{ED}^{\rm T}\Delta\boldsymbol{u}_n-\Delta t_n\left[\boldsymbol{I}+\Delta t_n\theta\boldsymbol{E}\boldsymbol{g}_{n,\sigma}\right]^{-1}\boldsymbol{E}\boldsymbol{g}_n \qquad (2.2.64)$$

After inserting into Eq. (2.2.63) we obtain

$$\begin{aligned} \boldsymbol{D}\left[\boldsymbol{E}-\boldsymbol{E}_n^*\right]\boldsymbol{D}^{\mathrm{T}}\Delta\boldsymbol{u}_n &= \Delta t_n \boldsymbol{D}\boldsymbol{E}\boldsymbol{g}_n, \\ \boldsymbol{E}_n^* &= \Delta t_n \theta \boldsymbol{E}\boldsymbol{g}_{n,\boldsymbol{\sigma}}\left[\boldsymbol{I}+\Delta t_n \theta \boldsymbol{E}\boldsymbol{g}_{n,\boldsymbol{\sigma}}\right]^{-1}\boldsymbol{E} \end{aligned} \tag{2.2.65}$$

Based on the derived equations, the following one-step scheme can be formulated:

set $n=0,\ \boldsymbol{\varepsilon}_0^{\mathrm{pl}}=\boldsymbol{0},\quad \boldsymbol{\xi}_0=\boldsymbol{0}$

solve BVP $\boldsymbol{D}\boldsymbol{E}\boldsymbol{D}^{\mathrm{T}}\boldsymbol{u}_0=-\bar{\boldsymbol{f}}$, calculate $\boldsymbol{\sigma}_0=\boldsymbol{E}\boldsymbol{D}^{\mathrm{T}}\boldsymbol{u}_0$

1: calculate

$\Delta\boldsymbol{\varepsilon}_n^{\mathrm{pl}}=\Delta t_n(\boldsymbol{g}_n+\theta\boldsymbol{g}_{n,\boldsymbol{\sigma}}\Delta t\boldsymbol{\sigma}_n)$,

$\Delta\boldsymbol{\xi}_n=\Delta t_n\left[\boldsymbol{I}-\Delta t_n\theta\boldsymbol{h}_{n,\boldsymbol{\xi}}\right]^{-1}\left[\boldsymbol{h}_n+\theta\boldsymbol{h}_{n,\boldsymbol{\sigma}}\Delta\boldsymbol{\sigma}_n\right]$,

$\boldsymbol{E}_n^*=\Delta t_n\theta\boldsymbol{E}\boldsymbol{g}_{n,\boldsymbol{\sigma}}\left[\boldsymbol{I}+\Delta t_n\theta\boldsymbol{E}\boldsymbol{g}_{n,\boldsymbol{\sigma}}\right]^{-1}\boldsymbol{E}$

solve BVP

$$\begin{aligned} \boldsymbol{D}\left[\boldsymbol{E}-\boldsymbol{E}_n^*\right]\boldsymbol{D}^{\mathrm{T}}\Delta\boldsymbol{u}_n &= \Delta t_n\boldsymbol{D}\boldsymbol{E}\boldsymbol{g}_n, \\ \Delta\boldsymbol{\sigma}_n &= \boldsymbol{E}(\boldsymbol{D}^{\mathrm{T}}\Delta\boldsymbol{u}_n-\Delta\boldsymbol{\varepsilon}_n^{\mathrm{pl}}) \end{aligned} \tag{2.2.66}$$

calculate

$$\begin{aligned} \boldsymbol{\varepsilon}_{n+1}^{\mathrm{pl}} &= \boldsymbol{\varepsilon}_n^{\mathrm{pl}}+\Delta\boldsymbol{\varepsilon}_n^{\mathrm{pl}},\ \boldsymbol{\xi}_{n+1}=\boldsymbol{\xi}_n+\Delta\boldsymbol{\xi}_n, \\ \boldsymbol{u}_{n+1} &= \boldsymbol{u}_n+\Delta\boldsymbol{u}_n,\ \boldsymbol{\sigma}_{n+1}=\boldsymbol{\sigma}_n+\Delta\boldsymbol{\sigma}_n \end{aligned}$$

if $t_{n+1}<t_N$ and $\omega<\omega_*$,

then set $n:=n+1$ go to 1

else finish

For $\theta>0$ this method provides an accuracy of higher order if compared with that for the explicit one-step Euler method. For example, for $\theta=1/2$ the method has a second order accuracy, see Subsubsect. 1.3.3.4 for elementary examples. Following the considered algorithm, both the fictitious force vector $\Delta t\boldsymbol{D}\boldsymbol{E}\boldsymbol{g}_n$ and the stiffness matrix $\boldsymbol{E}-\boldsymbol{E}_n^*$ must be updated at each time step. The modified stiffness leads to an additional effort in solving the boundary value problem (2.2.66).

2.2.2 Solution of Boundary Value Problems

According to the discussed time integration algorithms, linearized boundary value problems have to be solved at each time or iteration step. These problems include second-order partial differential equations with respect to the unknown displacements $\boldsymbol{u}(\boldsymbol{x},t_n)$ or displacement increments $\Delta\boldsymbol{u}(\boldsymbol{x},t_n)$. The effect of the accumulated inelastic strain is considered by means of fictitious force vectors and/or complementary stiffness matrices. The accumulated inelastic strains are determined by the entire deformation history. Therefore, the analytical methods developed within the theory of elasticity, e.g. the Fourier series approach (Altenbach et al, 2016) and the complex stress functions approach (Hahn, 1985), are not applicable in the general case of inelastic behavior with internal state variables. Only for some elementary problems, e.g. for the bar system or Bernoulli-Euler-type beam, analytical closed form solutions can be obtained (Boyle and Spence, 1983; Malinin, 1981; Odqvist and Hult, 1962). These solutions are helpful for the verification of the general computational methods or general purpose solvers.

In what follows let us briefly discuss the numerical methods applied in connection with inelastic structural analysis including:

- the finite difference method,
- the direct variational methods and
- the boundary element method.

Applying the finite difference method, the partial differential operators are replaced by finite differences leading to the solution of algebraic equations instead of the partial differential ones. The utilization is mostly efficient for problems leading to ordinary differential equations. Examples include creep of axi-symmetrically loaded shells of revolution and circular plates (Altenbach et al, 1996, 1997a,b; Betten and Borrmann, 1987; Betten et al, 1989; Burlakov et al, 1977; Byrne and Mackenzie, 1966; Murakami and Suzuki, 1973; Podgorny et al, 1984; Skrzypek and Ganczarski, 1998).

The widely used approach is based on the variational formulations. Starting from appropriate variational functionals, the following direct variational methods can be applied: the Ritz method, the Galerkin method and the Vlasov-Kantorovich method. Variational formulations and the classical direct variational methods will be discussed in the next subsection. The most powerful variational method for the structural analysis is the finite element method (Bathe, 1996; Zienkiewicz and Taylor, 1991) which is the basis of commercial general purpose solvers, e.g., ABAQUS, ADINA, ANSYS, COSMOS, etc. The possibility to incorporate a creep material model with internal state variables is available in commercial codes. The implementation can be performed by writing a user-defined material subroutine.

The boundary element method is based on the transformation of the partial differential equations into boundary integral equations. In order to solve these equations the boundary of the domain is divided into finite elements. As a result, a set of algebraic equations with respect to the vector of displacements (tractions) in the discretization points of the boundary can be obtained. In the case of inelastic analysis an additional domain discretization is necessary in order to store the components of the inelastic strain tensor (Becker et al, 1994). For details of the boundary element technique we refer to Brebbia et al (1983); Hartmann (1987); Katsikadelis (2002); Pilkey and Wunderlich (1994).

2.2.3 Variational Formulations and Procedures

Variational formulations are widely used in several problems of solid mechanics. They are the basis for direct variational methods, e.g. the Ritz method, the Galerkin method, the finite element method. With respect to the type of the BVP, different variational functionals have been proposed. Here let us consider a variational functional in terms of the displacement vector. Let $\boldsymbol{u}(q^i, t)$ be the solution of the BVP (2.1.4) - (2.1.9) under given $\boldsymbol{\varepsilon}^{\mathrm{pl}}$. Let $\delta\boldsymbol{u}$ be the vector of virtual displacements satisfying the kinematic boundary conditions (2.1.7). Starting from the equilibrium condition (2.1.6) we can write

$$\int_V (\nabla \cdot \boldsymbol{\sigma} + \rho \bar{\boldsymbol{f}}) \cdot \delta\boldsymbol{u} \mathrm{d}V = 0 \qquad (2.2.67)$$

With the identity

$$\int_V (\nabla \cdot \boldsymbol{\sigma}) \cdot \delta\boldsymbol{u} \mathrm{d}V = \int_V \left[\nabla \cdot (\boldsymbol{\sigma} \cdot \delta\boldsymbol{u}) - \boldsymbol{\sigma} \cdot\cdot (\nabla \delta\boldsymbol{u})^{\mathrm{T}}\right] \mathrm{d}V \qquad (2.2.68)$$

and applying the divergence theorem (Naumenko and Altenbach (2016, Sect. B.3)) by considering static boundary conditions (2.1.7) we obtain

$$\int_V \nabla \cdot (\boldsymbol{\sigma} \cdot \delta\boldsymbol{u}) \mathrm{d}V = \int_A (\boldsymbol{v} \cdot \boldsymbol{\sigma}) \cdot \delta\boldsymbol{u} \mathrm{d}A = \int_{A_p} \bar{\boldsymbol{p}} \cdot \delta\boldsymbol{u} \mathrm{d}A \qquad (2.2.69)$$

With $\boldsymbol{\sigma} \cdot\cdot (\nabla \delta\boldsymbol{u})^{\mathrm{T}} = \boldsymbol{\sigma} \cdot\cdot \delta(\nabla \boldsymbol{u})^{\mathrm{T}} = \boldsymbol{\sigma} \cdot\cdot \delta\boldsymbol{\varepsilon}$, Eqs (2.2.68) and (2.2.69) the variational Eq. (2.2.67) takes the following form

$$\int_V \boldsymbol{\sigma} \cdot\cdot \delta\boldsymbol{\varepsilon} \mathrm{d}V - \int_V \rho \bar{\boldsymbol{f}} \cdot \delta\boldsymbol{u} \mathrm{d}V - \int_{A_p} \bar{\boldsymbol{p}} \cdot \delta\boldsymbol{u} \mathrm{d}A = 0 \qquad (2.2.70)$$

or

$$\delta W_{\mathrm{i}} + \delta W_{\mathrm{e}} = 0,$$

$$\delta W_{\mathrm{i}} = -\int_V \boldsymbol{\sigma} \cdot\cdot \delta\boldsymbol{\varepsilon} \mathrm{d}V, \quad \delta W_{\mathrm{e}} = \int_V \bar{\boldsymbol{f}} \cdot \delta\boldsymbol{u} \mathrm{d}V + \int_{A_p} \bar{\boldsymbol{p}} \cdot \delta\boldsymbol{u} \mathrm{d}A \qquad (2.2.71)$$

The principle of virtual displacements (2.2.71) states that if a deformable system is in equilibrium, then the sum of the virtual work of external actions δW_{e} and the virtual work of internal forces δW_{i} is equal to zero (Altenbach et al, 2016; Pilkey and Wunderlich, 1994; Washizu, 1982). With the constitutive equation (2.1.9)

$$\begin{aligned}\boldsymbol{\sigma} \cdot\cdot \delta\boldsymbol{\varepsilon} &= \left({}^{(4)}\mathbf{C} \cdot\cdot (\boldsymbol{\varepsilon} - \boldsymbol{\varepsilon}^{\mathrm{pl}} - \boldsymbol{\varepsilon}^{\mathrm{th}})\right) \cdot\cdot \delta\boldsymbol{\varepsilon} \\ &= \frac{1}{2}\delta(\boldsymbol{\varepsilon} \cdot\cdot {}^{(4)}\mathbf{C} \cdot\cdot \boldsymbol{\varepsilon}) - (\boldsymbol{\varepsilon}^{\mathrm{pl}} + \boldsymbol{\varepsilon}^{\mathrm{th}}) \cdot\cdot {}^{(4)}\mathbf{C} \cdot\cdot \delta\boldsymbol{\varepsilon}\end{aligned}$$

the variational equation (2.2.71) can be rewritten as follows

$$\begin{aligned}\delta \Bigg[\frac{1}{2}\int_V \boldsymbol{\varepsilon} \cdot\cdot {}^{(4)}\mathbf{C} \cdot\cdot \boldsymbol{\varepsilon} \mathrm{d}V - \int_V \bar{\boldsymbol{f}} \cdot \boldsymbol{u} \mathrm{d}V - \int_{A_p} \bar{\boldsymbol{p}} \cdot \boldsymbol{u} \mathrm{d}A \\ - \int_V (\boldsymbol{\varepsilon}^{\mathrm{pl}} + \boldsymbol{\varepsilon}^{\mathrm{th}}) \cdot\cdot {}^{(4)}\mathbf{C} \cdot\cdot \boldsymbol{\varepsilon} \mathrm{d}V \Bigg] = 0\end{aligned}$$

or $\delta\Pi(\boldsymbol{u}) = 0$ with

$$\begin{aligned}\Pi(\boldsymbol{u}) = \frac{1}{2}\int_V \boldsymbol{\varepsilon} \cdot\cdot {}^{(4)}\mathbf{C} \cdot\cdot \boldsymbol{\varepsilon} \mathrm{d}V - \int_V \bar{\boldsymbol{f}} \cdot \boldsymbol{u} \mathrm{d}V - \int_{A_p} \bar{\boldsymbol{p}} \cdot \boldsymbol{u} \mathrm{d}A \\ - \int_V (\boldsymbol{\varepsilon}^{\mathrm{pl}} + \boldsymbol{\varepsilon}^{\mathrm{th}}) \cdot\cdot {}^{(4)}\mathbf{C} \cdot\cdot \boldsymbol{\varepsilon} \mathrm{d}V\end{aligned} \qquad (2.2.72)$$

Applying the matrix notation we can write

$$\begin{aligned}\Pi(\boldsymbol{u}) = \frac{1}{2}\int_V (\boldsymbol{D}^{\mathrm{T}}\boldsymbol{u})^{\mathrm{T}}\boldsymbol{E}\boldsymbol{D}^{\mathrm{T}}\boldsymbol{u} \mathrm{d}V - \int_V \bar{\boldsymbol{f}}^{\mathrm{T}}\boldsymbol{u} \mathrm{d}V - \int_{A_p} \bar{\boldsymbol{p}}^{\mathrm{T}}\boldsymbol{u} \mathrm{d}A \\ - \int_V \boldsymbol{\varepsilon}^{\mathrm{th}}\boldsymbol{E}\boldsymbol{D}^{\mathrm{T}}\boldsymbol{u} \mathrm{d}V - \int_V \boldsymbol{\varepsilon}^{\mathrm{pl}}\boldsymbol{E}\boldsymbol{D}^{\mathrm{T}}\boldsymbol{u} \mathrm{d}V\end{aligned} \qquad (2.2.73)$$

It is easy to verify that the condition $\delta\Pi(\boldsymbol{u}) = 0$ provides the partial differential equation with respect to the displacement vector and the static boundary condition (2.2.46).

The variational functional (2.2.73) has been derived from the principle of virtual displacements. By analogy, a variational functional in terms of stresses or stress functions can be formulated providing Eqs (2.2.50) as Euler equations. Furthermore, a mixed variational formulation in terms of displacements and stresses can be convenient for numerous structural mechanics problems. In Altenbach et al (1997a) a mixed variational functional has been utilized for the solution of the von Kármán type plate equations. In Altenbach et al (2001); Altenbach and Naumenko (2002) a mixed formulation has been applied to derive the first order shear deformation beam equations.

To solve the variational problem classical direct variational methods can be utilized. Let us illustrate the application of the Ritz method to the variational functional (2.2.73). The approximate solution for the displacement vector $\tilde{\boldsymbol{u}}$ is presented in the form of series

$$\tilde{u}_k = \sum_{i=1}^{N} a_{ki}\phi_{ki}(x_1,x_2,x_3), \quad k = 1,2,3 \tag{2.2.74}$$

(no summation over k) or

$$\tilde{\boldsymbol{u}} \equiv \begin{bmatrix} \tilde{u}_1 \\ \tilde{u}_2 \\ \tilde{u}_3 \end{bmatrix} = \begin{bmatrix} \boldsymbol{a}_1^{\mathrm{T}}\boldsymbol{\phi}_1 \\ \boldsymbol{a}_2^{\mathrm{T}}\boldsymbol{\phi}_2 \\ \boldsymbol{a}_3^{\mathrm{T}}\boldsymbol{\phi}_3 \end{bmatrix} = \begin{bmatrix} \boldsymbol{\phi}_1 & \boldsymbol{0} & \boldsymbol{0} \\ \boldsymbol{0} & \boldsymbol{\phi}_2 & \boldsymbol{0} \\ \boldsymbol{0} & \boldsymbol{0} & \boldsymbol{\phi}_3 \end{bmatrix}^{\mathrm{T}} \begin{bmatrix} \boldsymbol{a}_1 \\ \boldsymbol{a}_2 \\ \boldsymbol{a}_3 \end{bmatrix} = \boldsymbol{G}^{\mathrm{T}}\boldsymbol{a}, \tag{2.2.75}$$

where $\boldsymbol{\phi}_k$ are vectors of the trial (basis or shape) functions which should be specified a priori and $\boldsymbol{a}_k$ are vectors of unknown (free) parameters. The functions ϕ_{ki} in Eq. (2.2.74) must be linearly independent and satisfy the kinematical boundary conditions. Furthermore, the set of these functions must be complete in order to provide the convergence of $\tilde{\boldsymbol{u}}$ as $N \to \infty$. Inserting the approximate solution $\tilde{\boldsymbol{u}}$ into the variational functional (2.2.73) we can obtain for the time step t_n

$$\begin{aligned} \tilde{\Pi}_n(\tilde{\boldsymbol{u}}) &= \boldsymbol{a}^{\mathrm{T}}\left(\frac{1}{2}\int_V (\boldsymbol{D}^{\mathrm{T}}\boldsymbol{G})^{\mathrm{T}}\boldsymbol{E}\boldsymbol{D}^{\mathrm{T}}\boldsymbol{G}\,\mathrm{d}V\right)\boldsymbol{a} - \boldsymbol{a}^{\mathrm{T}}\int_V \boldsymbol{G}\bar{\boldsymbol{f}}\,\mathrm{d}V - \boldsymbol{a}^{\mathrm{T}}\int_{A_p} \boldsymbol{G}\bar{\boldsymbol{p}}\,\mathrm{d}A \\ &- \boldsymbol{a}^{\mathrm{T}}\left(\int_V (\boldsymbol{D}^{\mathrm{T}}\boldsymbol{G})^{\mathrm{T}}\boldsymbol{E}\boldsymbol{\varepsilon}^{\mathrm{th}^{\mathrm{T}}}\mathrm{d}V + \int_V (\boldsymbol{D}^{\mathrm{T}}\boldsymbol{G})^{\mathrm{T}}\boldsymbol{E}\boldsymbol{\varepsilon}_n^{\mathrm{pl}^{\mathrm{T}}}\,\mathrm{d}V\right) \\ &= \frac{1}{2}\boldsymbol{a}^{\mathrm{T}}\boldsymbol{K}\boldsymbol{a} - \boldsymbol{a}^{\mathrm{T}}(\boldsymbol{f} + \boldsymbol{f}^{\mathrm{th}} + \boldsymbol{f}_n^{\mathrm{pl}}) = \tilde{\Pi}_n(\boldsymbol{a}) \end{aligned} \tag{2.2.76}$$

with

$$\begin{aligned} \boldsymbol{K} &= \int_V (\boldsymbol{D}^{\mathrm{T}}\boldsymbol{G})^{\mathrm{T}}\boldsymbol{E}\boldsymbol{D}^{\mathrm{T}}\boldsymbol{G}\,\mathrm{d}V = \int_V \boldsymbol{B}^{\mathrm{T}}\boldsymbol{E}\boldsymbol{B}\,\mathrm{d}V, \quad \boldsymbol{B} = \boldsymbol{D}^{\mathrm{T}}\boldsymbol{G}, \\ \boldsymbol{f} &= \int_V \boldsymbol{G}\bar{\boldsymbol{f}}\,\mathrm{d}V - \int_{A_p} \boldsymbol{G}\bar{\boldsymbol{p}}\,\mathrm{d}A, \\ \boldsymbol{f}^{\mathrm{th}} &= \int_V \boldsymbol{B}^{\mathrm{T}}\boldsymbol{E}\boldsymbol{\varepsilon}^{\mathrm{th}^{\mathrm{T}}}\mathrm{d}V, \quad \boldsymbol{f}_n^{\mathrm{pl}} = \int_V \boldsymbol{B}^{\mathrm{T}}\boldsymbol{E}\boldsymbol{\varepsilon}_n^{\mathrm{pl}^{\mathrm{T}}}\,\mathrm{d}V \end{aligned}$$

From the condition $\delta\tilde{\Pi}_n(\boldsymbol{a}) = 0$, the following set of linear algebraic equations can be obtained

$$\boldsymbol{K}\boldsymbol{a} = \boldsymbol{f} + \boldsymbol{f}^{\mathrm{th}} + \boldsymbol{f}_n^{\mathrm{pl}} \tag{2.2.77}$$

After the solution of (2.2.77) the displacements can be computed from (2.2.75) and the stresses from (2.2.45). With the Ritz method and the explicit time integration procedure the step-by-step solution of a creep problem can be utilized as follows:

set $n = 0,\ \boldsymbol{\varepsilon}_0^{\mathrm{pl}} = \mathbf{0},\quad \boldsymbol{\zeta}_0 = \mathbf{0}$

calculate

$$\boldsymbol{K} = \int\limits_V \boldsymbol{B}^{\mathrm{T}}\boldsymbol{E}\boldsymbol{B}\mathrm{d}V,\quad \boldsymbol{f} = \int\limits_V \boldsymbol{G}\bar{\boldsymbol{f}}\mathrm{d}V - \int\limits_{A_p} \boldsymbol{G}\bar{\boldsymbol{p}}\mathrm{d}A,\quad \boldsymbol{f}^{\mathrm{th}} = \int\limits_V (\boldsymbol{D}^{\mathrm{T}}\boldsymbol{G})^{\mathrm{T}}\boldsymbol{E}\boldsymbol{\varepsilon}^{\mathrm{th}^{\mathrm{T}}}\mathrm{d}V$$

solve BVP $\boldsymbol{K}\boldsymbol{a}_0 = \boldsymbol{f} + \boldsymbol{f}^{\mathrm{th}}$ calculate $\tilde{\boldsymbol{u}}_0 = \boldsymbol{G}^{\mathrm{T}}\boldsymbol{a}_0,\ \boldsymbol{\sigma}_0 = \boldsymbol{E}\boldsymbol{D}^{\mathrm{T}}\tilde{\boldsymbol{u}}_0$

1: calculate

$$\Delta\boldsymbol{\varepsilon}_n^{\mathrm{pl}} = \Delta t_n \boldsymbol{g}(\boldsymbol{\sigma}_n, \boldsymbol{\zeta}_n, T_n), \qquad \Delta\boldsymbol{\zeta}_n = \Delta t_n \boldsymbol{h}(\boldsymbol{\sigma}_n, \boldsymbol{\zeta}_n, T_n)$$

$$\boldsymbol{\varepsilon}_{n+1}^{\mathrm{pl}} = \boldsymbol{\varepsilon}_n^{\mathrm{pl}} + \Delta\boldsymbol{\varepsilon}_n^{\mathrm{pl}}, \qquad \boldsymbol{\zeta}_{n+1} = \boldsymbol{\zeta}_n + \Delta\boldsymbol{\zeta}_n,$$

calculate

$$\boldsymbol{f}_{n+1}^{\mathrm{pl}} = \int\limits_V (\boldsymbol{D}^{\mathrm{T}}\boldsymbol{G})^{\mathrm{T}}\boldsymbol{E}\boldsymbol{\varepsilon}_{n+1}^{\mathrm{pl}^{\mathrm{T}}}\mathrm{d}V$$

solve $\boldsymbol{K}\boldsymbol{a}_{n+1} = \boldsymbol{f} + \boldsymbol{f}^{\mathrm{th}} + \boldsymbol{f}_{n+1}^{\mathrm{pl}}$

calculate $\tilde{\boldsymbol{u}}_{n+1} = \boldsymbol{G}^{\mathrm{T}}\boldsymbol{a}_{n+1},\ \boldsymbol{\sigma}_{n+1} = \boldsymbol{E}(\boldsymbol{D}^{\mathrm{T}}\tilde{\boldsymbol{u}}_{n+1} - \boldsymbol{\varepsilon}_{n+1}^{\mathrm{pl}})$

if $t_{n+1} < t_N$ and $\omega_l < \omega_{l*}, l = 1, \ldots, m$

then set $n := n+1$ go to 1

else finish

The vector $\boldsymbol{f}_n^{\mathrm{pl}}$ must be computed at each time step through a numerical integration. Therefore, the domain discretization is required to store the vectors $\boldsymbol{\varepsilon}^{\mathrm{pl}}$ and $\boldsymbol{\zeta}$. The accuracy of the solution by the Ritz method depends on the "quality" and the number of trial functions. For special problems with simple geometry, homogeneous boundary conditions, etc., trial functions can be formulated in terms of elementary functions (e.g. orthogonal polynomials, trigonometric or hyperbolic functions, etc.) defined within the whole domain, e.g. Altenbach et al (2016). Examples for such problems include beams (Altenbach et al, 2000b; Boyle and Spence, 1983) and plates (Altenbach et al, 1997a; Eisenträger et al, 2015; Naumenko et al, 2001). The Ritz method is simple in utilization and provides an approximate analytical solution.

In the general case of complex geometry, a powerful tool is the finite element method. The domain is subdivided into finite elements and the piecewise trial functions (polynomials) are defined within the elements. For details of finite element techniques we refer to the textbooks of Bathe (1996); Belytschko et al (2014); Betten (1998); Wriggers (2008); Zienkiewicz and Taylor (1991). By analogy with the Ritz method the finite element procedure results in a set of algebraic equations of the type

$$\boldsymbol{K}\boldsymbol{\delta}_n = \boldsymbol{f} + \boldsymbol{f}^{\mathrm{th}} + \boldsymbol{f}_n^{\mathrm{pl}}, \tag{2.2.78}$$

where $\boldsymbol{K}$ is the overall stiffness matrix, $\boldsymbol{\delta}_n$ is the vector of unknown nodal displacements and $\boldsymbol{f}$, $\boldsymbol{f}^{\text{th}}$ and $\boldsymbol{f}_n^{\text{pl}}$ are the nodal force vectors computed from given loads, thermal strains as well as creep strains at the time or iteration step. The commercial codes usually include more sophisticated time integration methods allowing the automatic time step size control. The vector $\boldsymbol{f}_n^{\text{pl}}$ depends on the distribution of creep strains at the current time step. The creep strains are determined by the constitutive model, and a variety of constitutive models can be applied depending on the material type, type of loading, available experimental data, etc. Therefore the possibility to incorporate a user-defined material law is usually available in commercial codes.

2.3 Temporal Scale Procedures

The aim of this section is the analysis of inelastic behavior under periodic cyclic loading regimes. The basic idea is to derive constitutive and evolution equations for rate of change of the mean variables observed over a long period of time and many loading cycles from a unified constitutive law which is available and valid for both monotonic and cyclic loading conditions.

Subsection 2.3.1 presents examples of inelastic material and structural behavior under cyclic loading. Basic time-scale approaches for the structural analysis are discussed in Subsect. 2.3.2. Basic ideas of asymptotic two-time-scales and time averaging procedures are discussed in Subsect. 2.3.3. Subsection 2.3.4 presents an example of the cyclic creep analysis based on the Frederick Armstrong type constitutive model and applying the two-time-scale asymptotic technique. In particular a technique to derive the constitutive equation for the mean inelastic strain rate and the evolution equation for the mean backstress variable is discussed. The constitutive models for the mean processes should contain only principal parameters of the cyclic loading, for example the mean stress, the stress amplitude, type of waveform, etc. and allow an efficient analysis of inelastic behavior under cyclic loading. Furthermore they are helpful in identifying material parameters and validation of a unified constitutive law for cyclic loadings.

2.3.1 Inelastic Behavior with Temporal Scale Effects

Many high-temperature components, for example turbine rotors, turbine housings, pipe systems and turbochargers, are subjected to varying thermal and mechanical loading paths. Changes of environment conditions may cause temperature gradients over the component section and lead to complex timely varying stress and strain states (Benaarbia et al, 2018; Nagode et al, 2011; Kostenko et al, 2013; Holdsworth et al, 2007; Wang et al, 2016; Zhu et al, 2017). In addition, high temperature creep processes (relatively slow time-dependent deformation, stress redistributions and changes in the microstructure observed over a long period of time) dominate for stationary loading regimes (Altenbach and Naumenko, 1997; Altenbach et al, 2001). The combination and interaction of rapidly varying and slowly varying processes can be crucial for the component life.

Analysis of structural behavior over many cycles of loading is crucial for life time estimations. Cyclic hardening , cyclic softening, creep ratcheting, fatigue damage evolution, etc. are examples of processes observable with respect to the global (slow) time scale after a certain number of loading cycles. Loading profiles and material responses within each loading cycle can be related to the (fast) local time scale. At elevated temperature not only the amplitudes of stresses/strains within a cycle, but also loading rates, hold times and many other factors exert an influence on the component life. Simulations of the mechanical behavior of components over many cycles of loading is numerically expensive. For an efficient analysis, time scale approaches and time averaging procedures can be applied. To this end two or more time scales (Altenbach et al, 2000a, 2018; Devulder et al, 2010a; Fish et al, 2012) can be introduced. A "slow or macroscopic" time scale can be used to capture the global cyclic behavior like cyclic hardening, softening or creep ratcheting. For the structural analysis within one loading cycle "fast or microscopic" time scales are useful. Figure 2.2 illustrates the inelastic strain vs. time response for the cyclic force profile with hold times. Two regimes can be recognized, the global one with the growth in the strain amplitude as a function of the "slow" time or cycle number and the rapid change of the inelastic strain within several cycles of loading. In Naumenko and Altenbach (2016, Sect. 1.2.) examples of structural analysis for components subjected to thermo-mechanical cyclic loading are presented. Figure 2.3a illustrates the local changes of normalized tangential strain and the normalized tangential stress components at point A of a rotor subjected to thermo-mechanical cyclic loading. Compressive regime during warm-up stage, creep and relaxation regimes during steady running stage and tensile regime during cool-down stage can be observed for one loading cycle. This "fast" time behavior determines the structural response over many cycles of loading. Figure 2.3b shows the normalized tangential stress as a function of the tangential mechanical strain. A change of the local hysteresis loop with respect to the cycle number can be observed. It shifts from cycle to cycle towards the compressive stress and strain axes. Prediction of such "slow" stress and stress changes over many loading cycles is crucial for life-time estimations of structural components.

Inelastic material behavior at high temperature is controlled by many microstructural processes having different characteristic times. For crystalline materials, inelastic flow is determined by dislocation glide and dislocation climb mechanisms (Frost and Ashby, 1982; Nabarro and de Villiers, 1995). The glide motion of dislocations dominates at lower homologous temperatures and higher stress levels. The climb of dislocations over obstacles is controlled by diffusion of lattice vacancies and becomes important in high-temperature regimes and moderate stress levels. As a thermally activated process, the diffusion of vacancies occurs over time scales that are much longer than the times required for glide steps. The difference in the time scales may be of many orders of magnitude depending on the stress and temperature levels.

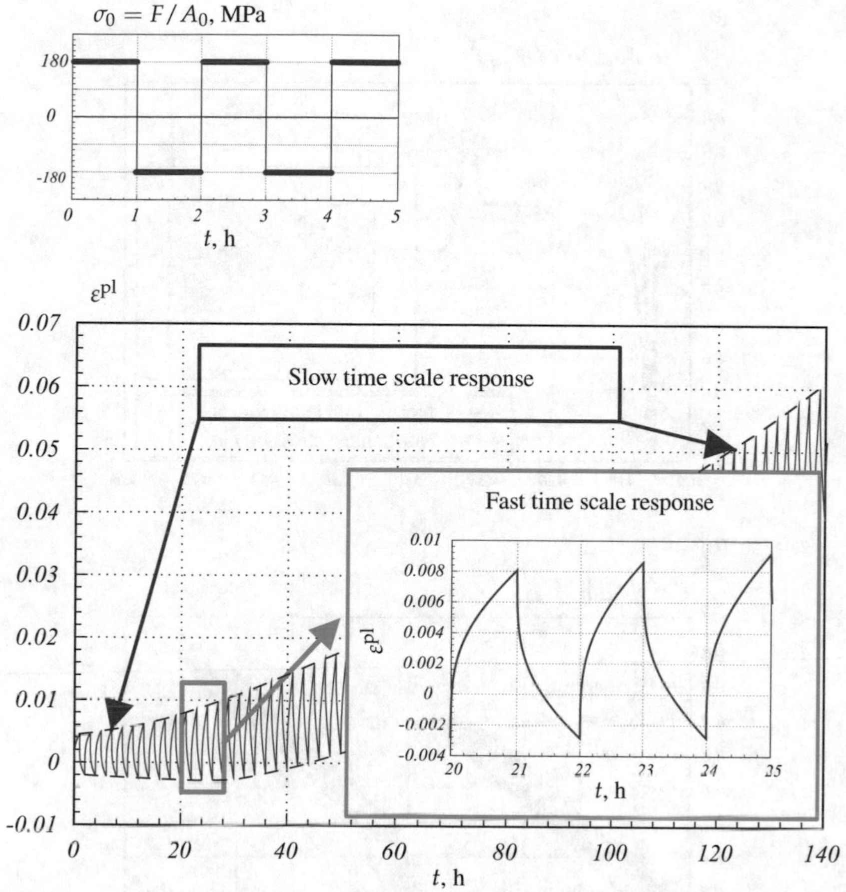

Fig. 2.2 Inelastic strain vs. time in a specimen from 12% Cr steel under cyclic force with hold times at 600 ° C (simulation)

2.3.2 Temporal scale Approaches

To analyze inelastic material behavior and structural responses, efficient numerical time step and iteration algorithms are developed and utilized (Belytschko et al, 2014; de Souza Neto et al, 2011; Shutov, 2016; Shutov et al, 2017). Examples of inelastic structural analysis of power plant components are presented in Kostenko et al (2013); Naumenko and Altenbach (2016); Naumenko and Kostenko (2009); Wang et al (2016); Zhu et al (2017) illustrating local changes of stress state for given external thermo-mechanical loading profiles. For the analysis of structural behavior over many loading/unloading cycles, a direct numerical solution would require computationally expensive cycle-by-cycle integration with relatively small time steps. Several approaches are available to decrease the computation time.

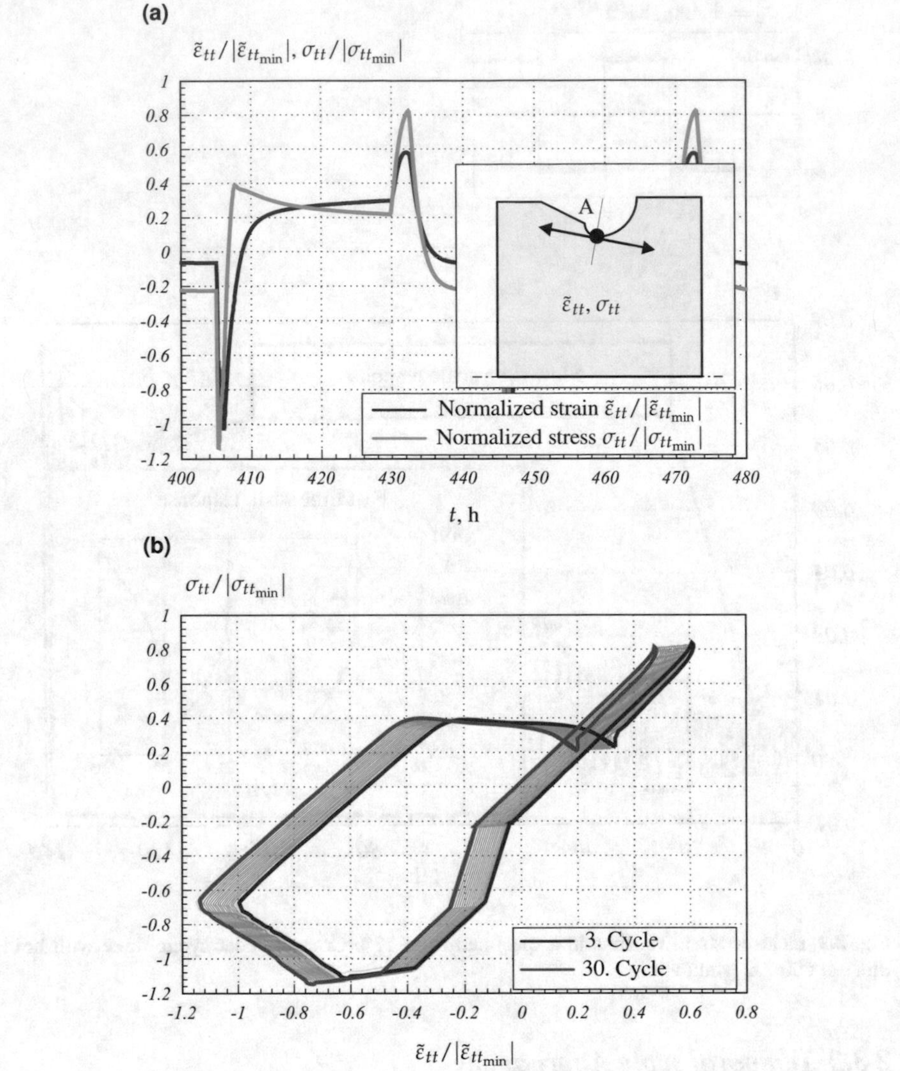

Fig. 2.3 Local stress and strain profiles at point A of a rotor with a groove subjected to thermo-mechanical cyclic loading. **(a)** Normalized tangential stress and tangential mechanical strain vs. time, **(b)** normalized tangential stress vs. tangential mechanical strain, after Naumenko et al (2011b)

The *empirical models* provide the rate of change of inelastic strains and internal state variables with respect to the cycle number as functions of stress state characteristics within one cycle, such as mean stress, stress amplitude, etc. To calibrate empirical models, experimental data for inelastic response under cyclic loading and/or simulations of material and structural behavior over several loading cycles are required. Examples include fatigue damage models of the type

$$\frac{\mathrm{d}\omega}{\mathrm{d}N} = f(\sigma_\mathrm{m}, \sigma_\mathrm{a}, \omega),$$

where N is the cycle number, σ_m is the mean stress and σ_a is the stress amplitude (Lemaitre and Desmorat, 2005), and the Paris law of fatigue crack growth (Paris and Erdogan, 1963)

$$\frac{\mathrm{d}a}{\mathrm{d}N} = C\Delta K^n,$$

where a is the crack length, C and n are material parameters and ΔK is the range of the stress intensity factor within the cycle

$$\Delta K = K_\mathrm{max} - K_\mathrm{min}$$

The *cycle jumping technique* is developed to enable inelastic strain tensor and internal variables to be calculated over many loading cycles, without the need to carry out the numerical integration of around all cycles (Lin et al, 1998; Labergere et al, 2015). This approach requires the solution of the initial-boundary value problem with a unified constitutive model for several cycles to achieve stabilization, from which the rate of change of unknown variables and, in turn, their rates of change with respect to loading cycle number can be evaluated.

Another method for analysis of inelastic responses under cyclic loading is based on *time-scales and time-averaging techniques*. Asymptotic time-scale methods were originally developed to solve differential equations for non-linear dynamical systems (Sanders and Verhulst, 1985; Bensoussan et al, 1978; Nayfeh, 1993). With regard to inelastic material response, the asymptotic approach was originally applied in Morachkovskii (1992); Moratschkowski and Naumenko (1995); Morachkovskii (1992); Altenbach et al (2000a) to analyze dynamic creep, i.e. the increase of inelastic strain under combined static and rapidly varying cyclic loading with a small amplitude. During the last decades the time-scale methods were adapted to analyze cyclic plasticity, visco-plasticity and fatigue. In Oskay and Fish (2004); Yu and Fish (2002); Devulder et al (2010b); Haouala and Doghri (2015) temporal asymptotic homogenisation is applied to solve non-linear initial-boundary value problems of visco-plasticity and damage for structural analysis of components subjected to cyclic loading. In Joseph et al (2010) the wavelet transformation based multi-time scaling algorithm is proposed for accelerated crystal plasticity finite element simulations of polycrystalline aggregates.

2.3.3 Two-time-scales and Time Averaging Procedures

To discuss the idea of time scales and time averaging let us consider a system of ordinary differential equations

$$\frac{\mathrm{d}\mathbf{x}}{\mathrm{d}t} = \mathbf{X}(t, \mathbf{x}) \qquad \mathbf{x}(0) = \mathbf{x}_0, \tag{2.3.79}$$

with

$$\mathbf{x} = (x_1, x_2, \ldots, x_r)^{\mathrm{T}}, \qquad \mathbf{X} = (X_1, X_2, \ldots, X_r)^{\mathrm{T}}, \tag{2.3.80}$$

where x_i is a set of unknown variables, for example, components of the inelastic strain tensor, the backstress tensor, damage parameters, etc. X_i are assumed to be continuous functions of time and x_i. The system (2.3.79) is non-autonomous as the external loading is a function of time. The loading profile usually consists of slow and fast parts. An example is the cyclic loading with the period of the cyclic component $\mathbb{T}$ much less than the time interval $[0, t_*]$, over which the material response is to be analyzed. An efficient method to solve Eqs (2.3.79) is based on the introduction of two or several time scales (Sanders and Verhulst, 1985; Bensoussan et al, 1978; Nayfeh, 1993). The basic idea of the two-time scales approach is to represent the time dependence of the solution $\mathbf{x}(t)$ in the form

$$\mathbf{x}(t) = \hat{\mathbf{x}}[\mathcal{T}_0(t), \mathcal{T}_1(t)], \tag{2.3.81}$$

where $\mathcal{T}_0$ and $\mathcal{T}_1$ are time scale functions.

Introducing a small parameter $\mu = \mathbb{T}/t_* \ll 1$ the slow and fast time scales can be introduced as follows

$$\mathcal{T}_0(t) = t, \quad \mathcal{T}_1(t) \equiv \tau(t) = \frac{t}{\mu}$$

With the slow time variable t and fast time variable τ the total time derivative takes the form

$$\frac{\mathrm{d}}{\mathrm{d}t} = \frac{\partial}{\partial t} + \frac{1}{\mu}\frac{\partial}{\partial \tau} \tag{2.3.82}$$

With Eq. (2.3.82) the ordinary differential equations (2.3.79) can be transformed to the following system of partial differential equations

$$\frac{\partial \mathbf{x}}{\partial t} + \frac{1}{\mu}\frac{\partial \mathbf{x}}{\partial \tau} = \mathbf{X}(t, \tau, \mathbf{x}) \tag{2.3.83}$$

One way to find an approximate solution to Eqs (2.3.83) is to apply the following asymptotic series expansion with respect to the small parameter μ

$$\mathbf{x}(t, \tau) = \mathbf{x}^{(0)}(t, \tau) + \mu \mathbf{x}^{(1)}(t, \tau) + \mu^2 \mathbf{x}^{(2)}(t, \tau) + \ldots, \tag{2.3.84}$$

where $\mathbf{x}^{(k)}(t, \tau)$ are unknown functions of two time variables. The right-hand side of Eqs (2.3.83) can be expanded as follows

$$\begin{aligned} \mathbf{X}(t, \tau, \mathbf{x}) &= \mathbf{X}(t, \tau, \mathbf{x}^{(0)} + \mu \mathbf{x}^{(1)} + \ldots) \\ &= \mathbf{X}(t, \tau, \mathbf{x}^{(0)}) + \mu \frac{\partial \mathbf{X}}{\partial \mathbf{x}}(t, \tau, \mathbf{x}^{(0)})\mathbf{x}^{(1)} + \ldots \end{aligned} \tag{2.3.85}$$

After inserting Eqs (2.3.84) and (2.3.85) into Eqs (2.3.83) and collecting terms with the same order of magnitude, the following set of differential equations can be obtained

$$\begin{aligned}
0(\mu^{-1}) &: \frac{\partial \mathbf{x}^{(0)}}{\partial \tau} = 0, \\
0(\mu^{0}) &: \quad \frac{\partial \mathbf{x}^{(0)}}{\partial t} + \frac{\partial \mathbf{x}^{(1)}}{\partial \tau} = \mathbf{X}(t,\tau,\mathbf{x}^{(0)}), \\
0(\mu^{1}) &: \quad \frac{\partial \mathbf{x}^{(1)}}{\partial t} + \frac{\partial \mathbf{x}^{(2)}}{\partial \tau} = \frac{\partial \mathbf{X}}{\partial \mathbf{x}}(t,\tau,\mathbf{x}^{(0)})\mathbf{x}^{(1)}, \\
&\ldots \quad \ldots
\end{aligned} \tag{2.3.86}$$

From Eq. $(2.3.86)_1$ it follows that the mean solution $\mathbf{x}^{(0)}$ is a function of the slow time t only. Following Sanders and Verhulst (1985) we introduce the following time average operator

$$\underset{\mathcal{T}}{M}\langle f(t,\tau)\rangle = \lim_{\mathcal{T}\to\infty} \frac{1}{\mathcal{T}} \int_0^{\mathcal{T}} f(t,\tau)\mathrm{d}\tau \tag{2.3.87}$$

Let us assume that the time average of the vector function $\mathbf{X}(t,\tau,\mathbf{x}^0)$ exists such that

$$\bar{\mathbf{X}}(t,\mathbf{x}^0) = \underset{\mathcal{T}}{M}\left\langle \mathbf{X}(t,\tau,\mathbf{x}^0)\right\rangle$$

and

$$\underset{\mathcal{T}}{M}\left\langle \frac{\partial \mathbf{x}^{(1)}}{\partial \tau}\right\rangle = \lim_{\mathcal{T}\to\infty} \frac{1}{\mathcal{T}}\left(\mathbf{x}^{(1)}(t,\mathcal{T}) - \mathbf{x}^{(1)}(t,0)\right) = \mathbf{0}$$

Then applying the average operator (2.3.87) to Eqs $(2.3.86)_2$ the following system of ordinary differential equations can be obtained

$$\frac{\mathrm{d}\mathbf{x}^{(0)}}{\mathrm{d}t} = \bar{\mathbf{X}}(t,\mathbf{x}^0), \tag{2.3.88}$$

Integration of Eq. (2.3.88) provides the solution for the mean variable $\mathbf{x}^{(0)}$. It is worth to note that $\mathbf{x}^{(1)}, \mathbf{x}^{(2)}, \ldots$ can be obtained by solving (2.3.86). The corresponding solution procedures are discussed in Sanders and Verhulst (1985); Bensoussan et al (1978); Nayfeh (1993). In many cases it is enough to find the mean solution $\mathbf{x}^{(0)}$. Let us note, that for periodic processes with the period $\mathbb{T}$ instead of (2.3.87) the following time averaging operator can be applied

$$\langle f(t,\tau)\rangle = \frac{1}{\mathbb{T}} \int_0^{\mathbb{T}} f(t,\tau)\mathrm{d}\tau = \int_0^1 f(t,\xi)\mathrm{d}\xi, \quad \xi = \frac{\tau}{\mathbb{T}} \tag{2.3.89}$$

2.3.4 Analysis of Cyclic Creep

In this subsection the two-time-scales asymptotic approach is applied to the Frederick-Armstrong type model in order to find the mean (zeroth order) approximation to the inelastic strain rate tensor. For several uni-axial loading profiles, constitutive and evolution equations for the mean inelastic strain rate and mean rate of the backstress will be derived. The influence of the mean stress, the stress amplitude as well as the type of the waveform on the mean inelastic strain rate will be analyzed. To validate the derived equations, the original constitutive model will be integrated numerically applying the cycle-by-cycle time-step procedure for various cyclic loading profiles. The results for the mean inelastic strain and the mean backstress will be compared and analyzed. Numerical examples will be presented for X20CrMoV12-1 steel in the cyclic creep range.

2.3.4.1 Constitutive Equations

Time averaging procedures discussed in Subsect. 2.3.3 can be applied to any unified constitutive model to derive equations for slow cyclic processes. To illustrate the basic ideas let us consider the Frederick-Armstrong-type constitutive model, see Subsect. 2.1.2. The model includes the constitutive equation for the inelastic strain rate tensor

$$\dot{\boldsymbol{\varepsilon}}^{\mathrm{pl}} = \frac{3}{2}\frac{\dot{\varepsilon}_{\mathrm{vM}}}{\overline{\sigma}_{\mathrm{vM}}}\overline{\boldsymbol{s}}, \quad \overline{\boldsymbol{s}} = \boldsymbol{s} - \boldsymbol{\beta},$$
$$\overline{\sigma}_{\mathrm{vM}} = \sqrt{\frac{3}{2}\mathrm{tr}\left(\overline{\boldsymbol{s}}^2\right)}, \quad \dot{\varepsilon}_{\mathrm{vM}} = R(T) f\left(\overline{\sigma}_{\mathrm{vM}}\right) \tag{2.3.90}$$

and the evolution equation for the backstress deviator

$$\dot{\boldsymbol{\beta}} = \frac{2}{3}C_h(T)\left[\dot{\boldsymbol{\varepsilon}}^{\mathrm{pl}} - \frac{3}{2}\dot{\varepsilon}_{\mathrm{vM}}\frac{\boldsymbol{\beta}}{h\left(\sigma_{\mathrm{vM}}\right)}\right], \quad \sigma_{\mathrm{vM}} = \sqrt{\frac{3}{2}\mathrm{tr}\left(\boldsymbol{s}^2\right)} \tag{2.3.91}$$

where $C_h(T)$ is a function of temperature and $h\left(\sigma_{\mathrm{vM}}\right)$ is a function of the von Mises equivalent stress.

For the uni-axial stress state the stress and the backstress deviators take the form

$$\boldsymbol{s} = \sigma\left(\boldsymbol{e}\otimes\boldsymbol{e} - \frac{1}{3}\boldsymbol{I}\right), \quad \boldsymbol{\beta} = \beta\left(\boldsymbol{e}\otimes\boldsymbol{e} - \frac{1}{3}\boldsymbol{I}\right), \tag{2.3.92}$$

where σ is the uni-axial stress, β is the uni-axial backstress and the unit vector $\boldsymbol{e}$ stands for the loading direction. Equations (2.3.90) and (2.3.91) take the following form

$$\dot{\varepsilon}^{\mathrm{pl}} = R(T) f\left(|\sigma - \beta|\right)\mathrm{sgn}(\sigma - \beta), \quad \dot{\beta} = C_h(T)\left[\dot{\varepsilon}^{\mathrm{pl}} - |\dot{\varepsilon}^{\mathrm{pl}}|\frac{\beta}{h(|\sigma|)}\right] \tag{2.3.93}$$

2.3.4.2 Constitutive Equations for Slow Process

Let us apply the two-time scales approach presented in Subsect. 2.3.3 to the Frederick-Armstrong-type constitutive model in order to evaluate the mean inelastic strain tensor and the mean backstress deviator. Assume that the stress tensor is given as the following function of time

$$\boldsymbol{\sigma}(t,\tau) = \boldsymbol{\sigma}^{(0)}(t) + \boldsymbol{\sigma}^{(1)}(\tau),$$

where $\boldsymbol{\sigma}^{(0)}(t)$ is the mean component and $\boldsymbol{\sigma}^{(1)}(\tau)$ is the periodic one. The deviatoric part is defined as follows

$$\boldsymbol{s}(t,\tau) = \boldsymbol{s}^{(0)}(t) + \boldsymbol{s}^{(1)}(\tau), \quad \boldsymbol{s}^{(0)} = \boldsymbol{\sigma}^{(0)} - \frac{1}{3}\mathrm{tr}\,\boldsymbol{\sigma}^{(0)}\boldsymbol{I}, \quad \boldsymbol{s}^{(1)} = \boldsymbol{\sigma}^{(1)} - \frac{1}{3}\mathrm{tr}\,\boldsymbol{\sigma}^{(1)}\boldsymbol{I}$$

The solution to Eqs. (2.3.90) and (2.3.91) is assumed as follows

$$\begin{aligned} \boldsymbol{\varepsilon}^{\mathrm{pl}}(t,\tau) &= \boldsymbol{\varepsilon}^{\mathrm{pl}^{(0)}}(t,\tau) + \mu\boldsymbol{\varepsilon}^{\mathrm{pl}^{(1)}}(t,\tau) + \ldots, \\ \boldsymbol{\beta}(t,\tau) &= \boldsymbol{\beta}^{(0)}(t,\tau) + \mu\boldsymbol{\beta}^{(1)}(t,\tau) + \ldots \end{aligned} \tag{2.3.94}$$

Taking into account Eqs (2.3.88) equations for the mean processes can be derived from Eqs (2.3.90) and (2.3.91). As a result we obtain

$$\begin{aligned} \dot{\boldsymbol{\varepsilon}}^{\mathrm{pl}^{(0)}} &= \frac{3}{2}R(T)\left\langle \frac{f(\tilde{\sigma}_{\mathrm{vM}})}{\tilde{\sigma}_{\mathrm{vM}}}\left(\boldsymbol{s}^{(0)} - \boldsymbol{\beta}^{(0)} + \boldsymbol{s}^{(1)}\right)\right\rangle, \\ \dot{\boldsymbol{\beta}}^{(0)} &= \frac{2}{3}C_h(T)\left(\dot{\boldsymbol{\varepsilon}}^{\mathrm{pl}^{(0)}} - \frac{3}{2}\boldsymbol{\beta}^{(0)}\left\langle\frac{\dot{\varepsilon}^{(0)}_{\mathrm{vM}}}{h(\sigma_{\mathrm{vM}})}\right\rangle\right), \end{aligned} \tag{2.3.95}$$

where

$$\tilde{\sigma}_{\mathrm{vM}} = \sqrt{\frac{3}{2}\mathrm{tr}\,\left(\boldsymbol{s}^{(0)} - \boldsymbol{\beta}^{(0)} + \boldsymbol{s}^{(1)}\right)^2}, \quad \dot{\varepsilon}^{(0)}_{\mathrm{vM}} = \sqrt{\frac{2}{3}\mathrm{tr}\,\left(\dot{\boldsymbol{\varepsilon}}^{\mathrm{pl}^{(0)2}}\right)}$$

Equations (2.3.95) can be solved providing $\boldsymbol{\varepsilon}^{\mathrm{pl}^{(0)}}(t)$ and $\boldsymbol{\beta}^{(0)}(t)$ for the given stress tensor profile.

To illustrate some analytical estimates based on Eqs. (2.3.95) let us assume a combined uni-axial cyclic load $\sigma = \sigma^{(0)} + \sigma^{(1)}(\tau)$ with the constant part $\sigma^{(0)}$ and the periodic part $\sigma^{(1)}(\tau)$. Equations (2.3.95) take the following form

$$\begin{aligned} \dot{\varepsilon}^{\mathrm{pl}^{(0)}} &= R(T)\left\langle f\left(\left|\sigma^{(0)} - \beta^{(0)} + \sigma^{(1)}\right|\right)\mathrm{sgn}\left(\sigma^{(0)} - \beta^{(0)} + \sigma^{(1)}\right)\right\rangle, \\ \dot{\beta}^{(0)} &= C_h(T)\left(\dot{\varepsilon}^{\mathrm{pl}^{(0)}} - \beta^{(0)}\left\langle\frac{|\dot{\varepsilon}^{\mathrm{pl}^{(0)}}|}{h(|\sigma|)}\right\rangle\right) \end{aligned} \tag{2.3.96}$$

In the first example we assume a rectangular waveform type stress profile, as shown in Fig. 2.4, with the mean stress $\sigma_\mathrm{m} > 0$ and the stress amplitude $0 < \sigma_\mathrm{a} < \sigma_\mathrm{m}$. Performing the integrations over the period of loading in Eqs. (2.3.96) we obtain

$$\begin{aligned} \dot{\varepsilon}^{\mathrm{pl}(0)} &= \frac{1}{2}\left(\dot{\varepsilon}^{\mathrm{pl}}_{\max} + \dot{\varepsilon}^{\mathrm{pl}}_{\min}\right), \\ \dot{\beta}^{(0)} &= C_h(T)\left[\dot{\varepsilon}^{\mathrm{pl}(0)} - \frac{1}{2}\left(\frac{\dot{\varepsilon}^{\mathrm{pl}}_{\max}}{h_{\max}} + \frac{\dot{\varepsilon}^{\mathrm{pl}}_{\min}}{h_{\min}}\right)\beta^{(0)}\right], \end{aligned} \tag{2.3.97}$$

where

$$\dot{\varepsilon}^{\mathrm{pl}}_{\max} = R(T) f(\sigma_\mathrm{m} + \sigma_\mathrm{a} - \beta^{(0)}), \quad \dot{\varepsilon}^{\mathrm{pl}}_{\min} = R(T) f(\sigma_\mathrm{m} - \sigma_\mathrm{a} - \beta^{(0)})$$
$$h_{\max} = h(\sigma_\mathrm{m} + \sigma_\mathrm{a}, T), \quad h_{\min} = h(\sigma_\mathrm{m} - \sigma_\mathrm{a}, T)$$

The constitutive equations (2.3.97) contain only the mean stress and the stress amplitude as loading parameters. One feature is that the mean creep rate and the mean backstress rate do not depend on the period of the fast loading. This is the result of the assumption $\mu \ll 1$.

2.3.4.3 Examples

To illustrate several analytical and numerical solutions, let us consider response functions and material parameters given in Naumenko et al (2011a,b) for the steel X20CrMoV12-1 as follows

$$R(T) = a_0 \mathrm{e}^{-\frac{\alpha}{T}}, \quad f(\sigma) = \sinh(B\sigma), \quad h(|\sigma|) = H_*|\sigma|, \tag{2.3.98}$$

where a_0, α, B, H_* are material parameters. Let us note that the response functions (2.3.98) are identified based on experimental data of creep for temperatures and stresses within the range $773\mathrm{K} \le T \le 873\mathrm{K}$ and $80\mathrm{MPa} \le |\sigma| \le 400\mathrm{MPa}$, respectively. For higher stress levels and/or lower absolute temperatures, advanced function of stress and temperature are required to capture stress-strain and creep curves accurately, as discussed in Eisenträger et al (2018) for steel X20CrMoV12-1. Below we limit our analysis to the typical creep regime characterized by low and moderate stress levels at high temperature. Introducing the following abbreviation

$$A = a_0 \mathrm{e}^{-\frac{\alpha}{T}}, \quad C = C_h(T), \quad M = \frac{\sigma_\mathrm{a}}{\sigma_\mathrm{m}}$$

the constitutive model (2.3.97) reduces to

$$
\begin{aligned}
\dot{\varepsilon}^{\mathrm{pl}^{(0)}} &= A\cosh(B\sigma_{\mathrm{a}})\sinh[B(\sigma_{\mathrm{m}}-\beta^{(0)})],\\
\dot{\beta}^{(0)} &= C\dot{\varepsilon}^{\mathrm{pl}^{(0)}}\left[1-\frac{\beta^{(0)}}{H_*\sigma_{\mathrm{m}}(1-M^2)}\left(1+M\tanh(B\sigma_{\mathrm{a}})\coth[B(\sigma_{\mathrm{m}}-\beta^{(0)})]\right)\right]
\end{aligned}
\qquad (2.3.99)
$$

Equations (2.3.99) can be integrated with respect to the slow time variable providing the mean backstress and the mean creep strain. The numerical solution can be performed by standard time integration techniques developed for creep problems, for example the explicit Euler method, and does not require advanced cycle-by-cycle time step algorithms or cycle jumping techniques. For several loading cases analytical solutions can be derived illustrating cycle-by-cycle evolution of mean inelastic strains. Such solutions are helpful for the identification of material parameters in Frederick-Armstrong type constitutive equations based on experimental data for cyclic loading. For example for $M^2 \ll 1$ the evolution equation $(2.3.99)_2$ can be integrated with elementary functions as follows

$$
\beta^{(0)}(\varepsilon^{\mathrm{pl}^{(0)}}) = H_*\sigma_{\mathrm{m}}\left[1-\exp\left(-\frac{C}{H_*\sigma_{\mathrm{m}}}\varepsilon^{\mathrm{pl}^{(0)}}\right)\right] \qquad (2.3.100)
$$

Inserting Eq. (2.3.100) into Eq. $(2.3.99)_1$ provides the following constitutive equation for the mean creep rate

$$
\dot{\varepsilon}^{\mathrm{pl}^{(0)}} = A\cosh(B\sigma_{\mathrm{a}})\sinh[B\sigma_{\mathrm{m}}g(\varepsilon^{\mathrm{pl}^{(0)}})], \qquad (2.3.101)
$$

where

$$
g(\varepsilon^{\mathrm{pl}^{(0)}}) = 1 - H_*\left[1-\exp\left(-\frac{C}{H_*\sigma_{\mathrm{m}}}\varepsilon^{\mathrm{pl}^{(0)}}\right)\right]
$$

We observe that the mean creep rate decreases with an increase of the mean creep strain towards the steady-state value

$$
\dot{\varepsilon}^{\mathrm{pl}^{(0)}} = A\cosh(B\sigma_{\mathrm{a}})\sinh[B\sigma_{\mathrm{m}}(1-H_*)] \qquad (2.3.102)
$$

An increase of the stress amplitude would result in an increase of the mean creep rate.

Let us note that the two-time scale asymptotic approach provides approximate values of the mean inelastic strain and the mean backstress since only the leading term in the asymptotic expansion (2.3.84) is considered. To analyze the accuracy let us compare the results of asymptotic procedure with the results of direct cycle-by-cycle numerical integration of Eqs (2.3.93). To this end the values of material parameters in Eqs (2.3.98) are taken from Naumenko et al (2011a) for $T = 823$ K as follows

$$
a_0 = 4.64\cdot 10^{23}\frac{1}{\mathrm{h}},\ \alpha = 6.12\cdot 10^{4}\frac{1}{\mathrm{K}},
$$

$$
c_{\mathrm{h}} = 8.84,\ B = 7.74\cdot 10^{-2}\frac{1}{\mathrm{MPa}},\ H_* = 0.46
$$

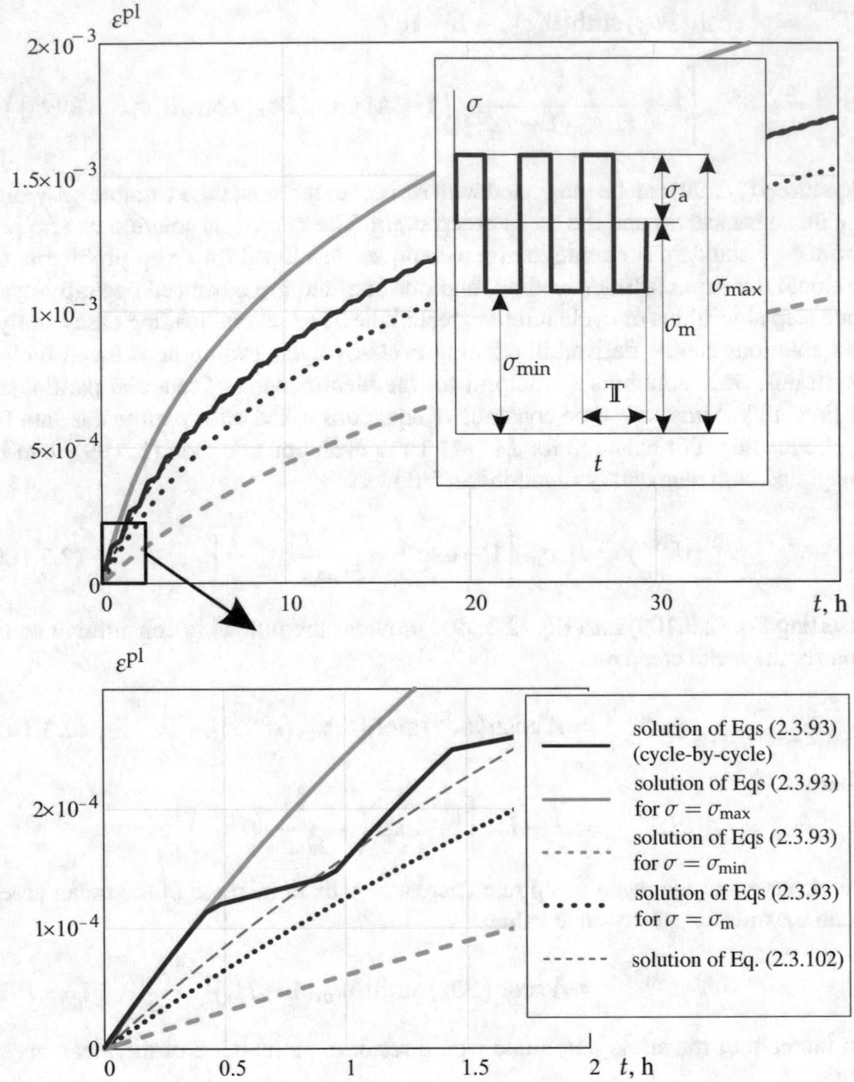

Fig. 2.4 Inelastic responses under uni-axial cyclic stress with rectangular waveform as well as constant maximum, minimum and mean stress levels compared with two-time-scales asymptotic solution

Figure 2.4 shows the result of numerical cycle-by-cycle integration of Eqs (2.3.93) for a cyclic stress profile with a rectangular waveform. For the analysis the loading parameters are assumed as follows $\sigma_m = 150$MPa, $\sigma_a = 10$MPa and $T = 1$h. Furthermore solutions of Eqs (2.3.93) for the constant minimum, mean and maximum stress levels are presented. The asymptotic creep curve obtained by Eq. (2.3.102) is located between the creep curves under constant minimum and maximum stress

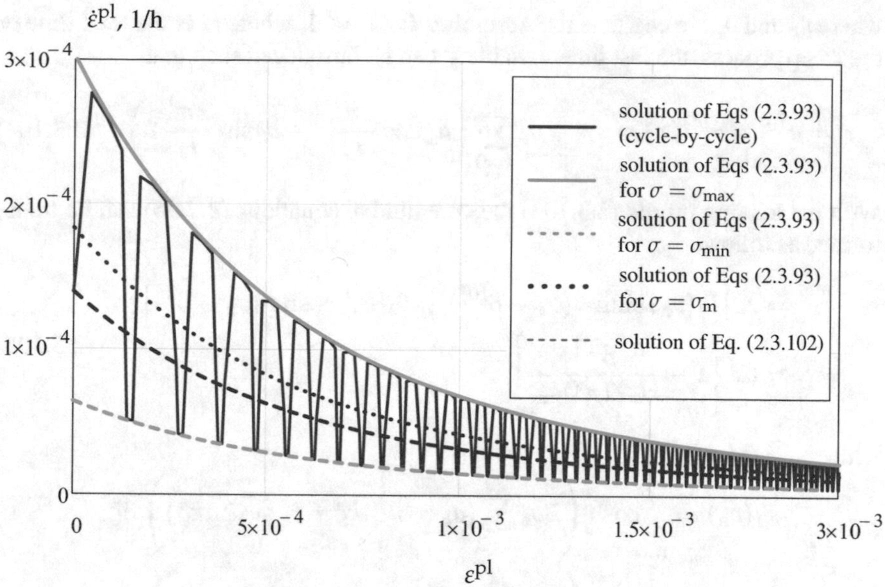

Fig. 2.5 Inelastic strain rate vs. inelastic strain for uni-axial cyclic stress with rectangular waveform as well as constant maximum, minimum and mean stress levels compared with two-time-scales asymptotic solution

level, respectively. Figure 2.5 illustrates the inelastic strain rate as a function of the inelastic strain computed by different methods. The asymptotic creep strain rate is significantly higher than the creep strain rate under constant mean stress. Within several loading cycles significant variations in the creep strain rate are observed. To predict this cyclic creep behavior, advanced time step integration procedures are required. On the other hand, for many cycles of loading over a long period of time, an asymptotic behavior is of primary interest. To evaluate the creep response for many loading cycles directly, an increased numerical effort is required. To avoid expensive step by step computations, the presented two-time-scales asymptotic approach is useful. Indeed, Eq. (2.3.102) predicts the mean inelastic response with satisfactory accuracy, as it is shown in Fig. 2.4. Furthermore, the dependence of the mean creep rate on the mean stress and the stress amplitude is expressed by the constitutive equation (2.3.102) and appropriate response function explicitly. This is an advantage of the two-time-scales approach over numerical techniques, for example, cycle jumping methods.

In the next examples let us consider cyclic loadings with different waveforms. Expanding the periodic part $\sigma^{(1)}$ in a Fourier series, the combined stress can be expressed as follows

$$\sigma = \sigma^{(0)} + \sigma^{(1)}(t) = \sigma_{\mathrm{m}} + \sigma_{\mathrm{a}} \sum_{k=0}^{\infty} \left(a_k \cos \frac{2\pi k}{T} t + b_k \sin \frac{2\pi k}{T} t \right) \qquad (2.3.103)$$

where a_k and b_k are coefficients. Assuming $t_*/T \ll 1$, where t_* is the total time of the creep process, the fast time variable τ can be introduced such that

$$\sigma = \sigma^{(0)} + \sigma^{(1)}(\tau) = \sigma_\mathrm{m} + \sigma_\mathrm{a} \sum_{k=0}^{\infty} \left(a_k \cos \frac{2\pi k}{t_*}\tau + b_k \sin \frac{2\pi k}{t_*}\tau \right) \qquad (2.3.104)$$

With the loading profile (2.3.104) the constitutive equations (2.3.96) can be transformed as follows

$$\begin{aligned} \dot{\varepsilon}^{\mathrm{pl}^{(0)}} &= A\left[F_1(\sigma_\mathrm{a}) \sinh B(\sigma_\mathrm{a} - \beta^{(0)}) + F_2(\sigma_\mathrm{a}) \cosh B(\sigma_\mathrm{a} - \beta^{0})\right], \\ \dot{\beta}^{(0)} &= C\dot{\varepsilon}\left[1 - \frac{\beta^{(0)}}{H_* F(M)\sigma_\mathrm{m}}\right] \end{aligned} \qquad (2.3.105)$$

with

$$\begin{aligned} F_1(\sigma_\mathrm{a}) &= \int_0^1 \cosh\left(B\sigma_\mathrm{a} \sum_{k=0}^{\infty} (a_k \cos 2\pi k\xi + b_k \sin 2\pi k\xi) \right) \mathrm{d}\xi, \\ F_2(\sigma_\mathrm{a}) &= \int_0^1 \sinh\left(B\sigma_\mathrm{a} \sum_{k=0}^{\infty} (a_k \cos 2\pi k\xi + b_k \sin 2\pi k\xi) \right) \mathrm{d}\xi, \\ F(M) &= \int_0^1 \left(1 + M \sum_{k=0}^{\infty} (a_k \cos 2\pi k\xi + b_k \sin 2\pi k\xi) \right) \mathrm{d}\xi \end{aligned}$$

The obtained system (2.3.105) provides the inelastic strain rate and the back stress rate for a slow time scale. The influence of the additional cyclic loading appears in the functions $F_1(\sigma_\mathrm{a}), F_2(\sigma_\mathrm{a})$ and $F(M)$. The second equation in (2.3.105) contains the function of the stress cycle asymmetry parameter $F(M)$. The structure of the function shows that an increase of M would result in an increase of $F(M)$ and an increase of the mean backstress saturation value. In Altenbach et al (2018) numerical examples are presented illustrating the efficiency of the two-time-scales approach for various types of cyclic waveforms.

The superposition of a cyclic component on a constant stress results in an increase of the creep strain rate. Such a behavior is known as dynamic creep and observed for many materials (Naumenko and Altenbach, 2016, Chapt. 1). Dynamic creep can also be described by power law creep constitutive equation modified by a function of the stress cycle asymmetry parameter (Morachkovskii, 1992; Altenbach et al, 2000a).

The benefit of the two-time-scale approach is the possibility to derive constitutive and evolution equations with respect to the slow time and containing principal parameters of cyclic loading, such as mean stress and stress amplitude, from unified constitutive laws which are applicable for both monotonic and cyclic loading regimes. The derived constitutive equations provide the explicit dependencies of the mean rates of the inelastic process, for example the mean creep rate, on the parameters of cyclic loading, such as mean stress and stress amplitude. This is an advantage of asymptotic methods over numerical approaches, such as cycle jumping technique. The constitutive laws for the mean quantities are useful in calibrating constitutive models with experimental data on creep under cyclic loading. Indeed, the equation for the mean deformation rate can be formulated in a closed analytical form and a least square method can be directly applied to estimate the values of material parameters.

The procedure can be generalized to multi-axial stress state for the use in structural analysis of components under combined static and cyclic loadings. The computational time can be significantly reduced since it is not necessary to integrate the constitutive equations within each loading cycle.

References

Abaqus User's Guide (2017) Abaqus Analysis User's Guide. Volume III: Materials

Altenbach H (2018) Kontinuumsmechanik: Einführung in die materialunabhängigen und materialabhängigen Gleichungen, 4th edn. Springer

Altenbach H, Naumenko K (1997) Creep bending of thin-walled shells and plates by consideration of finite deflections. Computational Mechanics 19:490 – 495

Altenbach H, Naumenko K (2002) Shear correction factors in creep-damage analysis of beams, plates and shells. JSME International Journal Series A, Solid Mechanics and Material Engineering 45:77 – 83

Altenbach H, Morachkovsky O, Naumenko K, Sichov A (1996) Zum Kriechen dünner Rotationsschalen unter Einbeziehung geometrischer Nichtlinearität sowie der Asymmetrie der Werkstoffeigenschaften. Forschung im Ingenieurwesen 62(6):47 – 57

Altenbach H, Morachkovsky O, Naumenko K, Sychov A (1997a) Geometrically nonlinear bending of thin-walled shells and plates under creep-damage conditions. Archive of Applied Mechanics 67:339 – 352

Altenbach H, Breslavsky D, Morachkovsky O, Naumenko K (2000a) Cyclic creep damage in thin-walled structures. The Journal of Strain Analysis for Engineering Design 35(1):1 – 11

Altenbach H, Kolarow G, Morachkovsky O, Naumenko K (2000b) On the accuracy of creep-damage predictions in thinwalled structures using the finite element method. Computational Mechanics 25:87 – 98

Altenbach H, Kushnevsky V, Naumenko K (2001) On the use of solid- and shell-type finite elements in creep-damage predictions of thinwalled structures. Archive of Applied Mechanics 71:164 – 181

Altenbach H, Naumenko K, Pylypenko S, Renner B (2007) Influence of rotary inertia on the fiber dynamics in homogeneous creeping flows. ZAMM-Journal of Applied Mathematics and Mechanics/Zeitschrift für Angewandte Mathematik und Mechanik 87(2):81 – 93

Altenbach H, Naumenko K, Gorash Y (2008) Creep analysis for a wide stress range based on stress relaxation experiments. International Journal of Modern Physics B 22:5413 – 5418

Altenbach H, Altenbach J, Naumenko K (2016) Ebene Flächentragwerke. Springer, Berlin

Altenbach H, Breslavsky D, Naumenko K, Tatarinova O (2018) Two-time-scales and time-averaging approaches for the analysis of cyclic creep based on armstrong-frederick type constitutive model. Proceedings of the Institution of Mechanical Engineers, Part C: Journal of Mechanical Engineering Science p 0954406218772609, DOI 10.1177/0954406218772609

Altenbach J, Altenbach H, Naumenko K (1997b) Lebensdauerabschätzung dünnwandiger Flächentragwerke auf der Grundlage phänomenologischer Materialmodelle für Kriechen und Schädigung. Technische Mechanik 17(4):353 – 364

Antman S (1995) Nonlinear Problems of Elasticity. Springer, Berlin

Backhaus G (1983) Deformationsgesetze. Akademie-Verlag, Berlin

Bassani JL, Hawk DE (1990) Influence of damage on crack-tip fields under small-scale-creep conditions. International Journal of Fracture 42:157 – 172

Bathe KJ (1996) Finite Element Rocedures. Prentice-Hall, Englewood Cliffs, New Jersey

Becker AA, Hyde TH, Xia L (1994) Numerical analysis of creep in components. The Journal of Strain Analysis for Engineering Design 29(3):185 – 192

Belytschko T, Liu WK, Moran B, Elkhodary K (2014) Nonlinear Finite Elements for Continua and Structures. Wiley

Benaarbia A, Rae Y, Sun W (2018) Unified viscoplasticity modelling and its application to fatigue-creep behaviour of gas turbine rotor. International Journal of Mechanical Sciences 136:36–49

Bensoussan A, Lions JL, Papanicolaou G (1978) Asymptotic Analysis for Periodic Structures. North-Holland, Amsterdam

Bertram A (2012) Elasticity and Plasticity of Large Deformations, 3rd edn. Springer, Berlin

Besseling JF (1958) A theory of elastic, plastic and creep deformation of an initially isotropic material showing anisotropic strain hardening, creep recovery and secondary creep. Trans of ASME J Appl Mech 25(1):529 – 536

Besseling JF, van der Giessen E (1994) Mathematical Modelling of Inelastic Deformation. Chapman & Hall, London

Betten J (1998) Anwendungen von Tensorfunktionen in der Kontinuumsmechanik anisotroper Materialien. ZAMM-Journal of Applied Mathematics and Mechanics/Zeitschrift für Angewandte Mathematik und Mechanik 78(8):507 – 521

Betten J (2001) Kontinuumsmechanik. Springer, Berlin

Betten J, Borrmann M (1987) Stationäres Kriechverhalten innendruckbelasteter dünnwandiger Kreiszylinderschalen unter Berücksichtigung des orthotropen Werkstoffverhaltens und des CSD - Effektes. Forschung im Ingenieurwesen 53(3):75 – 82

Betten J, Borrmann M, Butters T (1989) Materialgleichungen zur beschreibung des primären kriechverhaltens innendruckbeanspruchter zylinderschalen aus isotropem werkstoff. Ingenieur-Archiv 60(3):99 – 109

Blum W (2008) Mechanisms of creep deformation in steel. In: Abe F, Kern TU, Viswanathan R (eds) Creep-Resistant Steels, Woodhead Publishing, Cambridge, pp 365 – 402

Boyle JT, Spence J (1983) Stress Analysis for Creep. Butterworth, London

Brebbia CA, Telles JCT, Wrobel LC (1983) Boundary Element Techniques. Springer, Berlin

Burlakov AV, Lvov GI, Morachkovsky OK (1977) Polzuchest' tonkikh obolochek (Creep of thin shells, in Russ.). Kharkov State Univ. Publ., Kharkov

Byrne TP, Mackenzie AC (1966) Secondary creep of a cylindrical thin shell subject to axisymmetric loading. J Mech Eng Sci 8(2):215 – 225

Chaboche JL (2008) A review of some plasticity and viscoplasticity constitutive equations. International Journal of Plasticity 24:1642 – 1693

Chowdhury H, Naumenko K, Altenbach H, Krueger M (2017) Rate dependent tension-compression-asymmetry of Ti-61.8 at% Al alloy with long period superstructures at 1050°C. Materials Science and Engineering: A 700:503–511

Chowdhury H, Naumenko K, Altenbach H (2018) Aspects of power law flow rules in crystal plasticity with glide-climb driven hardening and recovery. International Journal of Mechanical Sciences 146-147:486 – 496

Courant R, Friedrichs K, Lewy H (1928) Über die partiellen Differenzengleichungen der mathematischen Physik. Mathematische Annalen 100:32–74

Curnier A (1994) Computational Methods in Solid Mechanics. Kluwer, Dordrect

Devulder A, Aubry D, Puel G (2010a) Two-time scale fatigue modelling: application to damage. Computational Mechanics 45(6):637 – 646

Devulder A, Aubry D, Puel G (2010b) Two-time scale fatigue modelling: application to damage. Computational Mechanics 45(6):637–646

Eisenträger J, Naumenko K, Altenbach H, Köppe H (2015) Application of the first-order shear deformation theory to the analysis of laminated glasses and photovoltaic panels. International Journal of Mechanical Sciences 96:163–171

Eisenträger J, Naumenko K, Altenbach H (2018) Calibration of a phase mixture model for hardening and softening regimes in tempered martensitic steel over wide stress and temperature ranges. The Journal of Strain Analysis for Engineering Design 53(3):156–177

Engeln-Müllges G, Reutter F (1991) Formelsammlung zur numerischen Mathematik mit QuickBASIC-Programmen. B.I. Wissenschaftsverlag, Mannheim

Eringen AC (1999) Microcontinuum Field Theories, vol I: Foundations and Solids. Springer, New York

Fish J, Bailakanavar M, Powers L, Cook T (2012) Multiscale fatigue life prediction model for heterogeneous materials. International Journal for Numerical Methods in Engineering 91(10):1087 – 1104

Frederick CO, Armstrong PJ (2007) A mathematical representation of the multiaxial Bauschinger effect. Materials at High Temperatures 24(1):1 – 26

Frost HJ, Ashby MF (1982) Deformation-Mechanism Maps. Pergamon, Oxford

Gariboldi E, Naumenko K, Ozhoga-Maslovskaja O, Zappa E (2016) Analysis of anisotropic damage in forged Al-Cu-Mg-Si alloy based on creep tests, micrographs of fractured specimen and digital image correlations. Materials Science and Engineering: A 652:175 – 185

Hahn HG (1985) Elastizitätstheorie. B.G. Teubner, Stuttgart

Hairer E, Wanner G (1996) Solving Ordinary Differential Equations. Stiff and Differential-Algebraic Problems II, vol. 14 of Springer Series in Computational Mathematics. Springer

Hairer E, Norset SP, Wanner G (1987) Solving ordinary differential equations, vol I: Nonstiff Problems. Springer, Berlin

Haouala S, Doghri I (2015) Modeling and algorithms for two-scale time homogenization of viscoelastic-viscoplastic solids under large numbers of cycles. International Journal of Plasticity 70:98–125

Hartmann F (1987) Methode der Randelemente. Springer, Berlin

Haupt P (2002) Continuum Mechanics and Theory of Materials. Springer, Berlin

Holdsworth S, Mazza E, Binda L, Ripamonti L (2007) Development of thermal fatigue damage in 1CrMoV rotor steel. Nuclear Engineering and Design 237:2292 – 2301

Hult JA (1966) Creep in Engineering Structures. Blaisdell Publishing Company, Waltham

Hyde T, Sun W, Hyde C (2013) Applied Creep Mechanics. McGraw-Hill Education

Joseph DS, Chakraborty P, Ghosh S (2010) Wavelet transformation based multi-time scaling method for crystal plasticity FE simulations under cyclic loading. Computer Methods in Applied Mechanics and Engineering 199(33 - 36):2177 – 2194

Katsikadelis J (2002) Boundary Elements: Theory and Applications. Elsevier Science

Kostenko Y, Almstedt H, Naumenko K, Linn S, Scholz A (2013) Robust methods for creep fatigue analysis of power plant components under cyclic transient thermal loading. In: ASME Turbo Expo 2013: Turbine Technical Conference and Exposition, American Society of Mechanical Engineers, pp V05BT25A040 – V05BT25A040

Labergere C, Saanouni K, Sun ZD, Dhifallah MA, Li Y, Duval JL (2015) Prediction of low cycle fatigue life using cycles jumping integration scheme. Applied Mechanics and Materials 784:308

Lai WM, Rubin D, Krempl E (1993) Introduction to Continuum Mechanics. Pergamon Press, Oxford

Längler F, Naumenko K, Altenbach H, Ievdokymov M (2014) A constitutive model for inelastic behavior of casting materials under thermo-mechanical loading. The Journal of Strain Analysis for Engineering Design 49:421 – 428

Lemaitre J, Chaboche JL (1990) Mechanics of Solid Materials. Cambridge University Press, Cambridge

Lemaitre J, Desmorat R (2005) Engineering Damage Mechanics: Ductile, Creep, Fatigue and Brittle Failures. Springer

Lin J, Dunne F, Hayhurst D (1998) Approximate method for the analysis of components undergoing ratchetting and failure. The Journal of Strain Analysis for Engineering Design 33(1):55–65

Lurie A (2005) Theory of Elasticity. Foundations of Engineering Mechanics, Springer

Lurie AI (1990) Nonlinear Theory of Elasticity. North-Holland, Dordrecht

Malinin NN (1981) Raschet na polzuchest' konstrukcionnykh elementov (Creep Calculations of Structural Elements, in Russ.). Mashinostroenie, Moskva

Maugin G (2013) Continuum Mechanics Through the Twentieth Century: A Concise Historical Perspective. Solid Mechanics and Its Applications, Springer

Morachkovskii OK (1992) Nonlinear creep problems of bodies under the action of fast field oscillations. International Applied Mechanics 28:489 – 495

Moratschkowski O, Naumenko K (1995) Analyse des Kriechverhaltens dünner Schalen und Platten unter zyklischen Belastungen. ZAMM-Journal of Applied Mathematics and Mechanics/Zeitschrift für Angewandte Mathematik und Mechanik 75(7):507 – 514

Murakami S, Suzuki K (1973) Application of the extended newton method to the creep analysis of shells of revolution. Ingenieur-Archiv 42:194 – 207

Nabarro FRN, de Villiers HL (1995) The Physics of Creep. Creep and Creep-resistant Alloys. Taylor & Francis, London

Nagode M, Längler F, Hack M (2011) Damage operator based lifetime calculation under thermomechanical fatigue for application on Ni-resist D-5S turbine housing of turbocharger. Engineering Failure Analysis 18(6):1565 – 1575

Naumenko K, Altenbach H (2005) A phenomenological model for anisotropic creep in a multi-pass weld metal. Archive of Applied Mechanics 74:808 – 819

Naumenko K, Altenbach H (2016) Modeling High Temperature Materials Behavior for Structural Analysis: Part I: Continuum Mechanics Foundations and Constitutive Models, Advanced Structured Materials, vol 28. Springer

Naumenko K, Gariboldi E (2014) A phase mixture model for anisotropic creep of forged Al-Cu-Mg-Si alloy. Materials Science and Engineering: A 618:368 – 376

Naumenko K, Kostenko Y (2009) Structural analysis of a power plant component using a stress-range-dependent creep-damage constitutive model. Materials Science and Engineering: A 510:169–174

Naumenko K, Altenbach J, Altenbach H, Naumenko VK (2001) Closed and approximate analytical solutions for rectangular Mindlin plates. Acta Mechanica 147:153 – 172

Naumenko K, Altenbach H, Kutschke A (2011a) A combined model for hardening, softening and damage processes in advanced heat resistant steels at elevated temperature. International Journal of Damage Mechanics 20:578 – 597

Naumenko K, Kutschke A, Kostenko Y, Rudolf T (2011b) Multi-axial thermo-mechanical analysis of power plant components from 9-12%Cr steels at high temperature. Engineering Fracture Mechanics 78:1657 – 1668

Nayfeh AH (1993) Introduction to Perturbation Methods. John Wiley and Sons, New York

Odqvist FKG (1974) Mathematical Theory of Creep and Creep Rupture. Oxford University Press, Oxford

Odqvist FKG, Hult J (1962) Kriechfestigkeit metallischer Werkstoffe. Springer, Berlin u.a.

Oskay C, Fish J (2004) Fatigue life prediction using 2-scale temporal asymptotic homogenization. International Journal for Numerical Methods in Engineering 61(3):329–359

Paris P, Erdogan F (1963) A critical analysis of crack propagation laws. Journal of basic engineering 85(4):528–533

Peirce D, Shih CF, Needleman A (1984) A tangent modulus method for rate dependent solids. Computers & Structures 18:875 – 887

Pilkey WD, Wunderlich W (1994) Mechanics of Structures: Variational and Computational Methods. CRC Press, Boca Raton

Podgorny AN, Bortovoj VV, Gontarovsky PP, Kolomak VD, Lvov GI, Matyukhin YJ, Morachkovsky OK (1984) Polzuchest' elementov mashinostroitel'nykh konstrykcij (Creep of mashinery structural members, in Russ.). Naukova dumka, Kiev

Sanders JA, Verhulst F (1985) Averaging Methods in Nonlinear Dynamical Systems. Springer, New York

Schmicker D, Naumenko K, Strackeljan J (2013) A robust simulation of Direct Drive Friction Welding with a modified Carreau fluid constitutive model. Computer Methods in Applied Mechanics and Engineering 265:186 – 194

Schwetlick H, Kretzschmar H (1991) Numerische Verfahren für Naturwissenschaftler und Ingenieure. Fachbuchverlag, Leipzig

Shutov A (2016) Efficient implicit integration for finite-strain viscoplasticity with a nested multiplicative split. Computer Methods in Applied Mechanics and Engineering 306:151–174

Shutov AV, Larichkin AY, Shutov VA (2017) Modelling of cyclic creep in the finite strain range using a nested split of the deformation gradient. ZAMM-Journal of Applied Mathematics and Mechanics/Zeitschrift für Angewandte Mathematik und Mechanik 97(9):1083–1099

Simo J, Hughes T (2000) Computational Inelasticity. Interdisciplinary Applied Mathematics, Springer

Skrzypek J, Ganczarski A (1998) Modelling of Material Damage and Failure of Structures. Foundation of Engineering Mechanics, Springer, Berlin

de Souza Neto EA, Peric D, Owen DR (2011) Computational methods for plasticity: theory and applications. John Wiley & Sons

Wang W, Buhl P, Klenk A, Liu Y (2016) The effect of in-service steam temperature transients on the damage behavior of a steam turbine rotor. International Journal of Fatigue 87:471–483

Washizu K (1982) Variational Methods in Elasticity and Plasticity. Pergamon Press, Oxford

Wriggers P (2008) Nonlinear Finite Element Methods. Springer, Berlin, Heidekberg

Yu Q, Fish J (2002) Temporal homogenization of viscoelastic and viscoplastic solids subjected to locally periodic loading. Computational Mechanics 29(3):199–211

Zhang SL, Xuan FZ (2017) Interaction of cyclic softening and stress relaxation of 9–12% Cr steel under strain-controlled fatigue-creep condition: Experimental and modeling. International Journal of Plasticity 98:45–64

Zhu X, Chen H, Xuan F, Chen X (2017) Cyclic plasticity behaviors of steam turbine rotor subjected to cyclic thermal and mechanical loads. European Journal of Mechanics-A/Solids 66:243–255

Zienkiewicz OC, Taylor RL (1991) The Finite Element Method. McGraw-Hill, London

Chapter 3
Beams

Beams are structural members that are designed to support lateral forces and bending moments. Beams can be also subjected to combined bending, torsion as well as axial tensile or compressive loads. In the case of linear elasticity the laterally loaded beams, rods subjected to torque as well as axially loaded rods can be analyzed separately and the superposition principle can be applied to establish the resultant stress and deformation states. For nonlinear material behavior such a superposition is not possible and combined loadings should be considered. Furthermore, inelastic material response may be different for tensile and compressive loadings leading to a shift of the neutral plane under pure bending.

Beams are also important in testing of materials. Three or four point bending tests are frequently used to analyze inelastic behavior experimentally. Examples are presented in Chuang (1986); Scholz et al (2008); Xu et al (2007) for homogeneous beams, in Weps et al (2013) for laminated beams and in Nordmann et al (2018) for beams with coatings.

Beams are discussed in monographs and textbooks on creep mechanics (Boyle and Spence, 1983; Hult, 1966; Kachanov, 1986; Kraus, 1980; Malinin, 1975, 1981; Odqvist, 1974; Penny and Mariott, 1995; Skrzypek, 1993), where the Bernoulli-Euler beam theory and elementary constitutive equations, such as the Norton-Bailey constitutive law for steady-state creep are applied.

Chapter 3 presents examples of inelastic structural analysis for beams. In Sect. 3.1 the classical Bernoulli-Euler beam theory is introduced. Governing equations and variational formulations for inelastic analysis are introduced. Closed-form solutions and approximate analytical solutions are derived for beams from materials that exhibit power law creep and stress regime dependent creep. Numerical solutions by the Ritz and finite element methods are discussed in detail. Creep and creep-damage constitutive models are applied to illustrate basic features of stress redistribution and damage evolution in beams. Furthermore, several benchmark problems for beams are introduced. The reference solutions for these problems obtained by the Ritz method are applied to verify user-defined creep-damage material subroutines and the general purpose finite element codes.

K. Naumenko and H. Altenbach, *Modeling High Temperature Materials Behavior for Structural Analysis*, Advanced Structured Materials 112,
https://doi.org/10.1007/978-3-030-20381-8_3

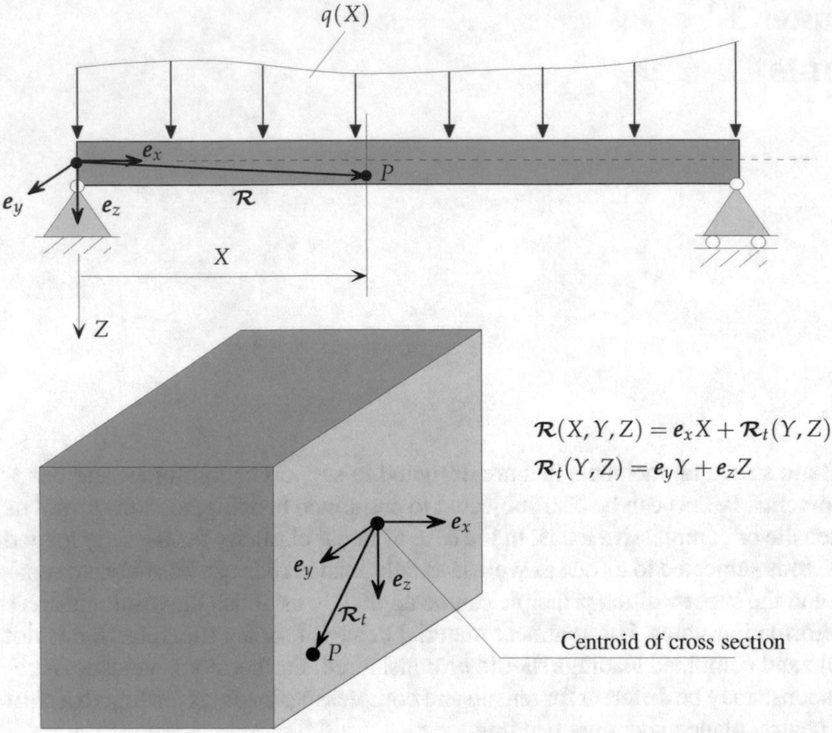

Fig. 3.1 Beam with a rectangular cross section. Geometry, loading and coordinates

For many materials inelastic behavior depends on the kind of stress state. Examples for stress state effects including different creep rates under tension, compression and torsion are discussed in Sect. 3.2. For such kind of material behavior, the classical beam theory may lead to errors in computed deformations and stresses. Section 3.3 presents a refined beam theory which includes the effect of transverse shear deformation (Timoshenko-type theory). Based on several examples, classical and refined theories are compared as they describe creep-damage processes in beams.

3.1 Classical Beam Theory

3.1.1 Governing Equations

Figure 3.1 illustrates a straight homogeneous beam with a rectangular cross section. The position of any point of the beam in the reference state can be defined by the vector

$$\boldsymbol{\mathcal{R}}(X,Y,Z) = \boldsymbol{e}_x X + \boldsymbol{\mathcal{R}}_t(Y,Z),$$

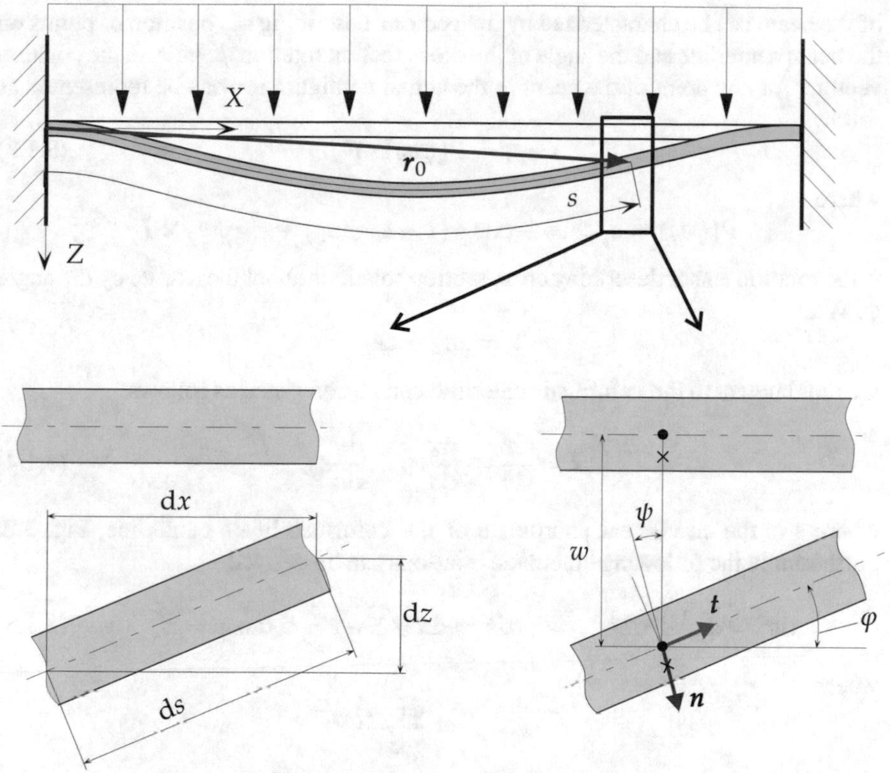

Fig. 3.2 Kinematics of a deformed beam

where X, Y, Z are Cartesian coordinates and $\boldsymbol{e}_x, \boldsymbol{e}_y, \boldsymbol{e}_z$ are basis vectors. The axial coordinate X gives the position of beam cross sections while the vector $\mathcal{R}_t(Y, Z)$ defines the position of a point in a cross section with respect to the centroid. Within the theory of rods and beams it is assumed that the cross sections remain plane during deformation, i.e. they behave like rigid bodies. To describe the deformation it is sufficient to define the actual position of the cross section centroid as well as to describe the cross section rotation. For the sake of brevity let us consider symmetrical bending in the plane spanned on the X and Z coordinate lines. In this special case any unit vector $\boldsymbol{m}$ connected with the actual cross section can be expressed as follows

$$\boldsymbol{m} = \sin\psi \boldsymbol{e}_x + \cos\psi \boldsymbol{e}_z, \tag{3.1.1}$$

where ψ is the angle of cross section rotation, Fig. 3.2. Specifying by $\boldsymbol{t}$ the unit tangent to the beam centerline, Fig. 3.2, we can write

$$\boldsymbol{t} = \cos\varphi \boldsymbol{e}_x - \sin\varphi \boldsymbol{e}_z, \tag{3.1.2}$$

where φ is the angle between the tangent and x directions. The actual configuration of the beam can be characterized by the vector $\boldsymbol{r}$ describing the position of points on the beam centerline and the angle of the cross section rotation ψ. Indeed, the position vector $\mathfrak{r}$ for any point of the beam in the actual configuration can be represented as follows

$$\mathfrak{r} = \boldsymbol{r} + \boldsymbol{P}(\psi \boldsymbol{e}_y) \cdot \mathcal{R}_t, \tag{3.1.3}$$

where

$$\boldsymbol{P}(\psi \boldsymbol{e}_y) = \boldsymbol{e}_y \otimes \boldsymbol{e}_y + \cos\psi(\boldsymbol{I} - \boldsymbol{e}_y \otimes \boldsymbol{e}_y) + \sin\psi\, \boldsymbol{e}_y \times \boldsymbol{I}$$

is the rotation tensor describing cross section rotations about the axis $\boldsymbol{e}_y$ by the angle ψ. With

$$\boldsymbol{r} = x\boldsymbol{e}_x + z\boldsymbol{e}_z$$

the unit tangent to the deformed centerline can be computed as follows

$$\boldsymbol{t} = \frac{\mathrm{d}\boldsymbol{r}}{\mathrm{d}s} = \frac{\mathrm{d}x}{\mathrm{d}s}\boldsymbol{e}_x + \frac{\mathrm{d}y}{\mathrm{d}s}\boldsymbol{e}_y, \tag{3.1.4}$$

where s is the curvilinear coordinate of the deformed beam centerline, Fig. 3.2. Furthermore the following kinematic relations can be derived

$$\mathrm{d}s^2 = \mathrm{d}x^2 + \mathrm{d}z^2 \quad \Rightarrow \quad \mathrm{d}s = \mathrm{d}x\sqrt{1 + z'^2}, \quad \tan\varphi = z', \tag{3.1.5}$$

where

$$(\ldots)' = \frac{\mathrm{d}(\ldots)}{\mathrm{d}x}$$

The derivative of the unit tangent with respect to s provides

$$\frac{\mathrm{d}\boldsymbol{t}}{\mathrm{d}s} = -\frac{1}{\rho}\boldsymbol{n} = -\chi\boldsymbol{n}, \tag{3.1.6}$$

where ρ is the local curvature radius of the beam centerline and χ is the corresponding curvature. Within the Bernoulli-Euler (transverse shear rigid) beam theory it is assumed that any cross section remains orthogonal to the beam centerline in the course of deformation. With $\boldsymbol{t} \cdot \boldsymbol{m} = 0$, Eqs (3.1.1) and (3.1.2) yield

$$\psi = -\varphi \tag{3.1.7}$$

In many cases the cross section rotations remain small such that the rotation tensor $\boldsymbol{P}(\psi \boldsymbol{e}_y)$ can be linearized as follows

$$\boldsymbol{P}(\psi \boldsymbol{e}_y) = \boldsymbol{I} + \psi\, \boldsymbol{e}_y \times \boldsymbol{I}$$

Furthermore with Eqs (3.1.5) and (3.1.7) we obtain

$$\psi = -\varphi = -z' \tag{3.1.8}$$

Within the geometrically-linear beam theory it is convenient to introduce the displacement vector $\boldsymbol{u}$ as follows

$$\boldsymbol{u} = \mathfrak{r} - \mathcal{R} = (u_0 - w'z)\boldsymbol{e}_x + w\boldsymbol{e}_z, \tag{3.1.9}$$

where $u_0 = \boldsymbol{r} \cdot \boldsymbol{e}_x$ is the axial displacement of the beam centerline and

$$w \approx z = \boldsymbol{r} \cdot \boldsymbol{e}_z$$

is the beam deflection. Furthermore all equations can be specified for initial (undeformed) configuration and the difference between the coordinates X, Y, Z and x, y, z can be neglected. The governing equations can be summarized as follows

- Kinematical equations

$$\begin{aligned} u(x,z) &= u_0(x) + \varphi(x)z, \quad \varphi(x) = -w'(x), \\ \varepsilon_x(x,z) &= u_0'(x) + \varphi'(x)z, \end{aligned} \tag{3.1.10}$$

  where $u(x,z)$ is the axial displacement and $\varepsilon_x(x,z)$ is the normal strain in the axial direction.
- Equilibrium conditions

$$N'(x) = 0, \quad Q'(x) + q(x) = 0, \quad M'(x) = Q(x), \tag{3.1.11}$$

  where $N(x)$ is the normal (axial) force, $Q(x)$ is the shear force, $M(x)$ is the bending moment and $q(x)$ is the given distributed lateral load.
- Constitutive equations
  - normal stress

$$\begin{aligned} \sigma_x(x,z) &= E[\varepsilon_x(x,z) - \alpha_T \Delta T(x,z) - \varepsilon_x^{\mathrm{pl}}(x,z)] \\ &= E[\varepsilon_{\mathrm{c}}(x) + \chi(x)z - \alpha_T \Delta T(x,z) - \varepsilon_x^{\mathrm{pl}}(x,z)], \end{aligned} \tag{3.1.12}$$

    where $\varepsilon_{\mathrm{c}} = u_0'$ is the strain of the beam centerline and $\chi = -w''$ is the curvature of the beam centerline.
  - stress resultants

$$\begin{aligned} N(x) &= \int_A \sigma_x \mathrm{d}A = EA\left[\varepsilon_{\mathrm{c}}(x) - \varepsilon_{\mathrm{c}}^{\mathrm{pl}}(x) - \varepsilon_{\mathrm{c}}^{\mathrm{th}}(x)\right], \\ M(x) &= \int_A \sigma_x z \mathrm{d}A = EI\left[\chi(x) - \chi^{\mathrm{pl}}(x) - \chi^{\mathrm{th}}(x)\right], \end{aligned} \tag{3.1.13}$$

    where A is the cross section area, I is the area moment of second degree and

$$\begin{aligned} \varepsilon_{\mathrm{c}}^{\mathrm{pl}}(x) &= \frac{1}{A}\int_A \varepsilon_x^{\mathrm{pl}}(x,z)\mathrm{d}A, \quad \varepsilon_{\mathrm{c}}^{\mathrm{th}}(x) = \alpha_T \frac{1}{A}\int_A \Delta T(x,z)\mathrm{d}A, \\ \chi^{\mathrm{pl}}(x) &= \frac{1}{I}\int_A \varepsilon_x^{\mathrm{pl}}(x,z)z\mathrm{d}A, \quad \chi^{\mathrm{th}}(x) = \alpha_T \frac{1}{I}\int_A \Delta T(x,z)z\mathrm{d}A \end{aligned} \tag{3.1.14}$$

are averages of thermal and inelastic strains. In terms of fictitious forces and moments Eqs (3.1.13) can be rewritten as follows

$$
\begin{aligned}
N(x) &= EA\varepsilon_c(x) - N^{pl}(x) - N^{th}(x), \\
M(x) &= EI\chi(x) - M^{pl}(x) - M^{th}(x)
\end{aligned}
\tag{3.1.15}
$$

with

$$
\begin{aligned}
N^{pl}(x) &= E\int_A \varepsilon_x^{pl}(x,z)\mathrm{d}A, \quad N^{th}(x) = E\alpha_T\int_A \Delta T(x,z)\mathrm{d}A, \\
M^{pl}(x) &= E\int_A \varepsilon_x^{pl}(x,z)z\mathrm{d}A, \; M^{th}(x) = E\alpha_T\int_A \Delta T(x,z)z\mathrm{d}A
\end{aligned}
\tag{3.1.16}
$$

- constitutive model for inelastic material behavior (2.1.12) can be specified for the beam as follows

$$
\begin{aligned}
\dot{\varepsilon}_x^{pl} &= f_\varepsilon(\sigma_x, \varepsilon_x^{pl}, \varsigma_k, \dot{\varsigma}_k, T), \quad k = 1, \ldots, n, \\
\dot{\varsigma}_k &= f_{\varsigma_k}(\sigma_x, \varepsilon_x^{pl}, \dot{\varepsilon}_x^{pl}, \varsigma_l, T), \quad l = 1, \ldots, n,
\end{aligned}
\tag{3.1.17}
$$

where ς_l are hardening, softening, damage or other internal state variables. For example the constitutive equation for steady-state inelastic flow (1.4.56) can be applied to analyze the behavior of beams as follows

$$
\dot{\varepsilon}_x^{pl} = \dot{\varepsilon}_0 f\left(\frac{|\sigma_x|}{\sigma_0}\right)\mathrm{sgn}(\sigma_x),
\tag{3.1.18}
$$

where $\dot{\varepsilon}_0$ and σ_0 are material parameters. $f(x)$ is the stress response function, see Subsect. 1.4.1. Another example is the Kachanov-Rabotnov-type creep-damage constitutive and evolution equations (Naumenko and Altenbach, 2016, Subsect. 5.7.1)

$$
\begin{aligned}
\dot{\varepsilon}_x^{pl} &= \frac{a|\sigma_x|^{n-1}\sigma_x}{(1-\omega)^n}, \quad \dot{\omega} = \frac{b\sigma_{eq}^k}{(1-\omega)^m}, \\
\sigma_{eq} &= \alpha\frac{|\sigma_x| + \sigma_x}{2} + (1-\alpha)|\sigma_x|,
\end{aligned}
\tag{3.1.19}
$$

where ω is the damage variable and a, b, n, k, m, α are material parameters.

The boundary conditions at $x = 0$ and $x = l$ (l is the beam length) can be formulated by two values from the set of the kinematic and static quantities w, φ, Q and M. The initial conditions at $t = 0$ are $\varepsilon_x^{pl} = 0$ and $\omega = 0$.

3.1.2 Variational Formulation and the Ritz Method

The variational formulations and the Ritz method were discussed in Sect. 2.2.3 for three-dimensional solids. To derive the variational functional for the beam we start from the principle of virtual displacements (2.2.71). Applying the kinematical relations (3.1.10) and the constitutive equations (3.1.12) we can write

$$\begin{aligned}\int\limits_V \sigma_x \delta\varepsilon_x \mathrm{d}V &= EI\int\limits_0^l w''\delta w''\mathrm{d}x + EA\int\limits_0^l u_0'\delta u_0' dx \\ &+ \int\limits_0^l M^{\mathrm{pl}}\delta w''\mathrm{d}x - \int\limits_0^l N^{\mathrm{pl}}\delta u_0'\mathrm{d}x \\ &= \int\limits_0^l q\delta w \mathrm{d}x\end{aligned} \tag{3.1.20}$$

To simplify the analysis, the fictitious force and moment associated with the thermal strain are neglected in Eq. (3.1.20). Assuming the inelastic strain to be known function of the coordinates x and z for the fixed time t, the following functional can be formulated

$$\begin{aligned}\Pi_t(w,u_0) &= \frac{1}{2}EI\int\limits_0^l w''^2\mathrm{d}x + \frac{1}{2}EA\int\limits_0^l u_0'^2\mathrm{d}x \\ &+ \int\limits_0^l M^{\mathrm{pl}}w''\mathrm{d}x - \int\limits_0^l N^{\mathrm{pl}}u_0' dx - \int\limits_0^l qw\mathrm{d}x\end{aligned} \tag{3.1.21}$$

The problem is to find such functions w and u_0 that yield an extremum of the functional for the given values of M^{pl}, N^{pl} and q. The approximate solutions can be represented in the form of series

$$w(x,t) = a_0^w(t)\phi_0^w(x) + \sum_{i=1}^{\mathcal{N}} a_i^w(t)\phi_i^w(x), \quad u_0(x) = \sum_{i=1}^{\mathcal{M}} a_i^u(t)\phi_i^u(x), \tag{3.1.22}$$

where $\phi_i^w(x)$ and $\phi_i^u(x)$ are trial functions to be specified a-priori. $a_i^w(t)$ and $a_i^u(t)$ are unknown coefficients to be computed by minimizing the functional (3.1.21). The trial functions should be formulated according to the types of constraints and the loading conditions.

As an example let us consider the simply supported and uniformly loaded beam. In this case the following trial functions can be applied. $\phi_0^w(x) = x(x-l)(x^2 - lx - l^2)$ is the first approximation which follows from the solution for the elastic deflection. $\phi_i^w(x)$ are the polynomials satisfying the boundary conditions for the deflection $w = 0$ and for the bending moment $M = 0$ at $x = 0$ and $x = l$

$$\phi_i^w(x) = x^{i+2}(l-x)^{i+2} \tag{3.1.23}$$

Assuming that $u_0 = 0$ at $x = 0$ the functions $\phi_i^u(x) = x^i$ can be utilized. After collecting the unknown constant coefficients into the vector $\boldsymbol{a}^{\mathrm{T}} = [\boldsymbol{a}^{w^{\mathrm{T}}} \quad \boldsymbol{a}^{u^{\mathrm{T}}}]$ with $\boldsymbol{a}^{w^{\mathrm{T}}} = [a_0^w \quad a_i^w]$, $i = 1,\ldots,\mathcal{N}$ and $\boldsymbol{a}^{u^{\mathrm{T}}} = [a_i^u]$, $i = 1,\ldots,\mathcal{M}$, the Ritz method yields a set of linear algebraic equations

$$\frac{\partial \Pi_t}{\partial a_k} = 0 \quad \Rightarrow \quad \boldsymbol{R}^{ww}\boldsymbol{a}^w = \boldsymbol{f}^w, \quad \boldsymbol{R}^{uu}\boldsymbol{a}^u = \boldsymbol{f}^u \tag{3.1.24}$$

with

$$\begin{aligned}
R_{kj}^{ww} &= EI\int_0^l \phi_k^{w\prime\prime}\phi_j^{w\prime\prime}dx, & R_{kj}^{uu} &= EA\int_0^l \phi_k^{u\prime}\phi_j^{u\prime}dx,\\
f_k^w &= q\int_0^l \phi_k^w dx - \int_0^l M^{\mathrm{pl}}\phi_k^{w\prime\prime}dx, & f_k^u &= \int_0^l N^{\mathrm{pl}}\phi_k^{u\prime}dx,\\
k &= 0,\ldots,\mathcal{M}, \quad j = 0,\ldots,\mathcal{M}
\end{aligned}$$

After solving Eqs (3.1.24) the displacements can be computed from (3.1.22). With (3.1.12) the stress $\sigma_x(x,z,t)$ can be calculated as follows

$$\sigma_x(x,z,t)) = E\left[u_0'(x,t) - zw''(x,t) - \varepsilon_x^{\mathrm{pl}}(x,z,t)\right]$$

For the known values of the stress and internal state variables the constitutive model (3.1.17) yields the rates of inelastic strain and internal state variables for the fixed time t. The corresponding values for the time $t+\Delta t$ are calculated using the implicit time integration procedure (see Subsubsect. 2.2.1.2)

$$\begin{aligned}
\varepsilon^{\mathrm{pl}}(x,z,t+\Delta t) &= \varepsilon^{\mathrm{pl}}(x,z,t) + \frac{\Delta t}{2}[\dot{\varepsilon}^{\mathrm{pl}}(x,z,t) + \dot{\varepsilon}^{\mathrm{pl}}(x,z,t+\Delta t)],\\
\varsigma_k(x,z,t+\Delta t) &= \varsigma_k(x,z,t) + \frac{\Delta t}{2}[\dot{\varsigma}_k(x,z,t) + \dot{\varsigma}_k(x,z,t+\Delta t)],\\
\varepsilon^{\mathrm{pl}}(x,z,0) &= 0, \quad \varsigma_k(x,z,0) = \varsigma_{k_0}
\end{aligned}$$

To evaluate the fictitious force N^{pl} and the moment M^{pl} defined in (3.1.16), the Simpson quadrature rule with $\mathcal{N}_h$ integration points in the thickness direction of the beam can be used. To compute the matrices $\boldsymbol{R}^{ww}$ and $\boldsymbol{R}^{uu}$ as well as the vectors $\boldsymbol{f}^w$ and $\boldsymbol{f}^u$ in (3.1.24) the Simpson quadrature rule with $\mathcal{N}_l$ integration points along the beam length axis x can be applied. The values of the inelastic strain and internal state variables at the current time step must be stored in all integration points. They are used for the calculations at the next time step.

3.1.3 Closed-form Solutions for Steady-state Creep

Consider a beam with a rectangular cross section $g \times h$ subjected to the constant bending moment M, Fig. 3.3. We seek for the rate of change of the beam curvature as well as for the distribution of the normal stress σ_x across the thickness direction as a consequence of the creep process under isothermal condition. This problem is discussed in many books on creep mechanics, e.g. Boyle and Spence (1983); Hult (1966); Malinin (1981); Naumenko and Altenbach (2007) under the assumption of the power law creep. In what follows we present this classical solution as well as solutions with advanced stress response functions.

Let us assume that the spontaneous deformation after the loading of the beam is elastic. Elasticity solutions for beams are presented within the elementary engineering mechanics courses. For the pure bending of the beam the normal stress is computed as follows

$$\sigma_x(z,0) = \frac{M}{W}\frac{2z}{h}, \tag{3.1.25}$$

where $W = gh^2/6$ is the section moment. Taking the time derivative of Eq (3.1.12) the rate of the normal stress is defined as follows

$$\dot{\sigma}_x = E(\dot{\chi}z - \dot{\varepsilon}_x^{\text{pl}}), \tag{3.1.26}$$

where $\dot{\varepsilon}_x^{\text{pl}}$ is given by Eq. (3.1.18). Equations (3.1.18) and (3.1.26) can be solved numerically to compute the stress redistribution in the beam. Numerical solutions will be discussed in Subsect. 3.1.4. When the bending moment M is kept constant, the steady creep state exists for which the normal stress does not change over time. The steady-state stress distribution can be derived in a closed analytical form. Setting $\dot{\sigma}_x = 0$, Eqs (3.1.18) and (3.1.26) yield

$$\dot{\chi}z = \dot{\varepsilon}_0 f\left(\frac{|\sigma_x|}{\sigma_0}\right)\text{sgn}(\sigma_x) \tag{3.1.27}$$

For the given normal stress the bending moment is computed as follows

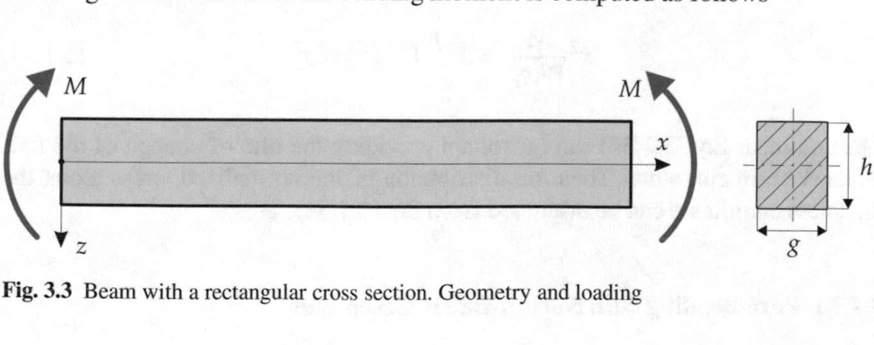

Fig. 3.3 Beam with a rectangular cross section. Geometry and loading

$$M = 2g \int_0^{h/2} \sigma_x z \mathrm{d}z \tag{3.1.28}$$

Let us introduce the new variables

$$\xi = \frac{2z}{h}, \quad \Sigma = \frac{\sigma_x}{\sigma_0}, \quad \dot{\kappa} = \frac{h}{2}\frac{\dot{\chi}}{\dot{\varepsilon}_0}$$

Equation (3.1.27) can be put in the following form

$$\dot{\kappa}\xi = f(|\Sigma|)\mathrm{sgn}(\Sigma) \tag{3.1.29}$$

For $0 \leq \xi \leq 1$ the normal stress is positive and

$$\dot{\kappa}\xi = f(\Sigma) \tag{3.1.30}$$

Since f is a monotonically increasing function an inverse function exists such that

$$\Sigma = f^{-1}(\dot{\kappa}\xi) \tag{3.1.31}$$

Equation (3.1.29) can be written as follows

$$\frac{M}{W\sigma_0} = 3\int_0^1 \Sigma\xi \mathrm{d}\xi \tag{3.1.32}$$

After inserting (3.1.31) we obtain

$$\frac{M}{W\sigma_0} = 3\int_0^1 f^{-1}(\dot{\kappa}\xi)\xi \mathrm{d}\xi \tag{3.1.33}$$

By changing the integration variable Eq. (3.1.33) can be put in the following form

$$\dot{\kappa}^2 \frac{M}{W\sigma_0} = 3\int_0^{\dot{\kappa}} f^{-1}(x) x \mathrm{d}x \tag{3.1.34}$$

The nonlinear Eq. (3.1.34) can be solved providing the rate of change of the normalized beam curvature. Then the distribution of the normalized stress along the thickness coordinate can be evaluated from Eq. (3.1.31).

3.1.3.1 Pure Bending with Norton-Bailey Creep Law

The classical solution to the steady-state creep of the beam is based on the power law stress function, e.g. Hult (1966); Malinin (1981). In this case $f(\Sigma) = \Sigma^n$ and

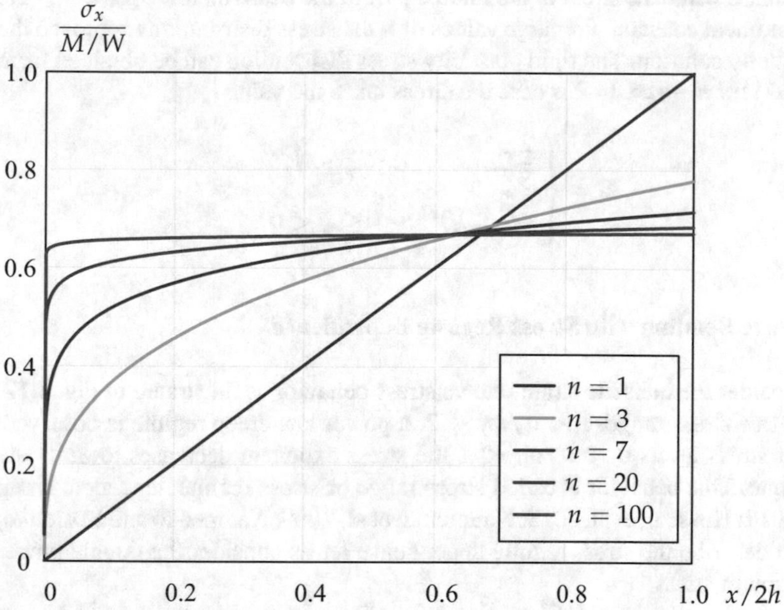

Fig. 3.4 Distribution of the normal stress across the thickness direction in a steady creep state for a beam with a rectangular cross section

$f^{-1}(x) = x^{1/n}$. For the power law creep, Eq. (3.1.34) is solved in the closed analytical form. The result is

$$\dot{\kappa}^{1/n} = \frac{2n+1}{3n}\frac{M}{W\sigma_0} \tag{3.1.35}$$

With Eq. (3.1.31) the normalized stress can be obtained as follows

$$\Sigma = \frac{2n+1}{3n}\frac{M}{W\sigma_0}|\xi|^{1/n}\mathrm{sgn}(\xi) \tag{3.1.36}$$

With the reference stress

$$\Sigma_{\mathrm{ref}} = \frac{\sigma_x(0, h/2)}{\sigma_0} = \frac{M}{W\sigma_0}$$

Eq. (3.1.36) takes the following form

$$\Sigma = \Sigma_{\mathrm{ref}}\frac{2n+1}{3n}|\xi|^{1/n}\mathrm{sgn}(\xi) \tag{3.1.37}$$

Figure 3.4 shows the distribution of the normal stress across the thickness coordinate in a steady creep state according to Eq. (3.1.37). For $n = 1$ the distribution is linear as in the case of linear elasticity. For nonlinear viscous behavior with $n > 1$ the stress redistribution takes place such that the stress at the outer layers of the beam decrease while the stress in the middle part of the beam increase providing the bending moment constant. For large values of n the stress distributions approach the rigid plasticity solution. The rigid plasticity stress distribution can be obtained from Eq. (3.1.37) for $n \to \infty$. In this case the stress takes the value

$$\Sigma = \begin{cases} \frac{2}{3}\Sigma_{\text{ref}}, & 0 \le \xi \le 1, \\ -\frac{2}{3}\Sigma_{\text{ref}}, & -1 \le \xi < 0 \end{cases}$$

3.1.3.2 Pure Bending with Stress Regime Dependence

Let us consider the inelastic strain rate vs stress behavior as illustrated in Fig. 1.12. For moderate stress range, $1 < \sigma/\sigma_0 < 2$, a power law creep regime is observed. For lower stress levels $0 < \sigma/\sigma_0 < 1$ the stress exponent decreases towards the value of one. This behavior is called stress range or stress regime dependent creep (Boyle, 2012; Hosseini et al, 2013; Naumenko et al, 2009; Naumenko and Kostenko, 2009). To describe the stress regime dependence let us consider the double power law function of stress

$$f(\Sigma) = \Sigma + \Sigma^n, \quad \Sigma = \frac{\sigma}{\sigma_0} \tag{3.1.38}$$

In this case two stress ranges or stress regimes can be captured. For $\sigma/\sigma_0 \ll 1$ the double power law stress function provides the linear creep behavior while for $\sigma/\sigma_0 \gg 1$ the power law creep is described, Fig. 1.12. With the double power law function, Eq. (3.1.34) can be solved to find the rate of the normalized beam curvature. With Eq. (3.1.31) the distribution of the normalized stress along the thickness direction can be computed.

Let us note that the stress value σ_0 corresponds to the cross point of the linear and the power law lines in the double logarithmic scale as shown in Fig. 1.12. To designate the validity range of the power law creep function in structural analysis applications, let us introduce the transition stress

$$\sigma_\epsilon = \sigma_0 \epsilon^{1/1-n}$$

such that for $\sigma > \sigma_\epsilon$ the relative deviation of the creep rate, defined by Eq. (3.1.38), from the creep rate, defined by the power law function $f(\Sigma) = \Sigma^n$ does not exceed the given accuracy $\epsilon < 1$.

The classical solution to the steady-state creep of the beam based on the power law stress function (3.1.36) follows from Eqs (3.1.29) – (3.1.34) by setting $\Sigma \gg 1$. On the other hand for $\Sigma \ll 1$ and $f(\Sigma) = \Sigma$ the linear creep solution can be obtained

$$\dot{\kappa} = \frac{M}{W\sigma_0}, \quad \Sigma = \frac{M}{W\sigma_0}\xi \tag{3.1.39}$$

To analyze the results for the wide stress range Eqs (3.1.31) and (3.1.34) were evaluated numerically for the case $n = 12$ and for different values of the normalized

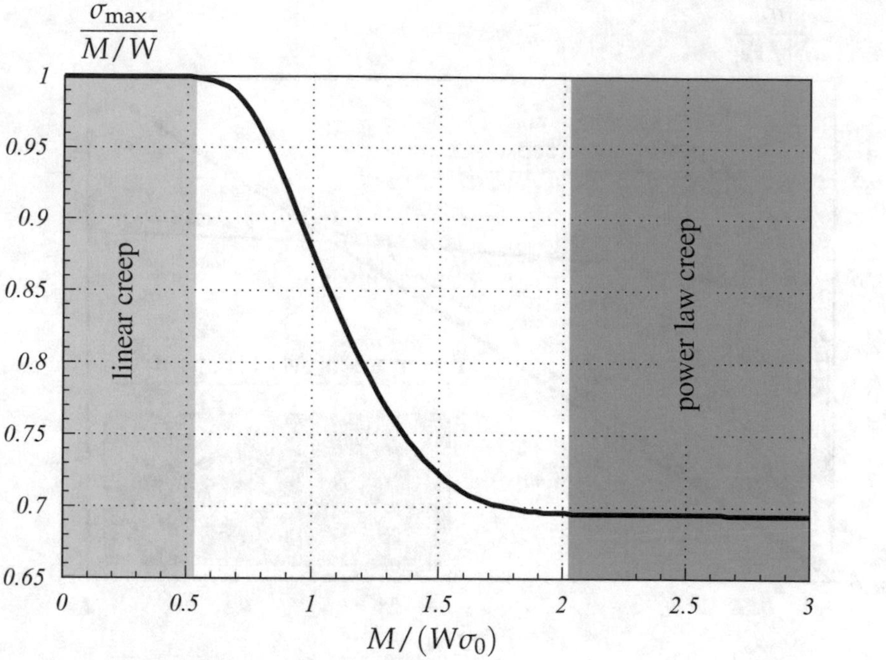

Fig. 3.5 Steady-state creep solutions for a beam, $n = 12$. Normalized maximum stress vs. normalized bending moment

reference stress

$$\Sigma_{\text{ref}} \equiv \frac{\sigma_x(0, h/2)}{\sigma_0} = \frac{M}{W\sigma_0}$$

Figure 3.5 shows the dependence of the normalized maximum stress in the steady creep state on the normalized reference stress. The transition from the linear to the power law creep range can be recognized clearly. In this transition range the power law underestimates the maximum stress value.

With the introduced transition stress σ_ϵ the classical solution for the maximum stress is valid when the following inequalities are satisfied

$$\sigma_{\text{ref}} > \sigma_{\text{ref}_\epsilon} = \frac{3n}{2n+1}\sigma_\epsilon \quad \Rightarrow \quad \Sigma_{\text{ref}} > \frac{3n}{2n+1}\epsilon^{1/1-n} \qquad (3.1.40)$$

Let us note that the upper bound for the power law creep solution also exists (power law breakdown).

Not only the maximum stress value but also the character of stress distribution depends on the normalized reference stress value, Fig. 3.6. For $\Sigma_{\text{ref}} = 2$ the solution according to the constitutive model (3.1.38) almost coincides with the power law solution (3.1.37) except a small region in the neighborhood of the neutral plane. For the given value of Σ_{ref} satisfying (3.1.40) the power law solution is valid for $\xi_\epsilon \leq \xi \leq 1$, where

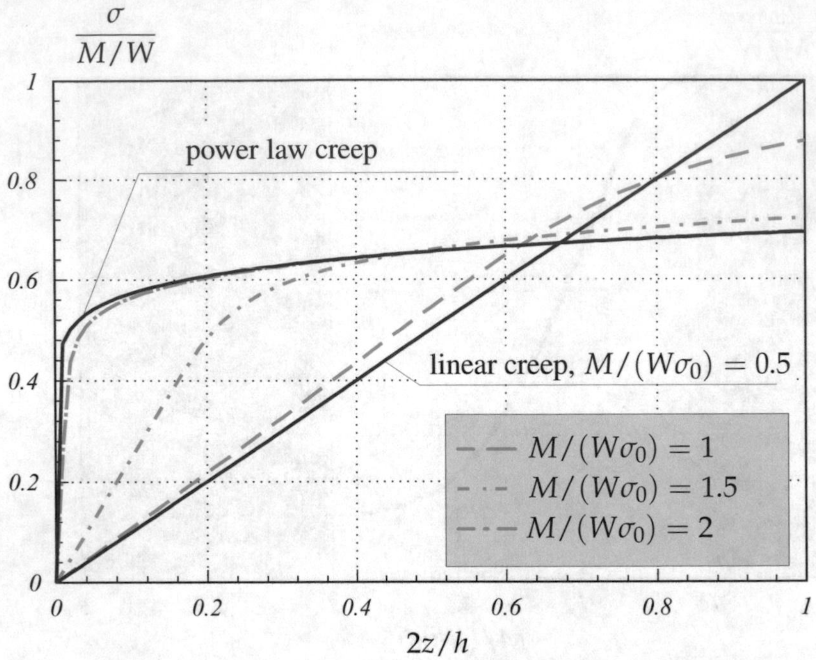

Fig. 3.6 Steady-state creep solutions for a beam, $n = 12$. Normalized stress vs normalized thickness coordinate

$$\xi_\epsilon = \frac{\epsilon^{1/1-n}}{\Sigma_{\text{ref}}} \frac{3n}{2n+1}$$

In the neighborhood of the neutral plane the normal stress distribution corresponds to the linear rather than the power law creep.

3.1.3.3 Bending Under Lateral Load

Let us present examples of a beam subjected to the constant lateral load in the steady creep state. We consider beams with arbitrary cross sections having the $x - z$ plane of symmetry. Assuming the power law stress function, the steady-state solution follows from (3.1.29)

$$\sigma_x = \left(\frac{1}{a}\right)^{\frac{1}{n}} |\dot{\chi} z|^{\frac{1}{n}-1} \dot{\chi} z, \quad a = \frac{\dot{\varepsilon}_0}{\sigma_0^n} \tag{3.1.41}$$

The bending moment is computed as follows

$$M = \int_A \sigma_x z \mathrm{d}A = \left(\frac{1}{a}\right)^{\frac{1}{n}} I_n |\dot{\chi}|^{\frac{1}{n}-1} \dot{\chi}, \tag{3.1.42}$$

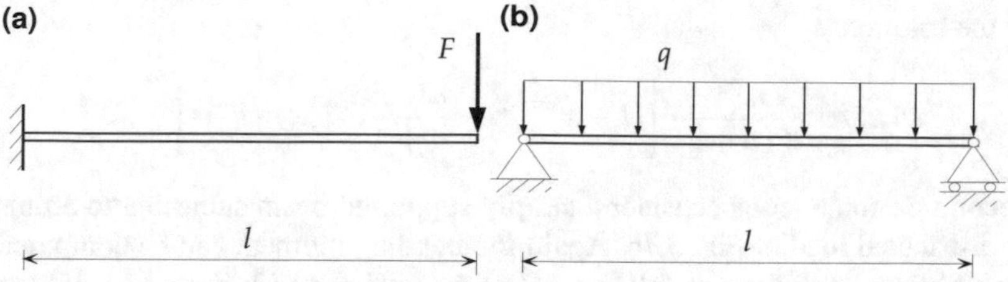

Fig. 3.7 Beams under lateral loading. **(a)** Cantilever beam, **(b)** simply-supported beam

where

$$I_n = \int_A |z|^{\frac{1}{n}-1} z^2 \mathrm{d}A \tag{3.1.43}$$

is the generalized area moment of degree $(n+1)/n$. From Eq. (3.1.42) the rate of change of the beam curvature in a steady creep state can be expressed as a power law function of the bending moment as follows

$$\dot{\chi} = \frac{a}{I_n^n} |M|^n \mathrm{sgn}(M) \tag{3.1.44}$$

With $\chi = -w''$ the following second order differential equation for the deflection rate can be obtained

$$\dot{w}'' = -\frac{a}{I_n^n} |M|^n \mathrm{sgn}(M) \tag{3.1.45}$$

If the bending moment is given as a function of the coordinate x, Eq. (3.1.45) can be integrated providing the deflection rate in a steady creep state.

As a first example let us consider a cantilever beam subjected to a single force F, Fig. 3.7a. In this statically determinate case the bending moment is

$$M(x) = F(x-l),$$

where l is the length of the beam. From Eq. (3.1.45) we obtain

$$\dot{w}(x)'' = \frac{a}{I_n^n} F^n (l-x)^n \tag{3.1.46}$$

The integration leads to

$$\dot{w}(x) = \frac{a}{I_n^n} \frac{F^n}{n+2} (l-x)^{n+2} + C_1 x + C_2 \tag{3.1.47}$$

The integration constants C_1 and C_2 can be determined from the boundary conditions of the clamped edge. With $\dot{w}(0) = 0$ and $\dot{w}'(0) = 0$ we obtain

$$C_1 = \frac{a}{I_n^n} \frac{F^n}{n+1} l^{n+1}, \quad C_2 = -\frac{a}{I_n^n} \frac{F^n}{n+2} l^{n+2}$$

Finally the solution is

$$\dot{w}(x) = \frac{aF^n}{I_n^n(n+2)}\left[(l-x)^{n+2} + \frac{n+2}{n+1}l^{n+1}x + l^{n+2}\right]$$

As a second example let us consider a simply supported beam subjected to a uniformly distributed load q, Fig. 3.7b. Again the bending moment can be computed from equilibrium conditions as follows $M(x) = qx(l-x)/2$. From (3.1.45) we obtain

$$\dot{w}(x)'' = -\frac{a}{I_n^n}\frac{q^n}{2^n}x^n(l-x)^n \tag{3.1.48}$$

Equation (3.1.48) can be integrated by taking into account boundary conditions $\dot{w}(0) = \dot{w}(l) = 0$. For integer values of the power n the solution is

$$\dot{w}(x) = \frac{a}{I_n^n}\frac{q^n}{2^n}x\sum_{k=0}^{n}\alpha_k(l^{n+k+1} - x^{n+k+1}) \tag{3.1.49}$$

with

$$\alpha_k = (-1)^k\frac{n!}{k!(n-k)!}\frac{l^{n-k}}{(n+k+1)(n+k+2)}$$

The reference elastic deflection is

$$w(x) = \frac{q}{24EI}x(x-l)(x^2 - lx - l^2)$$

Let us note that the closed-form solution for the steady-state deflection rate (3.1.49) is a polynomial of the order $2n+2$. Therefore, if the creep problem is numerically solved applying variational methods (see Sect. 3.1.4), trial functions for the deflection or deflection rate should contain polynomial terms of the order $2n+2$ instead of 4, as in the elastic case. The order of polynomial terms in the creep solution is material-dependent since n is the creep exponent in the Norton-Bailey creep law.

Many closed-form solutions for steady-state creep in beams with various types of boundary conditions and loading are presented in Boyle and Spence (1983); Malinin (1981); Rabotnov (1969); Odqvist (1974).

3.1.4 Solutions by Ritz Method

Let us consider a beam with the length l and the rectangular cross section $g \times h$, where g is the width and h is the height. The beam is simply supported and subjected to the uniformly distributed load q. Two cases of the material behavior will be discussed. In the first case we assume the idealized creep having the secondary stage only. The Norton-Bailey creep constitutive equation will be assumed. Although this equation describes stationary creep and the primary creep stage is neglected, the deflection vs. time curve of the beam exhibits the primary stage. The numerical results for the steady-state deflection rate will be compared with closed-form solution presented in Sect. 3.1.3.

In the second case the tertiary creep stage will be included by the introduction of the damage parameter. For the analysis the Kachanov-Rabotnov model will be applied. The reference numerical solution for the creep-damage behavior of the beam will be obtained by the Ritz method and the procedure presented in Sect. 3.3. This solution will be used to verify the capability of finite element codes for the creep-damage analysis. In particular, we present the results obtained by commercial finite element codes and user-defined creep-damage material subroutines.

The solution accuracy of the time dependent creep problem is primarily determined by the solution accuracy of the boundary value problem at each time or iteration step. By utilizing the Ritz method the solution accuracy depends on the type and the number of trial functions in (3.1.22), as shown in Altenbach et al (2000). By applying the finite element method the type of finite elements and the mesh density are most responsible for the accuracy. The following example is selected to examine the convergence of time dependent solutions with respect to different types of approximations including the series (3.1.22) and the finite element discretization.

For the numerical analysis we set

$$q = 60\ \text{N/mm}, \quad l = 10^3\ \text{mm}, \quad g = 30\ \text{mm}, \quad h = 80\ \text{mm} \tag{3.1.50}$$

Furthermore, we apply the creep-damage constitutive equations with the material parameters identified for the aluminium alloy BS 1472, see Naumenko and Altenbach (2016, Subsect. 6.2.4)

$$\dot{\varepsilon}_x = \frac{a|\sigma_x|^{n-1}}{(1-\omega)^m}\sigma_x, \quad \dot{\omega} = \frac{b|\sigma_x|^k}{(1-\omega)^p} \tag{3.1.51}$$

with the following set of material parameters

$$\begin{gathered} a = 1.35 \cdot 10^{-39}\ \text{MPa}^{-n}/\text{h}, \quad b = 3.029 \cdot 10^{-35}\ \text{MPa}^{-k}/\text{h}, \\ n = 14.37, \quad k = 12.895, \quad p = 12.5, \quad m = 10 \end{gathered} \tag{3.1.52}$$

The constitutive equation $(3.1.51)_1$ does not take into account primary creep.

3.1.4.1 Norton-Bailey Creep Law

First let us simplify the material behavior by neglecting the damage process. This can be accomplished by setting $b = 0$ in the damage evolution equation $(3.1.51)_2$. The creep constitutive equation $(3.1.51)_1$ simplifies to the Norton-Bailey law of steady-state creep. The creep rates are the same for the tensile and compressive loading. Therefore the fictitious normal force N^{pl} is zero. Consequently Eqs (3.1.24) can be simplified to

$$\boldsymbol{R}^{ww}\boldsymbol{a}^w = \boldsymbol{f}^w, \quad \boldsymbol{f}^u = \boldsymbol{0}, \quad \boldsymbol{a}^u = \boldsymbol{0} \tag{3.1.53}$$

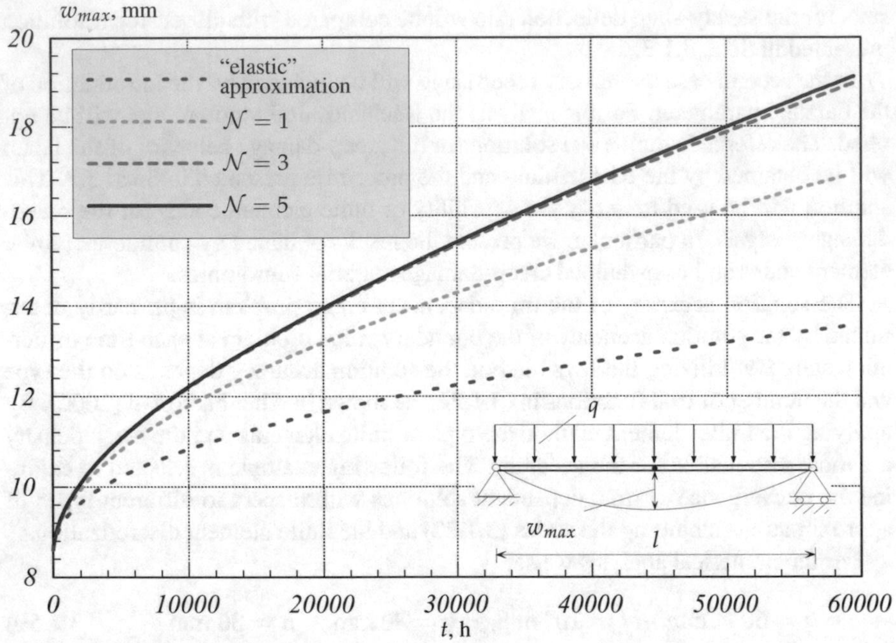

Fig. 3.8 Maximum deflection vs. time based on the Ritz method with different number of trial functions in (3.1.22)

The accuracy of the numerical solution depends on the number of trial functions in Eq. (3.1.22), and on the number of integration points $\mathcal{N}_l \times \mathcal{N}_h$. In the following example we fix the number of integration points to $N_l = 41$ and $N_h = 11$ and analyze the accuracy of the series approximation (3.1.22).

Figure 3.8 illustrates the time variation of the maximum deflection in the mid-point of the beam. We observe, that after a certain period of time the deflection rate approaches a constant, steady-state creep value. The steady-state deflection rate depends significantly on the number $\mathcal{N}$ of trial functions in (3.1.22). The first approximation for the deflection $w(x,t) = a_0^w(t)x(x-l)(x^2 - lx - l^2)$ is exact for the reference elastic state with $a_0^w(0) = q/(24EI)$. However, it is not sufficient to represent the creep behavior. The convergent solution can be obtained with $\mathcal{N} = 5$. Further increase of $\mathcal{N}$ did not result in better solutions.

Let us note, that the number of required series terms depends on the material behavior and, in particular, on the value of the creep exponent n (see Subsubsect. 3.1.3.3). Figure 3.9 shows the time variation of the normal stress σ_x in the bottom layer of the middle cross section of the beam. To verify the numerical results let us compare the steady-state stress value with the closed-form solution presented in Subsubsect. 3.1.3.3. From Eqs (3.1.41) and (3.1.42) the normal stress can be computed as follows

$$\sigma_x(x,z) = \frac{M(x)}{I_n}|z|^{\frac{1}{n}-1}z$$

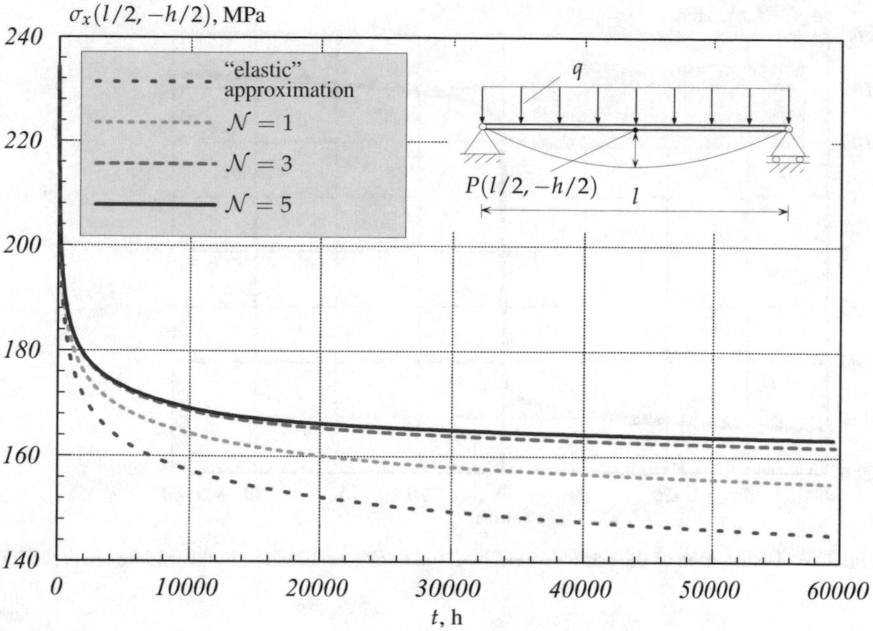

Fig. 3.9 Normal stress vs. time based on the Ritz method with different number of trial functions in (3.1.22)

For the rectangular cross section the generalized moment of inertia (3.1.43) reads

$$I_n = \frac{2gn}{n+1}\left(\frac{h}{2}\right)^{\frac{1}{n}+2}$$

With $M(l/2) = ql^2/8$ the normal stress in the middle cross section takes the form

$$\sigma(l/2, z) = \frac{ql^2}{16}\frac{2n+1}{gn}\left(\frac{2}{h}\right)^{\frac{1}{n}+2}|z|^{\frac{1}{n}-1}z \qquad (3.1.54)$$

Figure 3.10 shows the distribution of the normal stress over the thickness direction in the steady creep state. The solid line is the plot of Eq. (3.1.54) by taking into account (3.1.50) and $n = 14.37$. The circles denote the numerical stress values at the integration points. We observe that the numerical solution agrees well with the closed-form one. According to Eq. (3.1.54) the stress value in the bottom layer of the middle cross section is $\sigma_x(l/2, -h/2) = 161.7$ MPa, which is in good agreement with the obtained numerical solution.

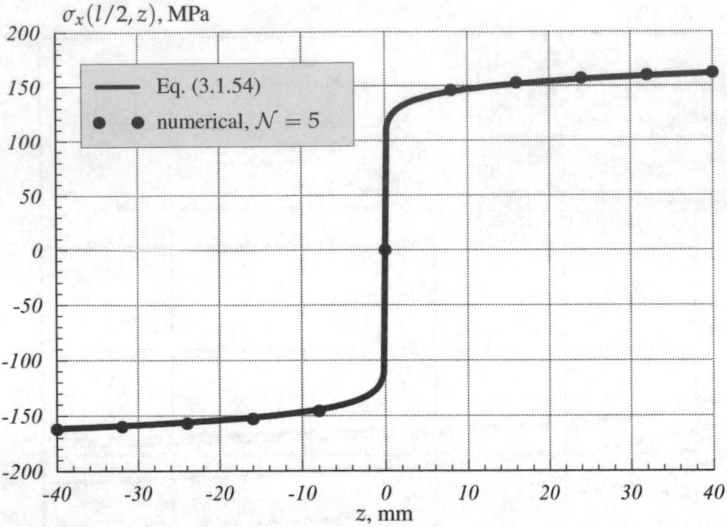

Fig. 3.10 Distribution of the normal stress over the thickness direction in the middle cross section

3.1.4.2 Kachanov-Rabotnov Creep-damage Law

In Eq. $(3.1.51)_2$ the damage rate is controlled by the absolute value of the normal stress. Therefore the damage state will be the same in tensile and compressive layers of the beam cross sections. The distribution of $|\sigma_x|$ will be symmetric with respect to the beam centerline. In this case one may apply the simplified equation (3.1.53) for the numerical analysis. The time step solutions were performed until the critical damage value is reached at one of the integration points. The condition of termination is $\omega(x_f, z_f, t_*) > 0.9$, where the integration point $P(x_f, z_f)$ can be specified as a point of failure initiation and the time t_* as the time to failure initiation. Let us note that more integration points over the thickness direction are required for the creep damage analysis if compared with the steady-state creep analysis. Our convergence studies suggest that 17 integration points provide enough accuracy for the deformations, stresses and the time to failure initiation.

Figures 3.11 and 3.12 illustrate the maximum deflection and the stress $\sigma_x(l/2, -h/2)$ as functions of time. The results have been obtained with different numbers of trial functions in (3.1.22). All applied approximations to the deflection function provide the same result for the reference elastic state. However, the results for creep are quite different and depend significantly on the number of polynomials. From Figs 3.11 and 3.12 we observe that the solutions converge against the accurate solution with increasing number of trial functions. By analogy with the uniaxial creep curve three creep stages of the beam can be recognized. The "primary" stage is characterized by the decrease in the deflection rate and significant stress relaxation. The "secondary" stage can be identified by slow changes in the rates of deflection growth and stress relaxation. During the "tertiary" stage the rates rapidly increase.

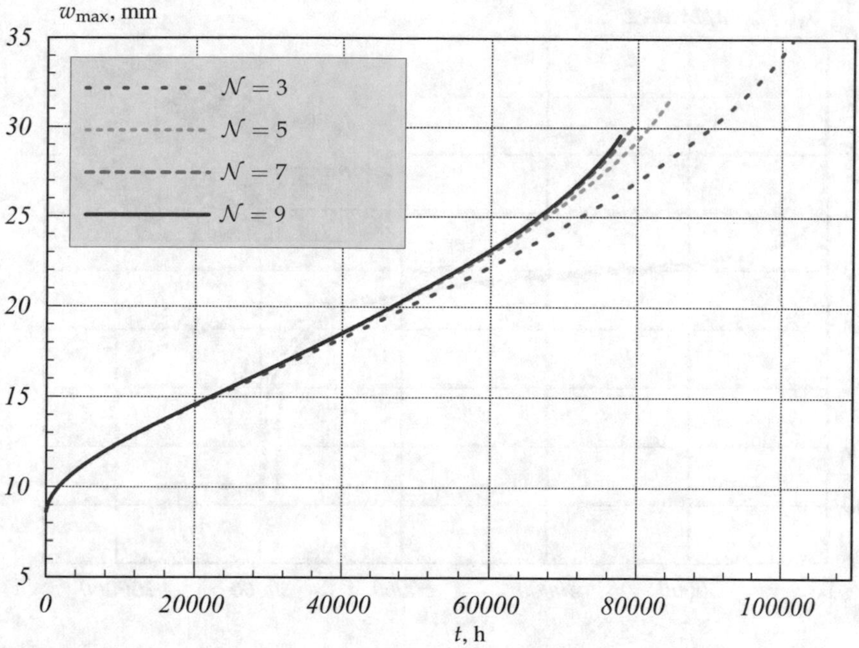

Fig. 3.11 Maximum deflection vs. time based on the Ritz method with different number of trial functions in (3.1.22)

The critical damage $\omega_* = 0.9$ appears in the bottom layer of the middle cross section. In all cases presented in Figs 3.11 and 3.12 the applied approximations provide almost the same solutions for the primary and secondary creep stages. The results differ only in the final stage. Therefore, we may conclude that the consideration of damage requires an increased order of approximation in comparison with the steady-state creep analysis.

3.1.5 Solutions by Finite Element Method

3.1.5.1 Norton-Bailey Creep Law

Structural creep analysis with the Norton-Bailey law can be performed with many commercial finite element codes. Many benchmarks exist which allow to verify the capability of the finite element method for the steady-state creep analysis, e.g. Abaqus Benchmarks (2017). They are based on the available closed-form solutions for elementary structures in the steady-state creep range. The following example is selected to examine the accuracy of the finite element solution for the simply supported beam. The beam is modeled as a plate strip with shell type elements. The solution is performed by the ABAQUS code. The midplane of the strip is divided

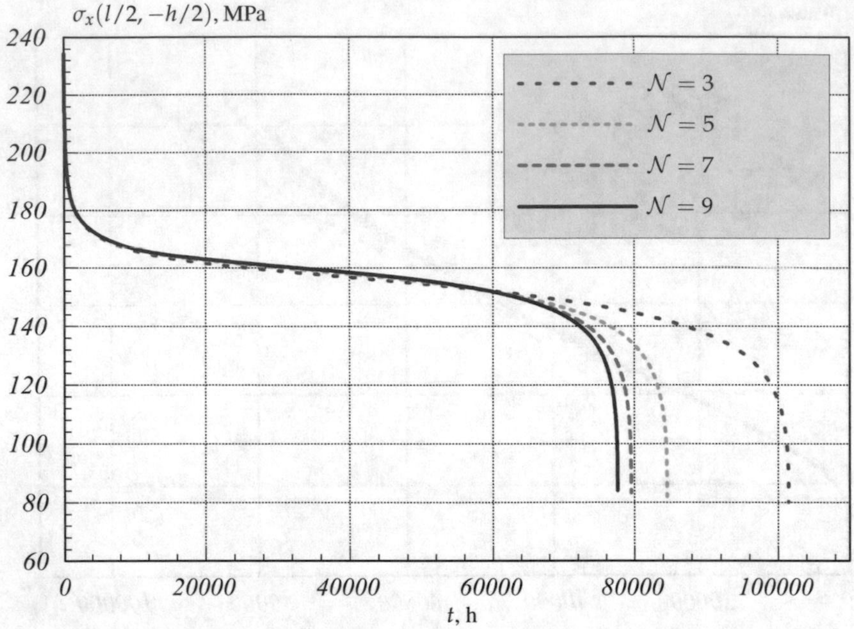

Fig. 3.12 Normal stress vs. time based on the Ritz method with different number of trial functions in (3.1.22)

by 4-node shell elements S4R5 as shown in Fig. 3.13. The Simpson quadrature rule with 11 integration points through the thickness of the cross section is selected. The automatic time step feature with the minimum time step size of 0.01 h, the maximum time step size of 1000 h and the creep strain error tolerance of 10^{-6} is applied for the time integration. Figures 3.13 and 3.14 illustrate the results for time variations of the maximum deflection and the normal stress $\sigma_x(l/2, -h/2)$ obtained with different number of finite elements. We observe that all of the used meshes provide the same result for the reference elastic state. However, time variations of the maximum deflection and the normal stress are sensitive to the mesh density. A mesh adjusted to the convergent solution of the linear elasticity problem (6 elements) is not fine enough for the creep analysis. The best result has been obtained with 30 elements. The corresponding plots for the time variations of the deflection and the stress agree well with those, previously obtained by the Ritz method (Figs 3.8 and 3.9). By analogy with the convergence of the series approximation (3.1.22) we may conclude that the number of finite elements required for an accurate solution of the creep problem depends on the material behavior, and in particular, on the value of the creep exponent n.

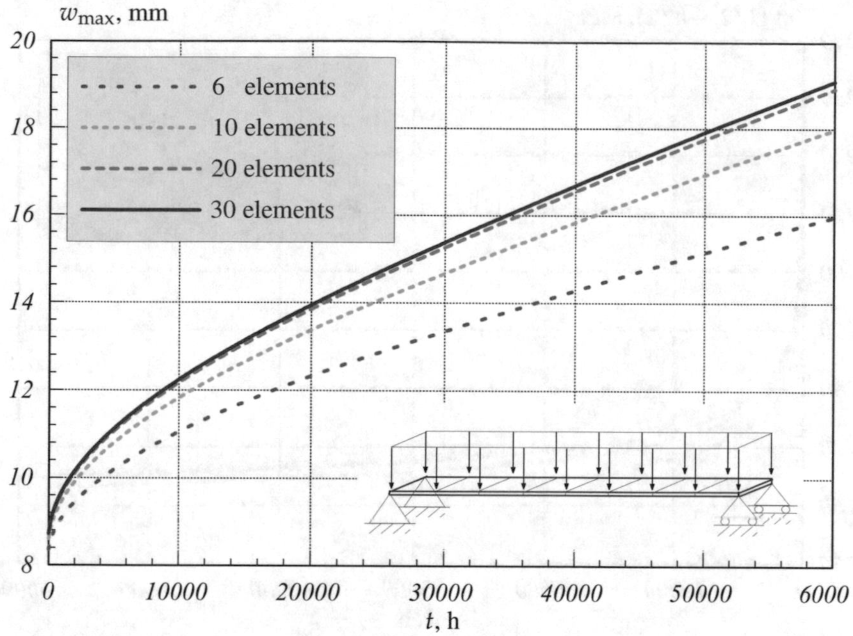

Fig. 3.13 Maximum deflection vs. time using the ABAQUS code with shell elements S4R5

3.1.5.2 Kachanov-Rabotnov Creep-damage Law

Creep constitutive models with damage state variables are not available in the general purpose finite element codes. However, a specific constitutive model with internal state variables can be incorporated in a general purpose code by writing a user-defined material subroutine. To verify the developed subroutine, the results of the finite element modeling must be compared with reference solutions of benchmark problems. Let us note that closed-form solutions to creep-damage problems are only available for the case of homogeneous stress (strain) states. Examples include a bar subjected to the uniform tension, a thin-walled tube subjected to the axial force and torque. Such benchmarks can be applied to assure that the developed subroutine is correctly coded and implemented. To analyze how the discretization parameters e.g., finite element type, mesh density, time step size, and time step control affect the solution, additional benchmark problems which involve inhomogeneous stress (strain) states are required.

The Kachanov-Rabotnov type constitutive model was incorporated into the ABAQUS and ANSYS finite element codes by means of the user-defined creep material subroutines. For details of the User Programmable Features provided by ANSYS and ABAQUS as well as the utilized time integration methods we refer to Abaqus User's Guide (2017); ANSYS (2001). To verify the subroutines we select the simply supported beam as a benchmark problem. This problem will be solved applying shell and plane stress finite elements. Furthermore, we perform the mesh

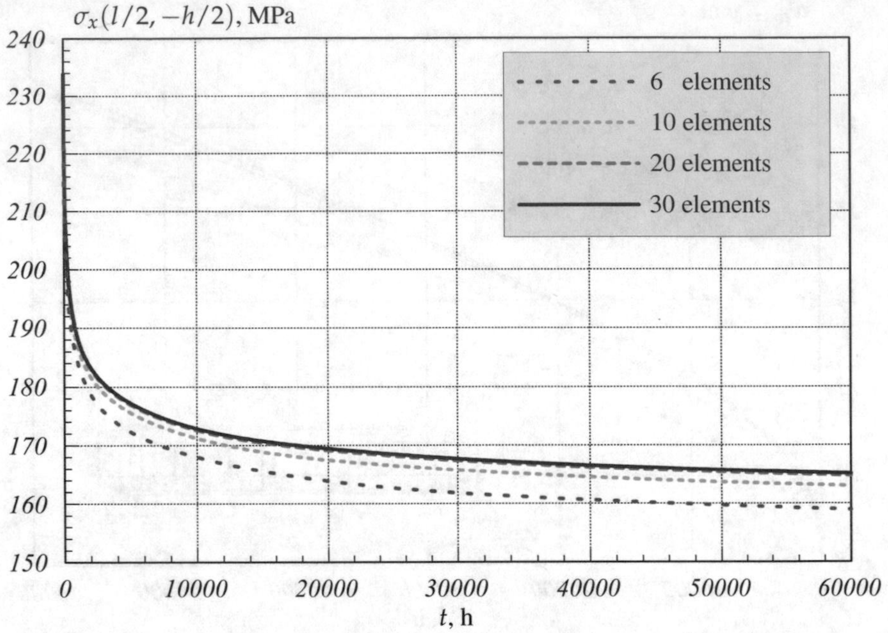

Fig. 3.14 Normal stress vs. time using the ABAQUS code with shell elements S4R5

convergence study illustrating basic features of the finite element solution to creep-damage problems. The results will be compared with those obtained by the Ritz method. In the case of shell type elements the beam will be considered as a plate strip. In addition, the beam will be modeled as a "wall" by using of the plane stress type finite elements.

In the next example we apply the 4-node shell element S4R5 with 17 integration points through the thickness direction. The settings for time step method are the same as in the previous example. Figures 3.15 and 3.16 illustrate time variations for the maximum deflection and the normal stress in the bottom layer of the middle cross section. The results have been obtained with different number of elements. We observe that all of the used meshes provide the same solutions for the reference elastic state. Furthermore, the meshes with 20, 30 and 40 elements lead to almost the same results in the "primary" and "secondary" creep ranges of the beam. The results differ only in the final stage before failure initiation. However, such a difference is not significant if we take into account the scatter of material data and the inaccuracy of the material behavior description. In this sense the mesh with 30 elements adjusted to the convergent solution in the steady-state creep range, cp. Figs 3.13 and 3.14, is fine enough for the numerical life-time predictions.

When studying creep-damage in structures with complex geometry it is difficult to perform convergence studies due to significant computational time. The presented results indicate that the optimal mesh size can be found based on the convergent solution in the steady-state creep range. With such meshes the accuracy of long term

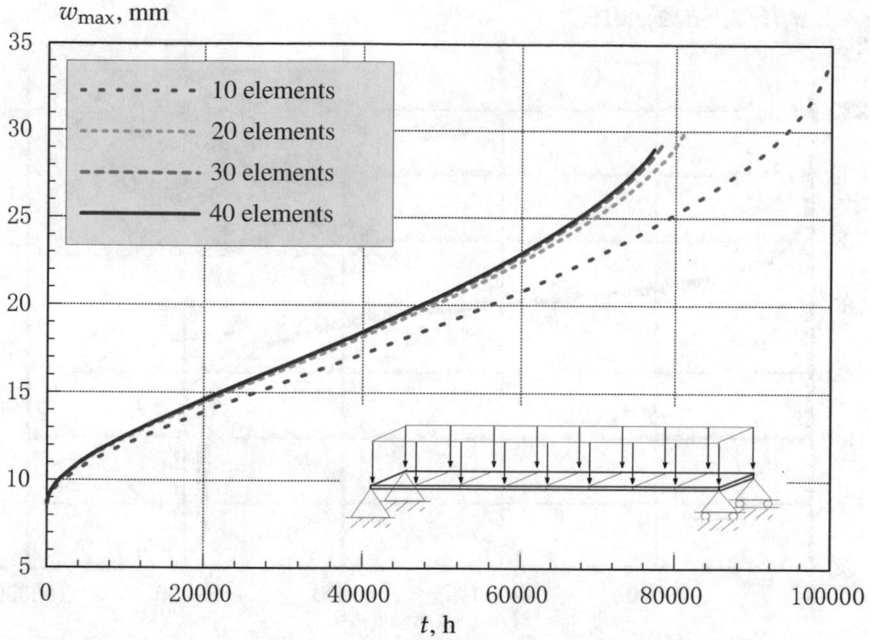

Fig. 3.15 Maximum deflection vs. time using the ABAQUS code with shell elements S4R5

Table 3.1 Element types and discretization parameters

Element Type	Element Abbreviation	Number of integration points	Mesh
Shell (ABAQUS)	S4R5	17 (Simpson)	40
Shell (ANSYS)	SHELL 43	5 (Gauss)	40
Plane Stress (ABAQUS)	CPS4R	-	100×8
Plane Stress (ANSYS)	PLANE 42	-	100×8

predictions is not less than the accuracy of the material data involved in computations.

In Altenbach and Naumenko (1997); Altenbach et al (2000, 2001) several benchmark problems for beams and rectangular plates solved by the Ritz method are presented. Finite element solutions for the same problems have been performed by ANSYS code applying shell, plane stress and solid type elements. The results illustrate the correctness of the developed subroutine over the wide range of element types as well as the convergence behavior of solutions.

To complete the analysis of the simply supported beam, let us compare the results obtained by the Ritz method with those of ABAQUS and ANSYS finite element codes. Shell and plane stress type finite elements elements were applied. Table 3.1 provides a summary of element types, the number of finite elements as well as the number of integration points through the thickness for shell-type elements. Figures

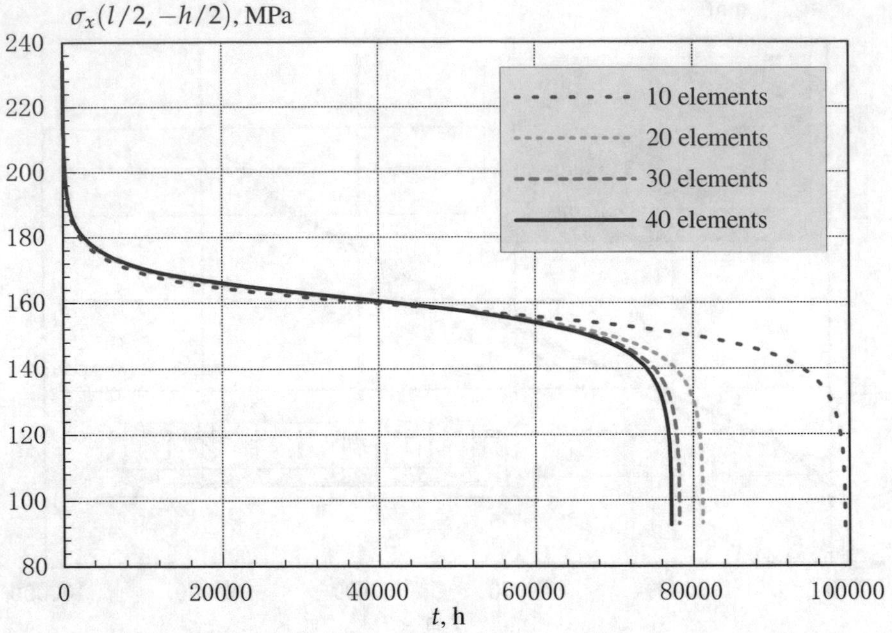

Fig. 3.16 Normal stress vs. time using the ABAQUS code with shell elements S4R5

3.17 and 3.18 illustrate the time variations of maximum deflection and the normal stress obtained by the Ritz method and the finite element method with shell and plane stress type elements. We observe that the results are in good agreement. The exception is the solution based on the shell element SHELL 43. The results are inaccurate due to insufficient number of integration points.

It is worth to mention that the results based on the elementary beam theory agree well with those based on the plane stress model. This confirms the kinematical assumptions of the elementary beam theory in creep-related analysis.

3.2 Stress State Effects and Cross Section Assumptions

For many materials stress state dependent material behavior has been observed in multiaxial tests, see Naumenko and Altenbach (2016, Subsect. 1.1.2). For example, the accelerated creep is significantly influenced by the kind of the stress state. Different tertiary creep rates and times to fracture can be obtained from creep tests under uniaxial tension with the stress σ and under torsion with the shear stress $\sqrt{3}\tau = \sigma$, e.g. (Kowalewski, 1996). As an example consider the creep behavior of type 316 stainless steel at 650°C. In Liu et al (1994) the following creep equations are applied

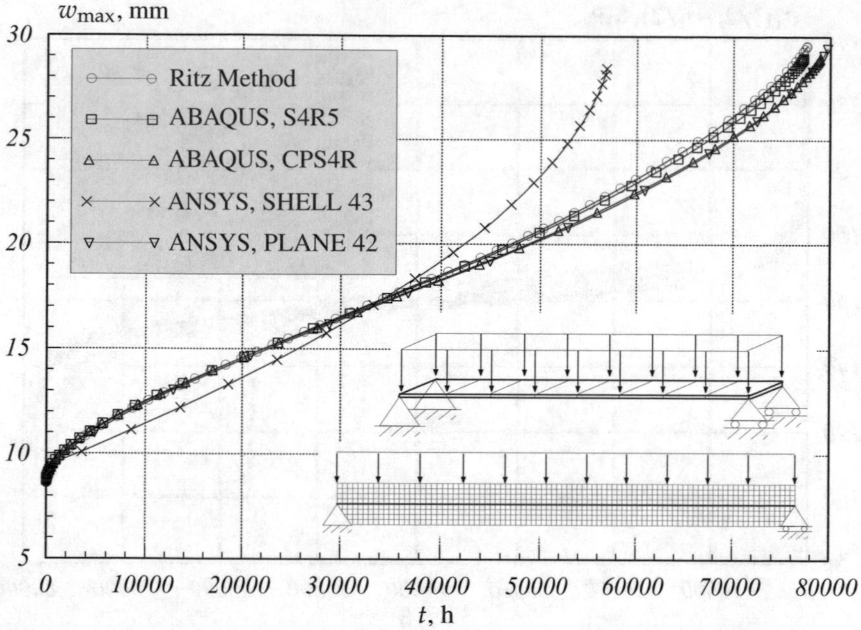

Fig. 3.17 Maximum deflection vs. time using the Ritz method and the finite element codes

$$
\begin{aligned}
\dot{\boldsymbol{\varepsilon}}^{\text{pl}} &= \frac{3}{2} f_1(\sigma_{\text{vM}}) g_1(\omega) \frac{\boldsymbol{s}}{\sigma_{\text{vM}}}, \quad \dot{\omega} = f_2\left[\sigma^{\omega}_{\text{eq}}(\boldsymbol{\sigma})\right] g_2(\omega), \\
\boldsymbol{\varepsilon}^{\text{pl}}|_{t=0} &= \mathbf{0}, \quad \omega|_{t=0} = 0, \quad 0 \leq \omega \leq \omega_*
\end{aligned}
\tag{3.2.55}
$$

The response functions f_1, f_2, g_1, and g_2 are

$$
\begin{aligned}
f_1(\sigma) &= a\sigma^n, \quad g_1(\omega) = (1-\omega)^{-n}, \\
f_2(\sigma) &= b\sigma^k, \quad g_2(\omega) = (1-\omega)^{-k}
\end{aligned}
\tag{3.2.56}
$$

The material parameters are given in Liu et al (1994) as follows

$$
\begin{aligned}
a &= 2.13 \cdot 10^{-13}\,\text{MPa}^{-n}/\text{h}, \quad b = 9.1 \cdot 10^{-10}\,\text{MPa}^{-k}/\text{h}, \\
n &= 3.5, \quad k = 2.8
\end{aligned}
\tag{3.2.57}
$$

The damage equivalent stress is assumed as follows

$$
\sigma^{\omega}_{\text{eq}}(\boldsymbol{\sigma}) = \frac{\sigma_I + |\sigma_I|}{2},
$$

where σ_I is the first principal stress.

Figure 3.19a shows the creep curves for tensile, compressive and shear stresses simulated according to Eqs (3.2.55) – (3.2.57). The normal and shear stress values are specified in order to provide the same constant value of the von Mises stress. It

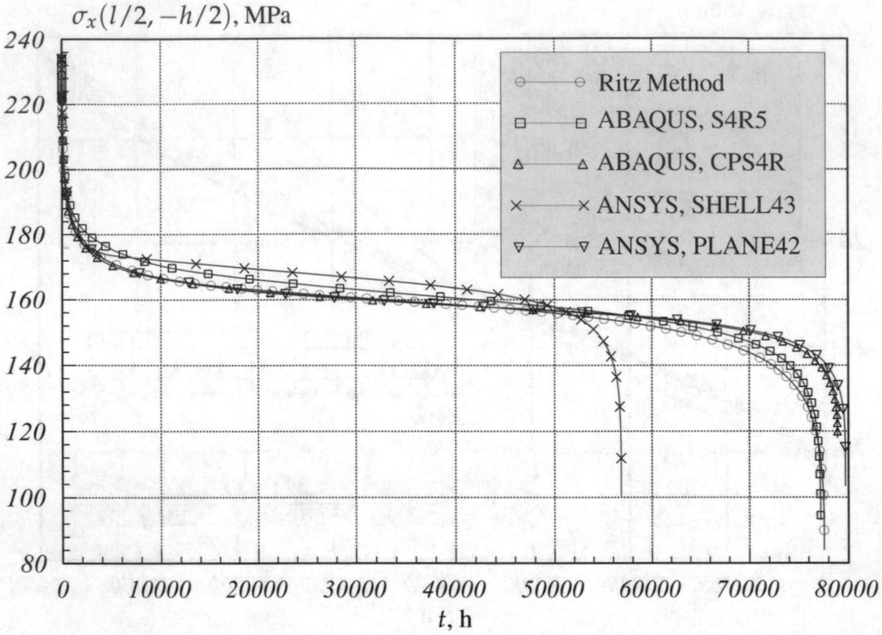

Fig. 3.18 Normal stress vs. time using the Ritz method and the finite element codes

is obvious that the tertiary creep rate is significantly dependent on the kind of loading. Figure 3.19b presents creep curves calculated by the combined action of the normal and shear stresses. We observe that even a small superposed shear stress can significantly influence the axial strain response and decrease the fracture time. Furthermore, combined tension-shear and compression-shear loadings with the same stress magnitudes lead to quite different creep responses. The change of the sign of the normal stress influences both the normal and the shear creep rates.

The considered loading case is typical for transversely loaded beams, plates and shells. For beams the local stress state is characterized by normal (bending) stress and small superposed transverse shear stress. Transverse shear strains are neglected within the classical theory of beams. The considered example indicates that small shear stress can significantly influence the material response and cause significant shear strains. Furthermore, the dependence of creep on the sign of the normal stress can lead to non-classical thickness distributions of the displacement, strain and stress fields. For example, the concept of the neutral stress-free plane fails and the distribution of the transverse shear stresses is non-symmetrical with respect to the midplane.

Cross section assumptions are usually the basis for different refined models of beams, plates and shells developed within the theory of elasticity, e.g. Eisenträger et al (2015); Reddy (1997); Schulze et al (2012); Naumenko and Eremeyev (2017). In what follows we apply the first order shear deformation theory (Timoshenko-type theory) to creep analysis. For a beam with a rectangular cross section we compare

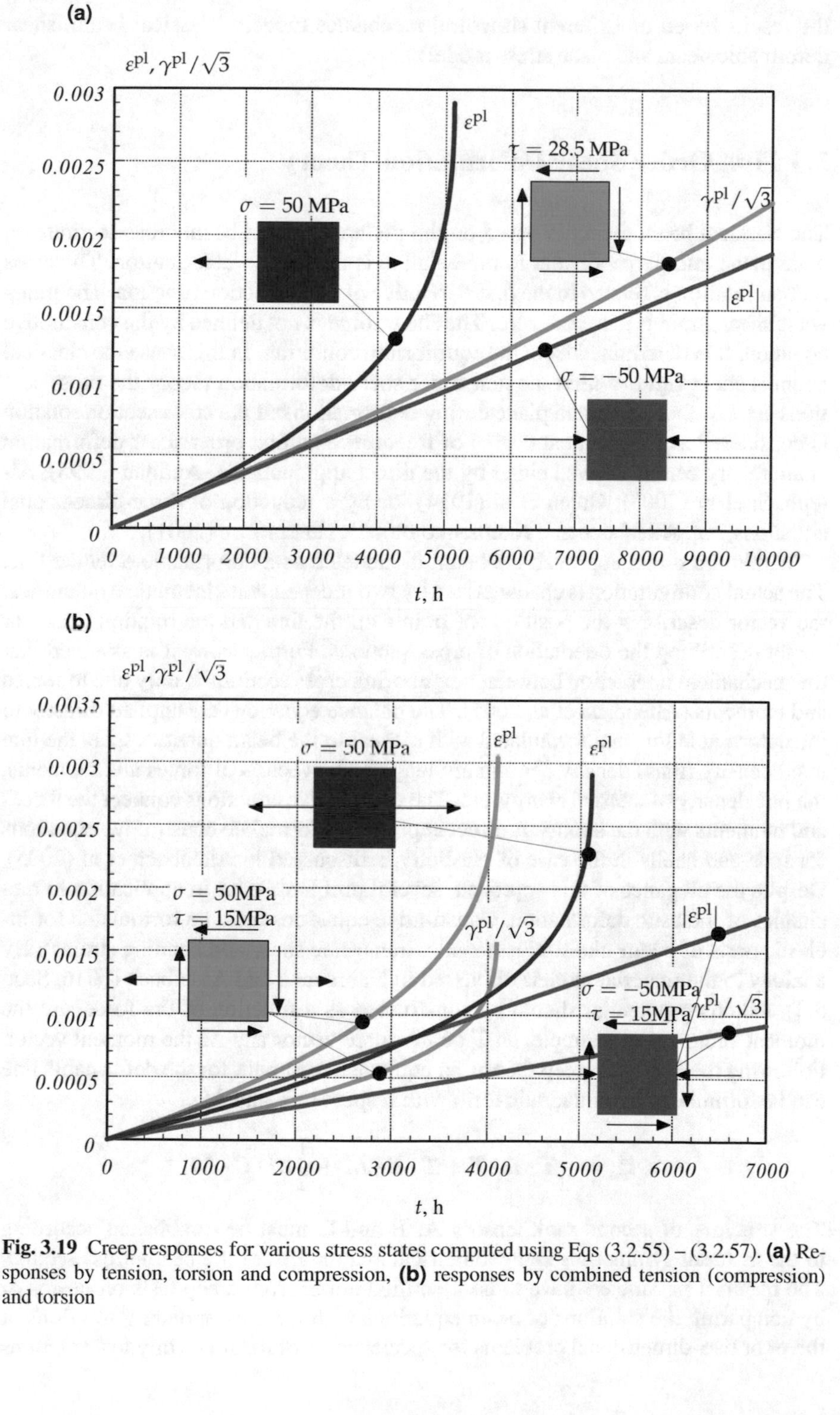

Fig. 3.19 Creep responses for various stress states computed using Eqs (3.2.55) – (3.2.57). **(a)** Responses by tension, torsion and compression, **(b)** responses by combined tension (compression) and torsion

the results based on different structural mechanics models (classical beam, shear deformable beam and plane stress model).

3.3 First Order Shear Deformation Theory

The classical beam theory is based on the assumption that beam cross sections remain plane and perpendicular to the beam axis during the deformation. The cross section rotation is related to the first derivative of the deflection function. The transverse shear strain is therefore zero. The shear force is not defined by the constitutive equation. It is determined from the equilibrium condition. In this sense the classical beam is shear rigid. Within the first order shear deformation theory the cross sections are assumed to remain plane during deformation but the cross section rotation is considered as independent degree of freedom. The first order shear deformation beam theory can be derived either by the direct approach, e.g. Antman (1995); Altenbach et al (2005); Green et al (1974), or by a reduction of three-dimensional equations, e.g. Altenbach and Naumenko (2002); Hutchinson (2001).

Within the direct approach the beam is modeled as a deformable oriented line. The actual configuration is characterized by two independent kinematical quantities: the vector describing the positions of points on the line and the rotation tensor or vector describing the orientation of cross sections. Furthermore, it is assumed that the mechanical interaction between neighboring cross sections is only due to forces and moments (Altenbach et al, 2005). The balance equations are applied directly to the deformable line and formulated with respect to the beam quantities, i.e. the line mass density (mass density per unit arc length), the vectors of forces and moments, the line density of internal energy, etc. The constitutive equations connect the forces and moments with the strains. A direct approach to formulate constitutive equations for rods and shells in the case of elasticity is discussed in Altenbach et al (2005). Despite the elegance of this approach, several problems arise in application to mechanics of inelastic deformation. Constitutive equations must be formulated for inelastic parts of beam-like strains (tensile, transverse shear and bending strains). By analogy to the material models discussed in Naumenko and Altenbach (2016, Sect. 5.4), the creep potential should be constructed as a function of the force and the moment vectors. For example, let $\boldsymbol{T}$ be the force vector and $\boldsymbol{M}$ the moment vector. Following the classical creep theory, an equivalent resultant for the deformable line can be formulated as a quadratic form with respect to $\boldsymbol{T}$ and $\boldsymbol{M}$

$$t_{\text{eq}}^2 = \frac{1}{2}\boldsymbol{T}\cdot\boldsymbol{A}\cdot\boldsymbol{T} + \boldsymbol{T}\cdot\boldsymbol{B}\cdot\boldsymbol{M} + \frac{1}{2}\boldsymbol{M}\cdot\boldsymbol{C}\cdot\boldsymbol{M}$$

The structure of second rank tensors $\boldsymbol{A}$, $\boldsymbol{B}$ and $\boldsymbol{C}$ must be established according to the material symmetries and geometrical symmetries of the beam cross section. The material parameters have to be identified either from creep tests on beams or by comparing the solutions of beam equations with the corresponding solutions of three- or two-dimensional problems for special cases of loading. Only few solutions

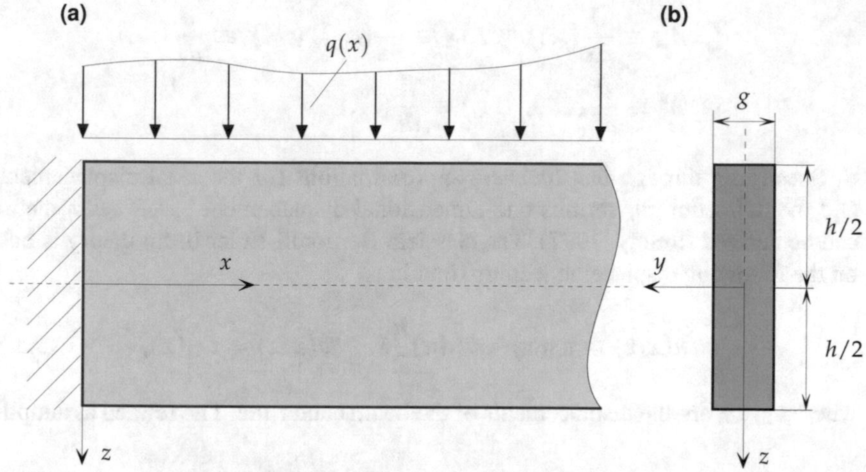

Fig. 3.20 Straight beam with a rectangular cross-section. **(a)** Loading, **(b)** geometry of the cross section

of this type are available in creep mechanics. An example is the pure bending of a beam under assumption of power law creep (see Subsubsect. 3.1.3.3). In this case the steady-state creep constitutive equation for the bending strain rate can be obtained from (3.1.44) as follows

$$\dot{\chi} = \frac{a}{I_n^n}|M|^{n-1}M$$

Alternatively the beam equations may be derived in the sense of approximate solution of two- or three-dimensional equations. To this end, through-the-thickness approximations of displacements and/or stresses are specified. Then, the two- or three-dimensional boundary value problem is reduced to ordinary differential equations by means of a variational principle. In order to discuss this approach let us consider a beam with a rectangular cross-section, Fig. 3.20. The governing two-dimensional equations for this case can be derived from (2.1.4) - (2.1.12) under the assumption of the plane stress state, i.e. $\boldsymbol{\sigma} \cdot \boldsymbol{e}_y = \boldsymbol{0}$. The principle of virtual displacements (2.2.71) yields

$$\frac{gh}{2}\int\limits_0^l\int\limits_{-1}^1(\sigma_x\delta\varepsilon_x + \tau_{xz}\delta\gamma_{xz} + \sigma_z\delta\varepsilon_z)\mathrm{d}\zeta\mathrm{d}x = \int\limits_0^l q(x)\delta w(x, -h/2)\mathrm{d}x \qquad (3.3.58)$$

For the sake of brevity, we assumed that the virtual work of tractions on the edges $x = 0$ and $x = l$ is zero. In (3.3.58) l denotes the beam length, $\sigma_x, \sigma_z, \tau_{xz}$ and $\varepsilon_x, \varepsilon_z, \gamma_{xz}$ are the Cartesian components of the stress and strain tensors, respectively, w is the beam deflection and $\zeta = 2z/h$ is the dimensionless thickness coordinate. In the following derivations we use the abbreviations

$$(\dots)_{,x} \equiv \frac{\partial}{\partial x}(\dots),\ (\dots)_{,z} \equiv \frac{\partial}{\partial z}(\dots),\ (\dots)' \equiv \frac{d}{dx}(\dots),$$
$$(\dots)^{\bullet} \equiv \frac{d}{d\zeta}(\dots),\ (\dot{\dots}) \equiv \frac{d}{dt}(\dots)$$

Specifying through-the-thickness approximations for the axial displacement u and the deflection w, various one-dimensional displacement based beam theories can be derived (Reddy, 1997). The classical Bernoulli-Euler beam theory is based on the following displacement approximations

$$u(x,z) = u_0(x) - w_0'(x)\frac{h}{2}\zeta, \quad w(x,z) = w_0(x), \tag{3.3.59}$$

where u_0, w_0 are the displacements of the beam centerline. The refined assumption

$$u(x,z) = u_0(x) + \varphi(x)\frac{h}{2}\zeta, \tag{3.3.60}$$

where φ denotes the independent cross-section rotation, provides the first order shear deformation (Timoshenko-type) beam theory. Another refined displacement-based beam model can be obtained with

$$\begin{aligned} u(x,\zeta) &= u_0(x) + \varphi(x)\frac{h}{2}\zeta + u_1(x)\Phi(\zeta), \\ w(x,\zeta) &= w_0(x) + w_1(x)\Omega(\zeta), \end{aligned} \tag{3.3.61}$$

where u_0 and w_0 are the displacements of the beam centerline, $\Phi(\zeta)$ and $\Omega(\zeta)$ are distribution functions, which should be specified, and $u_1(x)$ and $w_1(x)$ are unknown functions of the x-coordinate. The assumptions $\Phi(\zeta) = (\zeta h/2)^3, \Omega(z) = 0$ result in a Levinson-Reddy type theory (third order shear deformation theory) (Levinson, 1981; Reddy, 1984). From the boundary conditions $\gamma_{xz}(x, \pm 1) = 0$ it follows

$$\begin{aligned} u(x,\zeta) &= u_0(x) + \varphi(x)\frac{h}{2}\zeta - [w_0'(x) + \varphi(x)]\frac{h}{6}\zeta^3, \\ w(x,\zeta) &= w_0(x) \end{aligned}$$

and

$$\left.\frac{d\Phi}{d\zeta}\right|_{\zeta=-1} = \left.\frac{d\Phi}{d\zeta}\right|_{\zeta=1}$$

The next possibility is the use of stress-based approximations. For example, the solution of the Bernoulli-Euler beam equations in the linear-elastic range leads to the following stress distributions

$$\sigma_x = \frac{6M(x)}{gh^2}\zeta,$$
$$\tau_{xz} = \frac{3Q(x)}{2gh}\left(1-\zeta^2\right), \tag{3.3.62}$$
$$\sigma_z = \frac{3q(x)}{4g}\left(-\frac{2}{3}+\zeta-\frac{1}{3}\zeta^3\right)$$

Applying the stress approximations, equations for an elastic shear deformable plate have been derived by Reissner (1950) by means of a mixed variational principle. The displacement approximations (3.3.61) neglecting the terms $u_1\Phi$ and $w_1\Omega$ or the stress approximations (3.3.62) lead to the first order shear deformation beam theory. The stress approximations (3.3.62) are not applicable to inelastic beams because the normal stress σ_x is a nonlinear function of the thickness coordinate even in the case of steady-state creep (see Subsubsect. 3.1.3.3). To derive the beam equations for the creep analysis, the following approximations for the transverse shear and normal stresses were proposed in Altenbach and Naumenko (2002)

$$\tau_{xz} = \frac{2Q(x)}{gh}\frac{\psi^\bullet(\zeta)}{\psi_0},$$
$$\sigma_z = \frac{q(x)}{g}\frac{\psi(\zeta)-\psi(1)}{\psi_0}, \quad \psi_0 = \psi(1)-\psi(-1), \tag{3.3.63}$$

where $\psi(\zeta)$ is a function of distribution satisfying the boundary conditions $\psi^\bullet(\pm 1) = 0$. Furthermore, the linear through-the-thickness approximation of the axial displacement $u(x,\zeta) = u_0(x) + \zeta\varphi(x)h/2$ was assumed. Applying a mixed variational principle the following beam equations were derived in Altenbach and Naumenko (2002)

– equilibrium conditions

$$N' = 0, \quad M' - Q = 0, \quad Q' + q = 0, \tag{3.3.64}$$

– constitutive equation for the shear force

$$Q = GAk(\varphi + \tilde{w}' - \tilde{\gamma}^{\mathrm{pl}}), \tag{3.3.65}$$

where G is the shear modulus and

$$\frac{1}{k} = \frac{2}{\psi_0^2}\int\limits_{-1}^{1}\psi^{\bullet^2}(\zeta)d\zeta, \quad \tilde{w}(x) = \frac{1}{\psi_0}\int\limits_{-1}^{1} w(x,\zeta)\psi^\bullet(\zeta)d\zeta,$$
$$\tilde{\gamma}^{\mathrm{pl}}(x) = \frac{1}{\psi_0}\int\limits_{-1}^{1}\gamma_{xz}^{\mathrm{pl}}(x,\zeta)\psi^\bullet(\zeta)d\zeta \tag{3.3.66}$$

By setting $GAk \to \infty$ and $\tilde{\gamma}^{\mathrm{pl}} = 0$ in (3.3.65) the classical beam equations can be obtained. In this case $\varphi = -\tilde{w}'$ (the straight normal hypothesis). Let us note that Eqs (3.3.64) and (3.3.65) can be derived applying the direct approach. For plates and shells this way is shown in Altenbach and Zhilin (2004). However, in this case the meaning of the quantities GAk and $\tilde{\gamma}^{\mathrm{pl}}$ is different. The shear stiffness GAk plays the role of the beam-like material parameter and must be determined either from tests or by comparison of results according to the beam theory with solutions of three-dimensional equations of elasto-statics or -dynamics. For a review of different estimates of the shear correction factor k we refer to Altenbach et al (2015); Hutchinson (2001); Kaneko (1975). Furthermore, the direct approach would require a constitutive equation for the rate of transverse shear strain $\dot{\tilde{\gamma}}^{\mathrm{pl}}$. Within the applied variational procedure, Eqs (3.3.64) and (3.3.65) provide an approximate solution of the plane stress problem under special trial functions (3.3.63). Therefore k and $\tilde{\gamma}^{\mathrm{pl}}$ appear in (3.3.66) as numerical quantities and depend on the choice of the function $\psi(\zeta)$. For example, setting $\psi(\zeta) = \zeta$ we obtain

$$k = 1, \quad \tilde{\gamma}^{\mathrm{pl}}(x) = \frac{1}{2} \int_{-1}^{1} \gamma_{xz}^{\mathrm{pl}}(x, \zeta) d\zeta \tag{3.3.67}$$

With

$$\psi(\zeta) = \zeta - \zeta^3/3$$

we obtain the Reissner type approximation (3.3.62) and

$$k = 5/6, \quad \tilde{\gamma}^{\mathrm{pl}}(x) = \frac{3}{4} \int_{-1}^{1} \gamma_{xz}^{\mathrm{pl}}(x, \zeta)(1 - \zeta^2) d\zeta \tag{3.3.68}$$

As a next choice let us consider the steady-state creep solution of a Bernoulli-Euler beam (see Subsubsect. 3.1.3.3). According to Eqs (3.1.41) and (3.1.42) the bending stress σ_x can be expressed as follows

$$\sigma_x(x, \zeta) = \frac{M(x)}{gh^2} \frac{2(2n+1)}{n} |\zeta|^{(1/n)-1} \zeta$$

After inserting this equation into the equilibrium condition

$$\sigma_{x,x} + \frac{2}{h} \tau_{xz,\zeta} = 0 \tag{3.3.69}$$

and the integration with respect to ζ we obtain the transverse shear stress

$$\tau_{xz} = \frac{Q(x)}{gh} \frac{2n+1}{n+1} (1 - \zeta^2 |\zeta|^{(1/n)-1})$$

With the trial functions

$$\psi^{\bullet}(\zeta) = 1 - \zeta^2|\zeta|^{(1/n)-1}, \quad \psi(\zeta) = \zeta - \frac{n}{2n+1}\zeta|\zeta|^{\frac{1}{n}+1} \tag{3.3.70}$$

and from Eq. (3.3.66) it follows

$$k = \frac{3n+2}{4n+2}, \quad \tilde{\gamma}^{\text{pl}}(x) = \frac{2n+1}{2n+2}\int_{-1}^{1} \gamma_{xz}^{\text{pl}}(x,\zeta)(1 - \zeta^2|\zeta|^{(1/n)-1})d\zeta \tag{3.3.71}$$

By setting $n = 1$ in Eq. (3.3.71) we obtain Eq. (3.3.68). The value of n usually varies between 3 and 10 for metallic materials. For example, if $n = 3; 10$, $k = 11/14; 16/21$, respectively. It can be observed that with increasing creep exponent and consequently with increasing creep rate the value of k decreases (for $n \to \infty$ we obtain $k_{\infty} = 3/4$). The effect of damage is connected with the increase of the creep rate. Therefore a decrease of the value of k can be expected if damage evolution is taken into account. In addition, if the damage rate differs for tensile and compressive stresses, the thickness distribution of the transverse shear stress will be non-symmetrical. In this case the function $\psi^{\bullet}$ cannot be given a priori.

In Naumenko (2000); Altenbach and Naumenko (2002) the first order shear deformation equations are solved by the use of the Ritz method and a time step integration procedure. At a current time step the transverse shear stress is recovered by an approximate solution of Eq. (3.3.69). The proposed numerical procedure is applied in order to modify the trial functions as well as k and $\tilde{\gamma}^{\text{pl}}$ according to the time dependent redistribution of τ_{xz}.

Figure 3.21 presents the results for the uniformly loaded beam with clamped edges. The calculations have been performed with $l = 1000$ mm, $g = 50$ mm, $h = 100$ mm and $q_0 = 50$ N/mm. The constitutive model (3.2.55) and the material parameters for type 316 stainless steel at 650° C (3.2.57) were applied. Curve 1 in Fig. 3.21a is the time dependent maximum deflection calculated by the use of the Bernoulli-Euler beam theory. Curve 2 is obtained by the use of the first order shear deformation theory with the approximations (3.3.62) and Eq. (3.3.68). Curve 3 is the solution of the same equations but with the modified trial functions. Curve 4 is the solution of the plane stress problem with elements PLANE 42 by the ANSYS code. It is obvious that the Bernoulli-Euler beam theory cannot adequately predict the deflection growth. Furthermore, the first order shear deformation equations with the fixed trial functions underestimate the deflection, particularly in the tertiary creep regime. The best agreement with the plane stress solution is obtained if the trial functions are modified according to redistribution of the transverse shear stress. In this case the shear correction factor is time-dependent, Fig. 3.21b. With decreasing value of k we can conclude that the influence of the shear correction increases.

The results for the beam show that the modified shear stress influences the deflection growth in the creep-damage process. On the other hand if we neglect the damage evolution, the steady-state creep solution provides the shear stress distribution close to the parabolic one, see Eq. (3.3.70).

Figure 3.22 shows the distribution of the damage parameter at the last step of calculation. The damage evolution is controlled by the maximum tensile stress, see Eqs

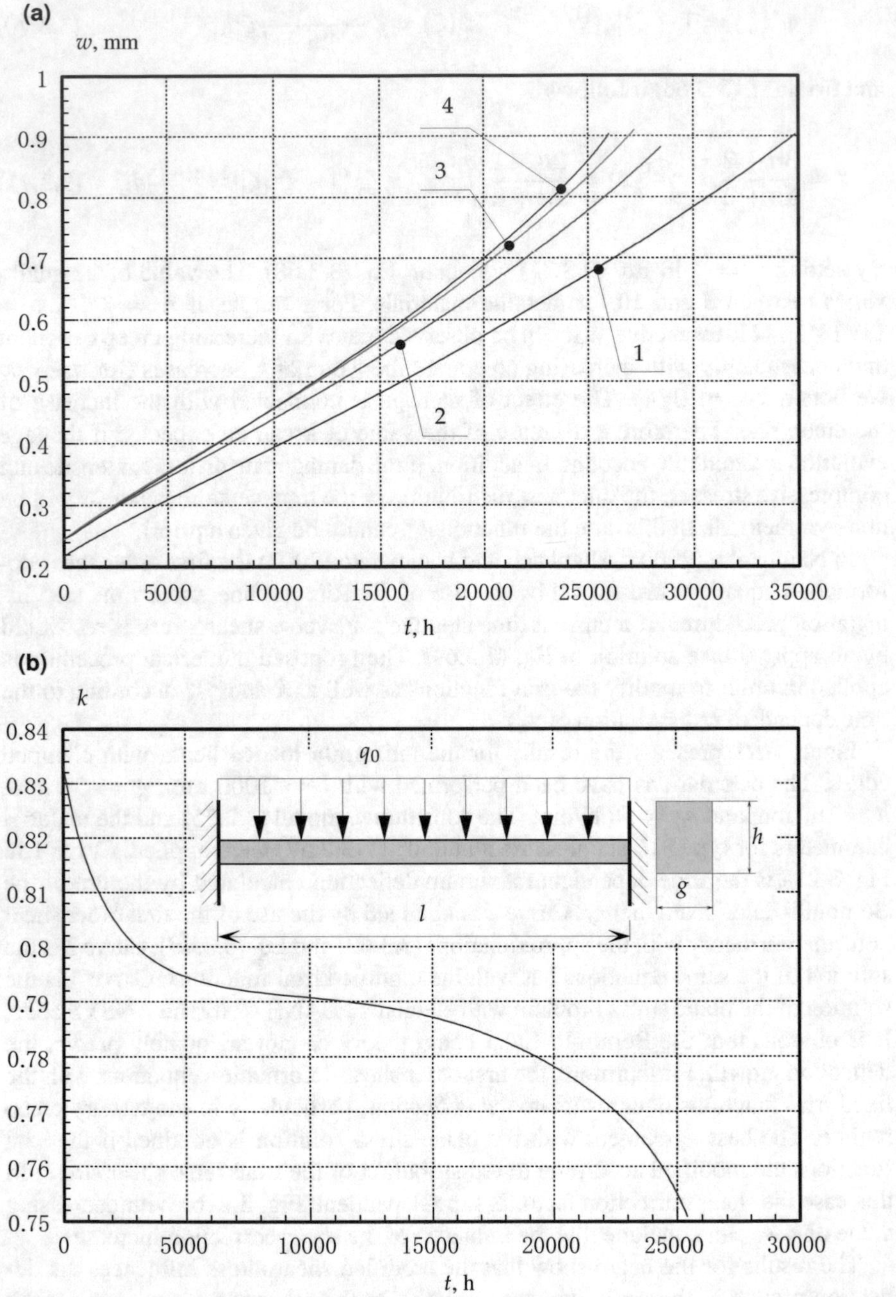

Fig. 3.21 Time-dependent solutions for a clamped beam. **(a)** Maximum deflection vs. time, **(b)** shear correction factor vs. time, 1 – Bernoulli–Euler beam theory, 2 – first order shear deformation theory with parabolic shear stress distribution, 3 – first order shear deformation theory with modified shear stress distribution, 4 – plane stress solution using the ANSYS code with PLANE 42 elements

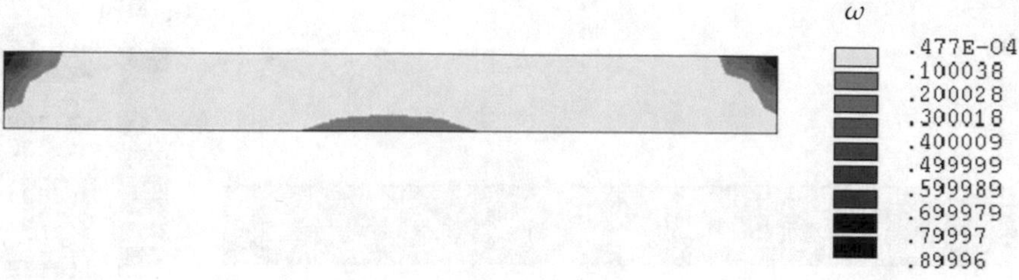

Fig. 3.22 Damage distribution in a beam at last time step

(3.2.55) and (3.2.57). Therefore the zones of the dominant damage are tensile layers of the clamped edges. Figure 3.23a presents the results for τ_{xz} obtained by ANSYS code with PLANE 42 elements. It can be observed that in the neighborhood of the beam edges, where the maximum damage occurs, the distribution of the transverse shear stress is non-symmetrical with respect to the beam midplane. Figures 3.23b and 3.23c show the solution for the transverse shear stress according to the derived beam equations. The transverse shear stress is calculated as a product of the shear force, the distribution function $\psi^\bullet$ and a constant factor. For the considered beam the shear force $Q(x) = q(l/2 - x)$ remains constant during the creep process. Therefore, the time redistribution of the transverse shear stress is only determined by the time-dependence of the function $\psi^\bullet$. Figure 3.23c illustrates $\psi^\bullet$ for different time steps.

The presented example shows that transverse shear deformation and transverse shear stress cannot be ignored in analysis of beams from materials with stress state effects of inelastic behavior. The first order shear deformation theory provides satisfactory results if compared to the results of the plane stress model. Additional investigations are required to establish the constitutive equations and material parameters for beams with arbitrary cross sections.

References

Abaqus Benchmarks (2017) Benchmarks Manual

Abaqus User's Guide (2017) Abaqus Analysis User's Guide. Volume III: Materials

Altenbach H, Naumenko K (1997) Creep bending of thin-walled shells and plates by consideration of finite deflections. Computational Mechanics 19:490 – 495

Altenbach H, Naumenko K (2002) Shear correction factors in creep-damage analysis of beams, plates and shells. JSME International Journal Series A, Solid Mechanics and Material Engineering 45:77 – 83

Altenbach H, Zhilin PA (2004) The theory of simple elastic shells. In: Kienzler R, Altenbach H, Ott I (eds) Theories of Plates and Shells. Critical Review and New Applications, Springer, Berlin, pp 1 – 12

Altenbach H, Kolarow G, Morachkovsky O, Naumenko K (2000) On the accuracy of creep-damage predictions in thinwalled structures using the finite element method. Computational Mechanics 25:87 – 98

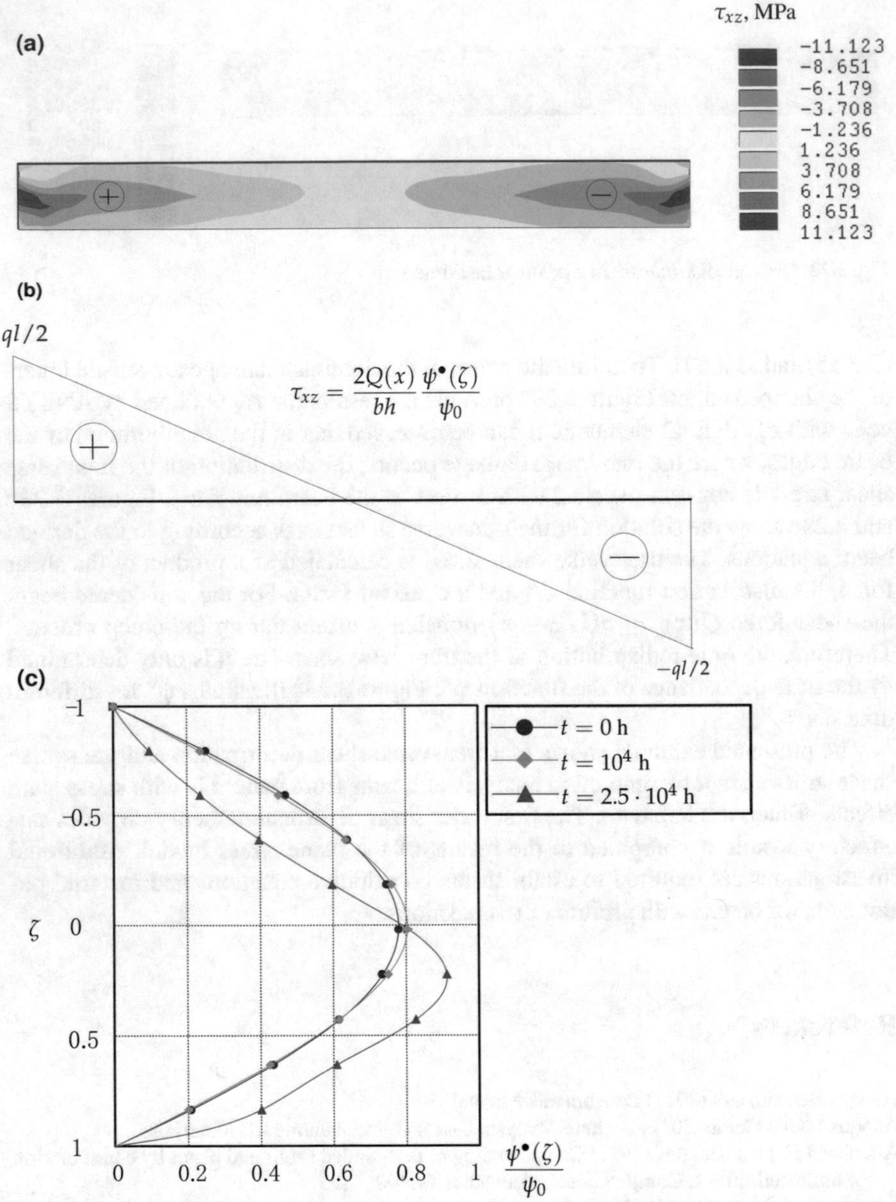

Fig. 3.23 Time-dependent solutions of a clamped beam. **(a)** Transverse shear stresses at last time step, solution with PLANE 42 elements, **(b)** shear force according to the beam equations, **(c)** function of the transverse shear stress distribution for different time steps

Altenbach H, Kushnevsky V, Naumenko K (2001) On the use of solid- and shell-type finite elements in creep-damage predictions of thinwalled structures. Archive of Applied Mechanics 71:164 – 181

Altenbach H, Naumenko K, Zhilin PA (2005) A direct approach to the formulation of constitutive equations for rods and shells. In: Pietraszkiewicz W, Szymczak C (eds) Shell Structures: Theory and Applications, Taylor & Francis, Leiden, pp 87 – 90

Altenbach H, Eremeyev VA, Naumenko K (2015) On the use of the first order shear deformation plate theory for the analysis of three-layer plates with thin soft core layer. ZAMM-Journal of Applied Mathematics and Mechanics/Zeitschrift für Angewandte Mathematik und Mechanik 95(10):1004–1011

ANSYS (2001) Theory Manual

Antman S (1995) Nonlinear Problems of Elasticity. Springer, Berlin

Boyle JT (2012) The creep behavior of simple structures with a stress range-dependent constitutive model. Archive of Applied Mechanics 82(4):495 – 514

Boyle JT, Spence J (1983) Stress Analysis for Creep. Butterworth, London

Chuang TJ (1986) Estimation of power-law creep parameters from bend test data. Journal of Materials Science 21(1):165–175

Eisenträger J, Naumenko K, Altenbach H, Köppe H (2015) Application of the first-order shear deformation theory to the analysis of laminated glasses and photovoltaic panels. International Journal of Mechanical Sciences 96:163–171

Green AE, Naghdi PM, Wenner ML (1974) On the theory of rods. II. Developments by direct approach. Proceedings of the Royal Society of London A Mathematical and Physical Sciences 337(1611):485 – 507

Hosseini E, Holdsworth SR, Mazza E (2013) Stress regime-dependent creep constitutive model considerations in finite element continuum damage mechanics. International Journal of Damage Mechanics 22(8):1186 – 1205

Hult JA (1966) Creep in Engineering Structures. Blaisdell Publishing Company, Waltham

Hutchinson JR (2001) Shear coefficients for timoshenko beam theory. Trans ASME J Appl Mech 68:87 – 92

Kachanov LM (1986) Introduction to Continuum Damage Mechanics. Martinus Nijhoff, Dordrecht

Kaneko T (1975) On Timoshenko's correction for shear in vibrating beams. J Phys D 8:1927 – 1936

Kowalewski ZL (1996) Creep rupture of copper under complex stress state at elevated temperature. In: Design and life assessment at high temperature, Mechanical Engineering Publ., London, pp 113 – 122

Kraus H (1980) Creep Analysis. John Wiley & Sons, New York

Levinson M (1981) A new rectangular beam theory. J Sound Vibr 74:81 – 87

Liu Y, Murakami S, Kageyama Y (1994) Mesh-dependence and stress singularity in finite element analysis of creep crack growth by continuum damage mechanics approach. European Journal of Mechanics A Solids 35(3):147 – 158

Malinin NN (1975) Prikladnaya teoriya plastichnosti i polzuchesti (Applied Theory of Plasticity and Creep, in Russ.). Mashinostroenie, Moskva

Malinin NN (1981) Raschet na polzuchest' konstrukcionnykh elementov (Creep Calculations of Structural Elements, in Russ.). Mashinostroenie, Moskva

Naumenko K (2000) On the use of the first order shear deformation models of beams, plates and shells in creep lifetime estimations. Technische Mechanik 20(3):215 – 226

Naumenko K, Altenbach H (2007) Modelling of Creep for Structural Analysis. Springer, Berlin et al.

Naumenko K, Altenbach H (2016) Modeling High Temperature Materials Behavior for Structural Analysis: Part I: Continuum Mechanics Foundations and Constitutive Models, Advanced Structured Materials, vol 28. Springer

Naumenko K, Eremeyev VA (2017) A layer-wise theory of shallow shells with thin soft core for laminated glass and photovoltaic applications. Composite Structures 178:434–446

Naumenko K, Kostenko Y (2009) Structural analysis of a power plant component using a stress-range-dependent creep-damage constitutive model. Materials Science and Engineering A510-A511:169 – 174

Naumenko K, Altenbach H, Gorash Y (2009) Creep analysis with a stress range dependent constitutive model. Archive of Applied Mechanics 79:619 – 630

Nordmann J, Thiem P, Cinca N, Naumenko K, Krüger M (2018) Analysis of iron aluminide coated beams under creep conditions in high-temperature four-point bending tests. The Journal of Strain Analysis for Engineering Design 53(4):255–265

Odqvist FKG (1974) Mathematical Theory of Creep and Creep Rupture. Oxford University Press, Oxford

Penny RK, Mariott DL (1995) Design for Creep. Chapman & Hall, London

Rabotnov YN (1969) Creep Problems in Structural Members. North-Holland, Amsterdam

Reddy JN (1984) A simple higher-order theory for laminated composite plate. Trans ASME J Appl Mech 51:745 – 752

Reddy JN (1997) Mechanics of Laminated Composite Plates: Theory and Analysis. CRC Press, Boca Raton

Reissner E (1950) A variational theorem in elasticity. J Math Phys 29:90 – 95

Scholz A, Schmidt A, Walther HC, Schein M, Schwienheer M (2008) Experiences in the determination of TMF, LCF and creep life of CMSX-4 in four-point bending experiments. International Journal of Fatigue 30(2):357–362

Schulze S, Pander M, Naumenko K, Altenbach H (2012) Analysis of laminated glass beams for photovoltaic applications. International Journal of Solids and Structures 49(15 - 16):2027 – 2036

Skrzypek JJ (1993) Plasticity and Creep. CRC Press, Boca Raton

Weps M, Naumenko K, Altenbach H (2013) Unsymmetric three-layer laminate with soft core for photovoltaic modules. Composite Structures 105:332–339

Xu B, Yue Z, Eggeler G (2007) A numerical procedure for retrieving material creep properties from bending creep tests. Acta Materialia 55(18):6275–6283

Chapter 4
Plane Stress and Plane Strain Problems

Many structural members can be analyzed applying simplifying assumptions of plane stress or plane strain state. Examples include plates under in-plane loading, thick pipes under internal pressure, rotating discs, etc. Although many problems of this type can be solved in a closed analytical form assuming linear-elastic material behavior (Altenbach et al, 2016; Lurie, 2005; Timoshenko and Goodier, 1951), only few solutions for elementary examples exist, where plasticity and/or creep are taken into account (Boyle and Spence, 1983; Malinin, 1975, 1981; Odqvist, 1974; Skrzypek, 1993).

Chapter 4 presents examples of inelastic structural analysis for plane stress and plane stress problems. In Sect. 4.1 basic assumptions are discussed and governing equations are introduced. Elementary structures including a pressurised thick cylinder, Sect. 4.2, a rotating disc, Sect. 4.3 and a plate with a circular hole, Sect. 4.4, are introduced to illustrate basic features of inelastic behavior under plane multi-axial stress and strain states. Classical results assuming the power law type creep as well as solutions with stress regime dependent inelastic behavior are presented.

4.1 Governing Equations

4.1.1 Assumptions and Preliminaries

As an example for the *plane stress* problem consider a flat plate, Fig. 4.1. The plate is thin-walled such that $h \ll \min(l_1, l_2)$, where h is the plate thickness and l_1, l_2 are plate lengths along $\boldsymbol{e}_1$ and $\boldsymbol{e}_2$ directions. The external force vectors have no components perpendicular to the plate mid-plane and are uniform with respect to the thickness direction. For such in-plane loading of a thin plate, the stress distribution can be assumed constant over the thickness coordinate and

$$\frac{\partial \boldsymbol{\sigma}}{\partial x_3} = \boldsymbol{0} \tag{4.1.1}$$

K. Naumenko and H. Altenbach, *Modeling High Temperature Materials Behavior for Structural Analysis*, Advanced Structured Materials 112,
https://doi.org/10.1007/978-3-030-20381-8_4

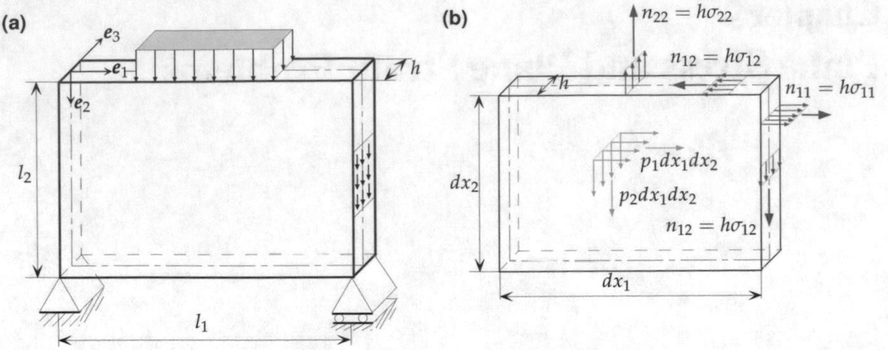

Fig. 4.1 Rectangular plate under in-plane loading. **(a)** Uniform distribution of boundary and reaction forces over the thickness h, **(b)** stress components σ_{ij}, force resultants n_{ij} [force/length] and body forces p_i per unit area of the midplane [force/area]

The surfaces $x_3 = \pm h/2$ are stress-free. With the boundary condition $\boldsymbol{e}_3 \cdot \boldsymbol{\sigma} = \boldsymbol{0}$ the out-of-plane stress components take zero values, i.e. $\sigma_{33} = \sigma_{13} = \sigma_{23} = 0$ on surfaces $x_3 = \pm h/2$. With Eq. (4.1.1) these out-of-plane stress components can be assumed zeros throughout the plate and the stress tensor takes the following simplified form

$$\boldsymbol{\sigma} = \boldsymbol{\sigma}_{\mathrm{p}} = \sigma_{11}\boldsymbol{e}_1 \otimes \boldsymbol{e}_1 + \sigma_{12}(\boldsymbol{e}_1 \otimes \boldsymbol{e}_2 + \boldsymbol{e}_2 \otimes \boldsymbol{e}_1) + \sigma_{22}\boldsymbol{e}_2 \otimes \boldsymbol{e}_2 \tag{4.1.2}$$

In Eq. (4.1.2) and in the sequel of this chapter, the index p will be used for the plane parts of tensors. As a result of the plane stress state $\boldsymbol{\sigma}_{\mathrm{p}}$, the strain state is defined by the following tensor

$$\begin{aligned}\boldsymbol{\varepsilon} &= \varepsilon_{11}\boldsymbol{e}_1 \otimes \boldsymbol{e}_1 + \varepsilon_{12}(\boldsymbol{e}_1 \otimes \boldsymbol{e}_2 + \boldsymbol{e}_2 \otimes \boldsymbol{e}_1) + \varepsilon_{22}\boldsymbol{e}_2 \otimes \boldsymbol{e}_2 + \varepsilon_{33}\boldsymbol{e}_3 \otimes \boldsymbol{e}_3 \\ &= \boldsymbol{\varepsilon}_{\mathrm{p}} + \varepsilon_{33}\boldsymbol{e}_3 \otimes \boldsymbol{e}_3,\end{aligned} \tag{4.1.3}$$

where the strain ε_{33} can be determined from constitutive equations. For circular plates it is convenient to use polar coordinates. Figure 4.2a illustrates the basis vectors $\boldsymbol{e}_r$, $\boldsymbol{e}_\varphi$ and the coordinates r, φ to describe positions of points. For the basis vectors the following relations are valid

$$\boldsymbol{e}_r(\varphi) = \cos\varphi \boldsymbol{e}_1 + \sin\varphi \boldsymbol{e}_2, \quad \boldsymbol{e}_\varphi(\varphi) = \frac{\mathrm{d}\boldsymbol{e}_r}{\mathrm{d}\varphi} = -\sin\varphi \boldsymbol{e}_1 + \cos\varphi \boldsymbol{e}_2$$

For an infinitesimal area element $dA = r\mathrm{d}\varphi\mathrm{d}r$, Fig. 4.2b one may draw a free-body diagram to introduce internal forces. The corresponding components of the stress tensor are shown in Fig. 4.2c. The plane stress state can be characterized by the following tensor

$$\boldsymbol{\sigma}_{\mathrm{p}} = \sigma_{rr}\boldsymbol{e}_r \otimes \boldsymbol{e}_r + \sigma_{r\varphi}(\boldsymbol{e}_r \otimes \boldsymbol{e}_\varphi + \boldsymbol{e}_\varphi \otimes \boldsymbol{e}_r) + \sigma_{\varphi\varphi}\boldsymbol{e}_\varphi \otimes \boldsymbol{e}_\varphi \tag{4.1.4}$$

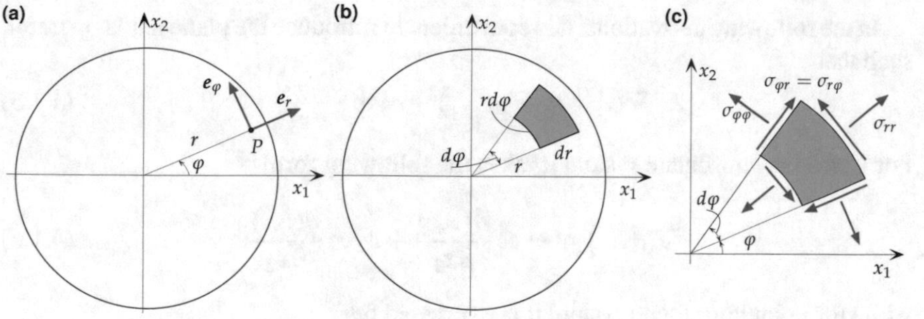

Fig. 4.2 Circular plate in a polar coordinate system. **(a)** Polar coordinates and basis vectors, **(b)** infinitesimal area element, **(c)** components of stress tensor

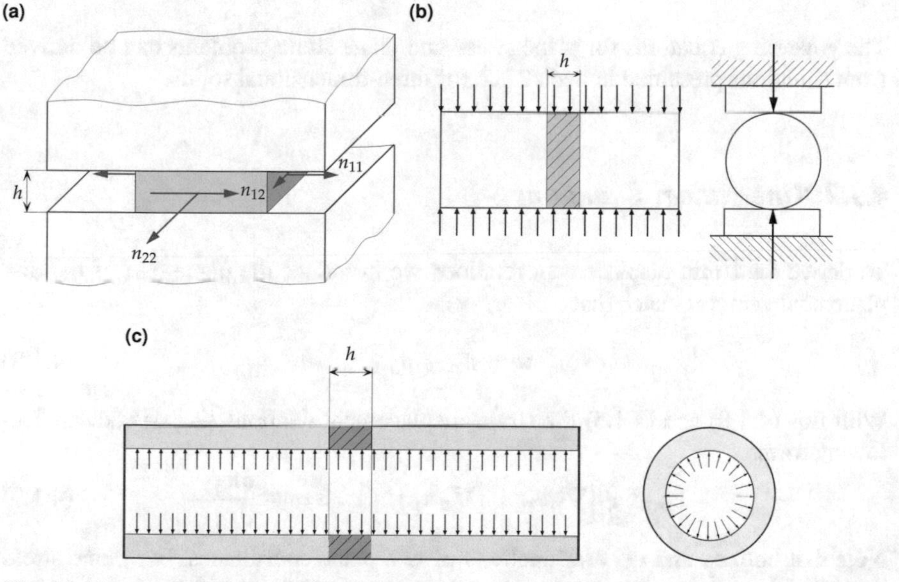

Fig. 4.3 Plane strain problems. **(a)** Thin plate between rigid blocks, **(b)** section of a long roller, **(c)** section of a thick-walled pipe

Figure 4.3 illustrates examples for problems which can be analyzed by plane strain state assumptions. For a thin plate placed between rigid blocks, Fig. 4.3a, the motion in the direction $\boldsymbol{e}_3$ is constrained and $\varepsilon_{33} = 0$ can be assumed. Because constraints the reaction forces act on the top and bottom surfaces of the plate leading to the stress σ_{33}. Similarly, for a thin section of a long roller, Fig. 4.3b or a thin section of a pipe, Fig. 4.3c, one may assume that the axial strain is negligible. For plane strain problems the stress and strain tensors have the following form

$$\boldsymbol{\varepsilon} = \boldsymbol{\varepsilon}_{\mathrm{p}}, \quad \boldsymbol{\sigma} = \boldsymbol{\sigma}_{\mathrm{p}} + \sigma_{33}\boldsymbol{e}_3 \otimes \boldsymbol{e}_3$$

In the following derivations, it is convenient to introduce the plane nabla operator, such that

$$\nabla(\dots) = \nabla_{\mathrm{p}}(\dots) + \boldsymbol{e}_3 \otimes \frac{\partial(\dots)}{\partial x_3} \qquad (4.1.5)$$

For Cartesian coordinate system it takes the following form

$$\nabla_{\mathrm{p}}(\dots) = \boldsymbol{e}_1 \otimes \frac{\partial(\dots)}{\partial x_1} + \boldsymbol{e}_2 \otimes \frac{\partial(\dots)}{\partial x_2}, \qquad (4.1.6)$$

while for polar coordinate system it is expressed by

$$\nabla_{\mathrm{p}}(\dots) = \boldsymbol{e}_r \otimes \frac{\partial(\dots)}{\partial r} + \boldsymbol{e}_\varphi \otimes \frac{1}{r}\frac{\partial(\dots)}{\partial \varphi} \qquad (4.1.7)$$

The governing equations for plane stress and plane strain problems can be derived from equations presented in Sect. 2.1.2 for three-dimensional solids.

4.1.2 Kinematical Equations

To derive the strain-displacement relations we introduce the plane part of the displacement vector $\boldsymbol{u}_{\mathrm{p}}$ such that

$$\boldsymbol{u} = \boldsymbol{u}_{\mathrm{p}} + u_3 \boldsymbol{e}_3 \qquad (4.1.8)$$

With Eqs (4.1.8) and (4.1.5) the strain-displacement relations (2.1.4) take the following form

$$\boldsymbol{\varepsilon}_{\mathrm{p}} = \frac{1}{2}\left[\nabla_{\mathrm{p}}\boldsymbol{u}_{\mathrm{p}} + (\nabla_{\mathrm{p}}\boldsymbol{u}_{\mathrm{p}})^{\mathrm{T}}\right], \quad \varepsilon_{33} = \frac{\partial u_3}{\partial x_3} \qquad (4.1.9)$$

Note that both $\boldsymbol{\varepsilon}_{\mathrm{p}}$ and ε_{33} are functions of two plane coordinates. For plane stress problems the transverse normal strain can be computed from constitutive equations and the displacement u_3 can be obtained by integrating Eq. $(4.1.9)_2$. Taking the plane divergence of Eq. $(4.1.9)_1$ we obtain

$$\nabla_{\mathrm{p}} \cdot \boldsymbol{\varepsilon}_{\mathrm{p}} = \frac{1}{2}\left(\Delta_{\mathrm{p}}\boldsymbol{u}_{\mathrm{p}} + \nabla_{\mathrm{p}}\nabla_{\mathrm{p}} \cdot \boldsymbol{u}_{\mathrm{p}}\right), \qquad (4.1.10)$$

where

$$\Delta_{\mathrm{p}}(\dots) = \nabla_{\mathrm{p}} \cdot \nabla_{\mathrm{p}}(\dots) \qquad (4.1.11)$$

is the two-dimensional Laplace operator. The divergence of Eq. (4.1.10) yields

$$\nabla_{\mathrm{p}} \cdot (\nabla_{\mathrm{p}} \cdot \boldsymbol{\varepsilon}_{\mathrm{p}}) = \Delta_{\mathrm{p}}\nabla_{\mathrm{p}} \cdot \boldsymbol{u}_{\mathrm{p}} \qquad (4.1.12)$$

On the other hand, by taking the trace of $(4.1.9)_1$ we obtain

$$\mathrm{tr}\,\boldsymbol{\varepsilon}_{\mathrm{P}} = \nabla_{\mathrm{P}} \cdot \boldsymbol{u}_{\mathrm{P}} \tag{4.1.13}$$

From Eqs (4.1.12) and (4.1.13) the following compatibility condition for the tensor $\boldsymbol{\varepsilon}_{\mathrm{P}}$ can be derived

$$\nabla_{\mathrm{P}} \cdot (\nabla_{\mathrm{P}} \cdot \boldsymbol{\varepsilon}_{\mathrm{P}}) = \Delta_{\mathrm{P}} \mathrm{tr}\,\boldsymbol{\varepsilon}_{\mathrm{P}} \tag{4.1.14}$$

4.1.3 Equilibrium Conditions

With Eq. (4.1.5) and the plane stress assumption, the equilibrium conditions (2.1.4) yield

$$\nabla_{\mathrm{P}} \cdot \boldsymbol{\sigma}_{\mathrm{P}} + \rho \bar{\boldsymbol{f}}_{\mathrm{P}} = \boldsymbol{0}, \tag{4.1.15}$$

where $\bar{\boldsymbol{f}}_{\mathrm{P}}$ is the plane part of the force density. For plane strain problems the stress tensor $\boldsymbol{\sigma}_{\mathrm{P}}$ can be computed from the equilibrium conditions (4.1.15) while the normal stress σ_{33} follows from constitutive equations.

Let us assume that the force $\rho \bar{\boldsymbol{f}}_{\mathrm{P}}$ has a potential such that

$$\rho \bar{\boldsymbol{f}}_{\mathrm{P}} = -\nabla_{\mathrm{P}} \Omega \tag{4.1.16}$$

Then the Airy stress function Φ can be introduced as follows

$$\boldsymbol{\sigma}_{\mathrm{P}} = (\Delta_{\mathrm{P}} \Phi + \Omega) \boldsymbol{P} - \nabla_{\mathrm{P}} \nabla_{\mathrm{P}} \Phi, \tag{4.1.17}$$

where $\boldsymbol{P} = \boldsymbol{I} - \boldsymbol{e}_3 \otimes \boldsymbol{e}_3$ is the projector. With Eq. (4.1.17) the equilibrium conditions (4.1.15) are identically satisfied.

4.1.4 Constitutive Equations

For isotropic materials the generalized Hooke's law (2.1.9) can be formulated as follows

$$\boldsymbol{\sigma} = \frac{E}{1+\nu} \left(\boldsymbol{\varepsilon}^{\mathrm{el}} + \frac{\nu}{1-2\nu} \mathrm{tr}\,\boldsymbol{\varepsilon}^{\mathrm{el}} \boldsymbol{I} \right) \tag{4.1.18}$$

or in the inverse form

$$\boldsymbol{\varepsilon}^{\mathrm{el}} = \frac{1}{E} [(1+\nu)\boldsymbol{\sigma} - \nu \mathrm{tr}\,\boldsymbol{\sigma} \boldsymbol{I}] \tag{4.1.19}$$

For the *plane stress state*, with $\sigma_{33} = 0$, Eq. (4.1.18) yields

$$\sigma_{33} = \varepsilon_{33}^{\mathrm{el}} + \frac{\nu}{1-2\nu} \mathrm{tr}\,\boldsymbol{\varepsilon}^{\mathrm{el}} = 0 \tag{4.1.20}$$

As a result we obtain

$$\varepsilon_{33}^{\mathrm{el}} = -\frac{\nu}{1-\nu} \mathrm{tr}\,\boldsymbol{\varepsilon}_{\mathrm{P}}^{\mathrm{el}} \tag{4.1.21}$$

Inserting Eq. (4.1.21) into Eq. (4.1.18) the following constitutive equation for the stress tensor can be obtained

$$\boldsymbol{\sigma} = \boldsymbol{\sigma}_{\mathrm{p}} = \frac{E}{1+\nu}\left(\boldsymbol{\varepsilon}^{\mathrm{el}}_{\mathrm{p}} + \frac{\nu}{1-\nu}\mathrm{tr}\,\boldsymbol{\varepsilon}^{\mathrm{el}}_{\mathrm{p}}\,\boldsymbol{P}\right) \tag{4.1.22}$$

In the case of *plane strain* state we have to set

$$\varepsilon_{33} = \varepsilon^{\mathrm{el}}_{33} + \varepsilon^{\mathrm{pl}}_{33} + \varepsilon^{\mathrm{th}}_{33} = 0 \quad \Rightarrow \quad \varepsilon^{\mathrm{el}}_{33} = -\varepsilon^{\mathrm{pl}}_{33} - \varepsilon^{\mathrm{th}}_{33} \tag{4.1.23}$$

Equation (4.1.19) yields

$$\varepsilon^{\mathrm{el}}_{33} = \frac{1}{E}\left[(1+\nu)\sigma_{33} - \nu\,\mathrm{tr}\,\boldsymbol{\sigma}\right] \tag{4.1.24}$$

With $\mathrm{tr}\,\boldsymbol{\sigma} = \mathrm{tr}\,\boldsymbol{\sigma}_{\mathrm{p}} + \sigma_{33}$, Eq. (4.1.24) provides the stress component σ_{33} as follows

$$\sigma_{33} = \nu\,\mathrm{tr}\,\boldsymbol{\sigma}_{\mathrm{p}} + E\varepsilon^{\mathrm{el}}_{33} \tag{4.1.25}$$

or

$$\sigma_{33} = \nu\,\mathrm{tr}\,\boldsymbol{\sigma}_{\mathrm{p}} - E(\varepsilon^{\mathrm{th}}_{33} + \varepsilon^{\mathrm{pl}}_{33}) \tag{4.1.26}$$

Inserting Eq. (4.1.26) into Eq. (4.1.19) we obtain

$$\boldsymbol{\varepsilon}^{\mathrm{el}}_{\mathrm{p}} = \frac{1+\nu}{E}(\boldsymbol{\sigma}_{\mathrm{p}} - \nu\,\mathrm{tr}\,\boldsymbol{\sigma}_{\mathrm{p}}\,\boldsymbol{P}) + \nu\left(\varepsilon^{\mathrm{th}}_{33} + \varepsilon^{\mathrm{pl}}_{33}\right)\boldsymbol{P} \tag{4.1.27}$$

The inelastic strain rate tensor must be defined by a constitutive equation. Constitutive models discussed in Naumenko and Altenbach (2016, Chapt. 5) can be applied with the assumption of plane stress or plane strain states. As an example consider the von Mises-Odqvist flow rule of steady-state creep (2.1.34). For plane strain problems it can be specified as follows

$$\begin{aligned}
\dot{\boldsymbol{\varepsilon}}^{\mathrm{pl}}_{\mathrm{p}} &= \frac{3}{2}\dot{\varepsilon}_0 f\left(\frac{\sigma_{\mathrm{vM}}}{\sigma_0}\right)\frac{1}{\sigma_{\mathrm{vM}}}\left(\boldsymbol{\sigma}_{\mathrm{p}} - \frac{1}{3}\mathrm{tr}\,\boldsymbol{\sigma}_{\mathrm{p}}\boldsymbol{P} - \frac{1}{3}\sigma_{33}\boldsymbol{P}\right),\\
\dot{\varepsilon}^{\mathrm{pl}}_{33} &= \frac{3}{2}\dot{\varepsilon}_0 f\left(\frac{\sigma_{\mathrm{vM}}}{\sigma_0}\right)\frac{1}{\sigma_{\mathrm{vM}}}\left(\frac{1}{3}\sigma_{33} - \frac{1}{3}\mathrm{tr}\,\boldsymbol{\sigma}_{\mathrm{p}}\right),\\
\dot{\boldsymbol{\varepsilon}}^{\mathrm{pl}}_{\mathrm{p}} &= \boldsymbol{P}\cdot\dot{\boldsymbol{\varepsilon}}^{\mathrm{pl}}\cdot\boldsymbol{P}
\end{aligned} \tag{4.1.28}$$

where the von Mises equivalent stress takes the form

$$\sigma^2_{\mathrm{vM}} = \frac{3}{2}\mathrm{tr}\,\boldsymbol{\sigma}^2_{\mathrm{p}} - \frac{1}{2}(\mathrm{tr}\,\boldsymbol{\sigma}_{\mathrm{p}})^2 - \mathrm{tr}\,\boldsymbol{\sigma}_{\mathrm{p}}\sigma_{33} + \sigma^2_{33},$$

$\dot{\varepsilon}_0$ and σ_0 are material parameters and f is a function of the von Mises equivalent stress, for example the power law. For plane stress problems Eqs (4.1.28) are simplified as follows

$$\dot{\varepsilon}^{\text{pl}}_{\text{p}} = \frac{3}{2}\dot{\varepsilon}_0 f\left(\frac{\sigma_{\text{vM}}}{\sigma_0}\right)\frac{1}{\sigma_{\text{vM}}}\left(\boldsymbol{\sigma}_{\text{p}} - \frac{1}{3}\text{tr}\,\boldsymbol{\sigma}_{\text{p}}\boldsymbol{P}\right),$$
$$\dot{\varepsilon}^{\text{pl}}_{33} = -\frac{3}{2}\dot{\varepsilon}_0 f\left(\frac{\sigma_{\text{vM}}}{\sigma_0}\right)\frac{1}{\sigma_{\text{vM}}}\frac{1}{3}\text{tr}\,\boldsymbol{\sigma}_{\text{p}}, \tag{4.1.29}$$
$$\sigma^2_{\text{vM}} = \frac{3}{2}\text{tr}\,\boldsymbol{\sigma}^2_{\text{p}} - \frac{1}{2}(\text{tr}\,\boldsymbol{\sigma}_{\text{p}})^2$$

4.2 Pressurized Thick Cylinder

Consider a thick cylinder section as shown in Fig. 4.4. We assume that the cylinder is uniformly heated and loaded by the internal pressure p. The problem of steady state stress distribution as a result of power law creep is discussed in textbooks on creep mechanics, e.g. Boyle and Spence (1983); Hult (1966); Malinin (1981) and is usually applied as a benchmark problem to verify finite element solutions, e.g. Abaqus Benchmarks (2017). However, the available results are only valid for either linear or the power law creep ranges. In Altenbach et al (2008); Boyle (2012); Naumenko et al (2009); Naumenko and Kostenko (2009) this problem is revised by taking into account the stress regime dependent creep behavior. Section 4.2 provides governing equations and solutions for the steady-state creep of a thick cylinder under internal pressure.

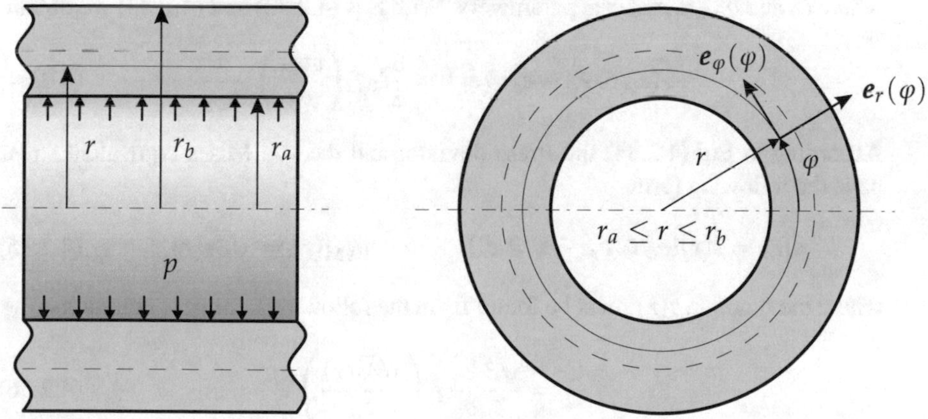

Fig. 4.4 Thick cylinder. Geometry and loading

4.2.1 Governing Equations for Steady-State Flow

Let $\boldsymbol{v}$ be the vector characterizing the steady-state flow velocity of the material point. Assuming the plane strain state we may write

$$\boldsymbol{v} = v_r(r)\boldsymbol{e}_r \quad \Rightarrow \quad \nabla\boldsymbol{v} = (\nabla\boldsymbol{v})^T = \frac{\partial v_r}{\partial r}\,\boldsymbol{e}_r \otimes \boldsymbol{e}_r + \frac{v_r}{r}\,\boldsymbol{e}_\varphi \otimes \boldsymbol{e}_\varphi \tag{4.2.30}$$

The condition of the volume constancy (2.1.35) yields

$$\nabla \cdot \boldsymbol{v} = \frac{\partial v_r}{\partial r} + \frac{v_r}{r} = 0 \quad \Rightarrow \quad v_r = \frac{\mathsf{C}}{r}, \tag{4.2.31}$$

where C is an integration constant. The symmetric part of the velocity gradient is the strain rate tensor

$$\dot{\boldsymbol{\varepsilon}} = \frac{\mathsf{C}}{r^2}(\boldsymbol{e}_\varphi \otimes \boldsymbol{e}_\varphi - \boldsymbol{e}_r \otimes \boldsymbol{e}_r), \tag{4.2.32}$$

and the latter can be additively decomposed into the elastic and the creep part. When damage processes are negligible and the pressure is constant over time, a steady-state exists for which $\dot{\boldsymbol{s}} = \boldsymbol{0}$. Let us apply the von Mises-Odqvist flow rule of steady-state creep (2.1.34). It can be rewritten as follows

$$\dot{\boldsymbol{\varepsilon}} = \frac{3}{2}\dot{\varepsilon}_0 f\left(\frac{\sigma_{\mathrm{vM}}}{\sigma_0}\right)\frac{\boldsymbol{s}}{\sigma_{\mathrm{vM}}}, \tag{4.2.33}$$

where $\dot{\varepsilon}_0$ and σ_0 are material parameters. With Eqs (4.2.32) and (4.2.33) we obtain

$$\frac{\mathsf{C}}{r^2}(\boldsymbol{e}_\varphi \otimes \boldsymbol{e}_\varphi - \boldsymbol{e}_r \otimes \boldsymbol{e}_r) = \frac{3}{2}\dot{\varepsilon}_0 f\left(\frac{\sigma_{\mathrm{vM}}}{\sigma_0}\right)\frac{\boldsymbol{s}}{\sigma_{\mathrm{vM}}} \tag{4.2.34}$$

According to Eq. (4.2.34) the stress deviator and the von Mises equivalent stress have the following form

$$\boldsymbol{s}(r) = s(r)(\boldsymbol{e}_\varphi \otimes \boldsymbol{e}_\varphi - \boldsymbol{e}_r \otimes \boldsymbol{e}_r) \quad \Rightarrow \quad \sigma_{\mathrm{vM}}(r) = \sqrt{3}s(r), \tag{4.2.35}$$

where the function $s(r)$ must be found from the following nonlinear equation

$$\frac{\mathsf{C}}{r^2} = \frac{\sqrt{3}}{2}\frac{\dot{\varepsilon}_0}{\sigma_0}f\left(\frac{\sqrt{3}s(r)}{\sigma_0}\right) \tag{4.2.36}$$

Applying the equilibrium condition (2.1.36) we can write

$$\nabla\sigma_{\mathrm{m}} + \nabla \cdot \boldsymbol{s} = \boldsymbol{0} \quad \Rightarrow \quad \frac{d\sigma_{\mathrm{m}}}{dr} = \frac{ds}{dr} + \frac{2}{r}\,s, \tag{4.2.37}$$

By integration of Eq. (4.2.37) the mean stress σ_{m} can be computed as follows

$$\sigma_{\mathrm{m}} = s + \int \frac{2}{r}\, s\mathrm{d}r + D, \tag{4.2.38}$$

where D is the integration constant. With

$$\boldsymbol{\sigma} = \sigma_{\mathrm{m}}\boldsymbol{I} + s(\boldsymbol{e}_{\varphi} \otimes \boldsymbol{e}_{\varphi} - \boldsymbol{e}_r \otimes \boldsymbol{e}_r)$$

the radial stress σ_{rr}, the hoop stress $\sigma_{\varphi\varphi}$ and the axial stress σ_{33} are defined as follows

$$\sigma_{rr} = \sigma_{\mathrm{m}} - s, \quad \sigma_{\varphi\varphi} = \sigma_{\mathrm{m}} + s, \quad \sigma_{33} = \sigma_{\mathrm{m}} \tag{4.2.39}$$

The integration constants C and D can be computed from the following boundary conditions

$$\sigma_{rr}(r_a) = -p, \quad \sigma_{rr}(r_b) = 0 \tag{4.2.40}$$

Let us introduce the normalized variables

$$\tilde{s} = \frac{s}{\sigma_0}\sqrt{3} = \frac{\sigma_{\mathrm{vM}}}{\sigma_0}, \quad \tilde{C} = \frac{2C}{\sqrt{3}\dot{\varepsilon}_0 r_a^2}, \quad \eta = \frac{r}{r_a}, \quad \zeta = \frac{r_a}{r_b} \tag{4.2.41}$$

With the boundary condition $(4.2.40)_2$, Eqs. (4.2.38) and (4.2.39) yield the following expressions for stress distributions along the normalized radial coordinate η

$$\begin{aligned}
\sigma_{rr}(\eta, \tilde{C}) &= \frac{2\sigma_0}{\sqrt{3}} \int\limits_{1/\zeta}^{\eta} \frac{1}{\eta} f^{-1}(\tilde{C}/\eta^2)\mathrm{d}\eta, \\
\sigma_{\varphi\varphi}(\eta, \tilde{C}) &= \frac{2\sigma_0}{\sqrt{3}} \left[\int\limits_{1/\zeta}^{\eta} \frac{1}{\eta} f^{-1}(\tilde{C}/\eta^2)\mathrm{d}\eta + f^{-1}(\tilde{C}/\eta^2) \right] \\
\sigma_{\mathrm{vM}}(\eta, \tilde{C}) &= \sigma_0 f^{-1}(\tilde{C}/\eta^2), \quad \sigma_{33} = \sigma_{\mathrm{m}} = \frac{1}{2}(\sigma_{rr} + \sigma_{\varphi\varphi})
\end{aligned} \tag{4.2.42}$$

With the boundary condition $\sigma_r(1) = -p$, Eq. $(4.2.42)_1$ yield

$$\int\limits_{1}^{1/\zeta} \frac{2}{\eta} f^{-1}(\tilde{C}/\eta^2)\mathrm{d}\eta = \frac{\sqrt{3}p}{\sigma_0} \tag{4.2.43}$$

By solving Eq. (4.2.43) the integration constant $\tilde{C}$ can be computed.

4.2.2 Solution with Norton-Bailey Creep law

The classical approach to the steady state creep of the thick cylinder assumes the power law creep constitutive equation, e.g. Betten (2008); Odqvist (1974); Skrzypek (1993). In this case Eq. (4.2.36) simplifies to

$$\frac{C}{r^2} = \frac{\sqrt{3}}{2}\frac{\dot{\varepsilon}_0}{\sigma_0}\left(\frac{\sqrt{3}s(r)}{\sigma_0}\right)^n$$

With the normalized variables (4.2.41) we obtain

$$f(\tilde{s}) = \tilde{s}^n \quad \Rightarrow \quad f^{-1}(\tilde{C}/\eta^2) = \frac{\tilde{C}^{1/n}}{\eta^{2/n}},$$

From Eq. (4.2.43) the integration constant $\tilde{C}$ is computed as follows

$$\tilde{C}^{1/n} = \frac{\sqrt{3}p}{\sigma_0}\frac{1}{n(1-\zeta^{2/n})}$$

Taking the integrals in Eqs (4.2.42) the following expressions for the radial and the hoop stress can be obtained

$$\begin{aligned} \sigma_{rr} &= \frac{p}{1-\zeta^{2/n}}\left(\zeta^{2/n} - \eta^{-2/n}\right), \\ \sigma_{\varphi\varphi} &= \frac{p}{1-\zeta^{2/n}}\left(\zeta^{2/n} - \frac{n-2}{n}\eta^{-2/n}\right) \end{aligned} \tag{4.2.44}$$

The von Mises equivalent stress can be computed from Eqs $(4.2.42)_3$ as follows

$$\sigma_{\text{vM}} = \frac{\sqrt{3}p}{n(1-\zeta^{2/n})}\frac{1}{\eta^{2/n}} \tag{4.2.45}$$

From Eqs $(4.2.42)_4$ and (4.2.44) the following expression for the axial stress (hydrostatic stress) can be obtained

$$\sigma_{33} = \sigma_m = \frac{p}{1-\zeta^{2/n}}\left(\zeta^{2/n} - \frac{n-1}{n}\eta^{-2/n}\right) \tag{4.2.46}$$

Figure 4.5 illustrates the distributions of the radial stress along the radial coordinate for $r_b/r_a = 2$ and different values of the creep exponent. The corresponding distributions of the hoop stress are shown in Fig. 4.6. For $n = 1$ (elasticity or linear creep) the hoop stress has the maximum value at the inner radius of the cylinder. For $n = 2$ the hoop stress is constant over the radial coordinate. For $n > 2$ the hoop stress takes the maximum value at the outer cylinder radius. The stress redistribution from initial elastic state to the steady creep state can be clearly observed by comparison of graphs for $n = 1$ and $n > 1$ shown in Fig. 4.6. The hoop stress relaxes at the inner radius and increases at the outer radius. For large values of n, the stress distributions approach the solutions for rigid plastic material behavior. The corresponding expressions can be obtained from (4.2.44) by setting $n \to \infty$ as follows

$$\sigma_{rr} = -p\frac{\ln\eta}{\ln\zeta} \quad \sigma_{\varphi\varphi} = -p\frac{1+\ln\eta}{\ln\zeta} \tag{4.2.47}$$

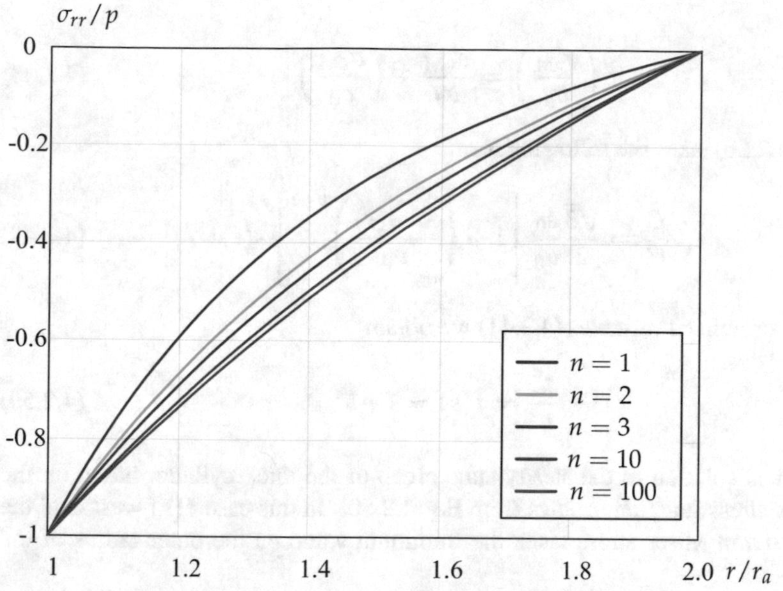

Fig. 4.5 Distribution of the radial stress along the radial coordinate for a pressurized cylinder in a steady creep state

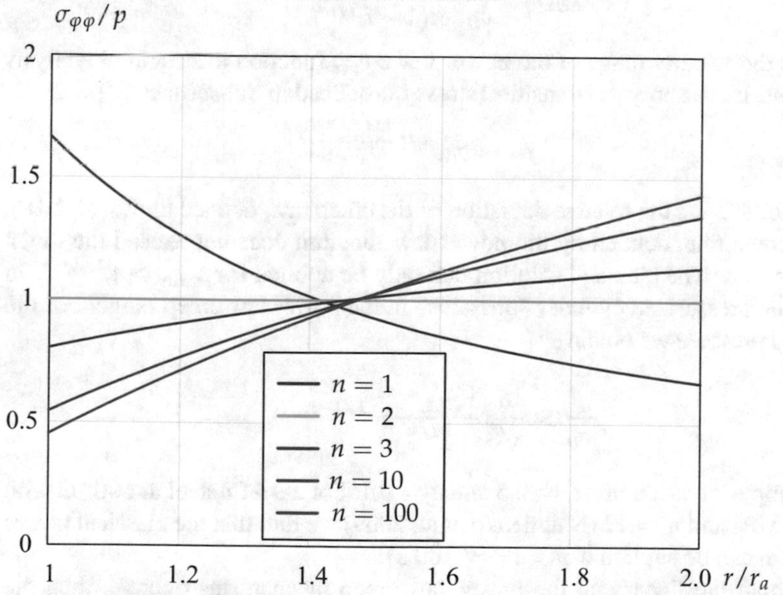

Fig. 4.6 Distribution of the hoop stress along the radial coordinate for a pressurized cylinder in a steady creep state

4.2.3 Solution with Stress Regime Dependent Creep law

Let us consider the inelastic strain rate vs stress behavior as illustrated in Fig. 1.12. To describe the stress regime dependence let us apply the double power law function of stress (3.1.38). In this case

$$f\left(\frac{\sigma_{\mathrm{vM}}}{\sigma_0}\right) = \frac{\sigma_{\mathrm{vM}}}{\sigma_0} + \left(\frac{\sigma_{\mathrm{vM}}}{\sigma_0}\right)^n, \tag{4.2.48}$$

and Eq. (4.2.36) takes the following form

$$\frac{\mathsf{C}}{r^2} = \frac{\sqrt{3}}{2}\frac{\dot{\varepsilon}_0}{\sigma_0}\left[1 + \left(\frac{\sqrt{3}s(r)}{\sigma_0}\right)^{n-1}\right]s(r) \tag{4.2.49}$$

With the normalized variables (4.2.41) we obtain

$$\frac{\tilde{\mathsf{C}}}{\eta^2} = f(\tilde{s}) = \tilde{s} + \tilde{s}^n \tag{4.2.50}$$

The classical solution to the steady-state creep of the thick cylinder based on the power law stress function follows from Eq (4.2.50). In this case $f(\tilde{s}) = \tilde{s}^n$ and the normalized von Mises stress takes the minimum value on the outer radius of the cylinder

$$\tilde{s}_{\min} = \frac{\sqrt{3}p}{\sigma_0}\frac{\zeta^{2/n}}{n(1-\zeta^{2/n})}$$

To discuss the validity range of the power law creep function in structural analysis applications, let us apply the transition stress introduced in Subsubsect. 3.1.3.2

$$\sigma_\epsilon = \sigma_0 \epsilon^{1/1-n}$$

such that for $\sigma > \sigma_\epsilon$ the relative deviation of the creep rate, defined by Eq. (4.2.48), from the creep rate, defined by the power law function does not exceed the given accuracy $\epsilon < 1$. The classical solution can only be applied for $s_{\min} > \epsilon^{1/1-n}$. In this case the pressurized cylinder "operates" in the power law creep range. For the normalized pressure we obtain

$$\frac{p}{\sigma_0} > \frac{n}{\sqrt{3}}\frac{1-\zeta^{2/n}}{\zeta^{2/n}}\epsilon^{1/1-n}$$

As an example let us assume $\zeta = 0.5$ and $\epsilon = 0.1$. For a 9%Cr steel at 600° C with $\sigma_0 = 100$ MPa and $n = 12$ (Naumenko et al, 2009) we find that the classical power law solution can be applied if $p > \sigma_0 = 100$ MPa.

When both the linear and the power law creep mechanisms operate, then the character of the stress distribution additionally depends on the applied pressure. To assess such a dependence, the transition from the linear to the power law creep must be taken into account. In this case the inverse function f^{-1} and consequently the

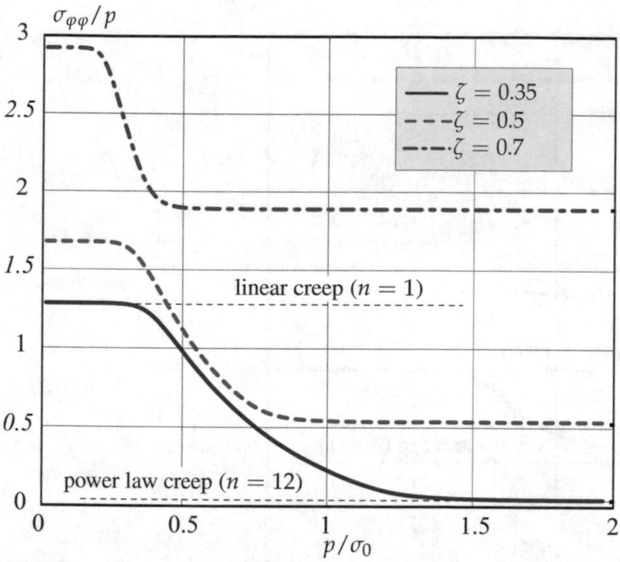

Fig. 4.7 Normalized hoop stress vs. normalized pressure at the inner radius

stress distributions cannot be given in a closed analytical form. To obtain an approximate solution, two numerical procedures including the numerical integration and finding the root of a non-linear algebraic equation were applied. First, the non-linear Eq. (4.2.43) was solved numerically to compute the constant $\tilde{C}$. Then the integrals (4.2.42) were evaluated to find the stress distributions along the radial coordinate. The input parameters are the ratio of the cylinder radii ζ, the normalized pressure $\tilde{p} = p/\sigma_0$ and the creep exponent n.

Figures 4.7 and 4.8 illustrate the dependence of the hoop stress on the normalized pressure for different values of ζ. The validity ranges of the classical solution can be recognized. For both the low and the high pressures the results do not depend on the normalized pressure $\tilde{p}$ and coincide with the classical solution (4.2.44) for $n = 1$ and $n = 12$. For "moderate" pressures usually encountered in practice, the use of the power law would lead to an underestimation of the hoop stress on the inner surface, Fig. 4.7, and overestimation on the outer surface, Fig. 4.8. The range of validity of the power law depends on ζ and extends with an increase of ζ (decrease of the wall thickness). Therefore for thin-walled pipes loaded by a moderate pressure, the power law creep is a reasonable approximation.

Figures 4.9 and 4.10 illustrate stress distributions along the radial coordinate for the case $\zeta = 0.5$. For small values of the normalized pressure, for example $\tilde{p} = 0.2$ the results agree with the classical solution assuming the linear creep behavior. For high values of the normalized pressure, for example $\tilde{p} = 0.9$ the results correspond to the power law creep regime. In this case, the hoop stress relaxes on the inner surface and increases on the outer surface, Fig. 4.10. For moderate values of the normalized pressure, for example $\tilde{p} = 0.7$, the distribution of the hoop stress deviates significantly from the classical solution. While the relaxation is still observed on the inner surface, the maximum shifts towards the core layer of the cylinder.

Figures 4.11 and 4.12 illustrate the dependence of solutions on the value of the stress exponent. The validity range of the classical solution depends on the value

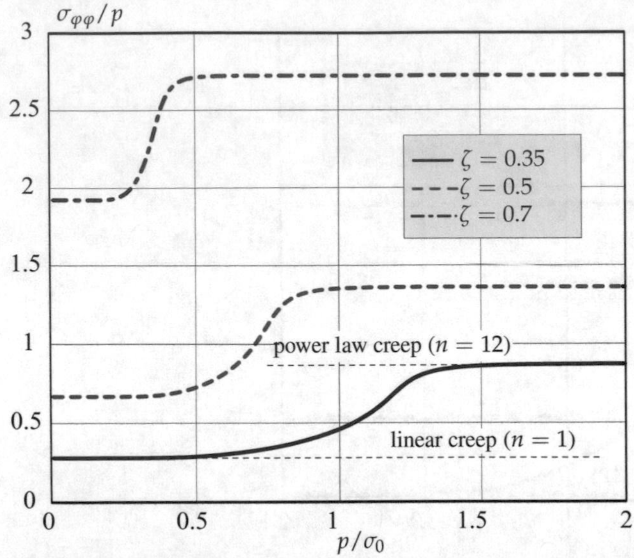

Fig. 4.8 Normalized hoop stress vs. normalized pressure. at the outer radius

of the normalized pressure and the creep exponent. For example, with $n = 3$ the classical solution is valid for $\tilde{p} > 2$, while with $n = 12$ it is applicable for $\tilde{p} > 1$. On the other hand, the transition range, where the double power law should be applied extends as the creep exponent decreases, Fig. 4.11. For example, for a moderate value of the pressure $\tilde{p} = 0.7$ within this transition range and for all considered values of the creep exponent except $n = 1$, the distributions of the hoop stress over the radial coordinate differ significantly from the classical solution (4.2.44) shown in Fig. 4.6.

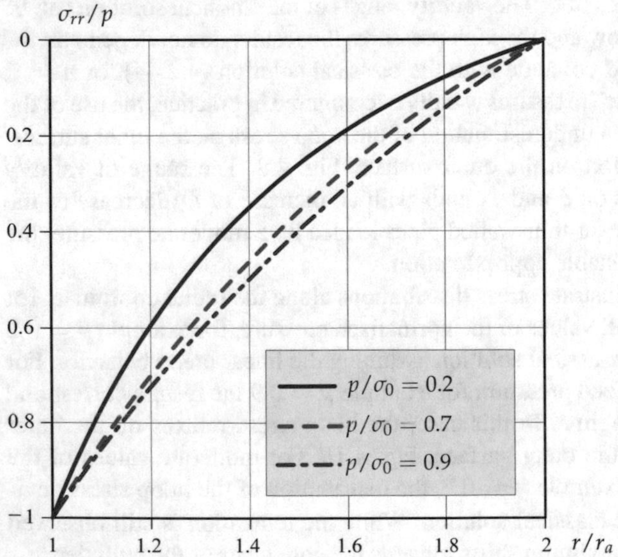

Fig. 4.9 Normalized radial stress vs. normalized radial coordinate for $\zeta = 0.5$ and different values of pressure

4.2.4 Finite Element Solution

The creep constitutive equation (4.2.33) with the double power las stress function (4.2.48) is utilized inside the ABAQUS finite element code by the user subroutine creep. The results obtained by the approximate solution of Eqs (4.2.42) are not based on the finite element method and can be applied to verify the finite element procedures. Let us note that the steady state creep problem of the thick cylinder in the power law creep range is the standard benchmark (Abaqus Benchmarks, 2017). It can be used to verify user material subroutines, time integration methods, etc. The geometrical input parameters and the finite element model are assumed as given in the benchmark manual (Abaqus Benchmarks, 2017). In the analysis we set $r_a = 25.4$ mm, $r_b = 50.8$ mm. The pipe section has been meshed by 10 CAX8R elements. The following material parameters in Eqs (4.2.33) and (4.2.48) are assumed

$$\dot{\varepsilon}_0 = 2.5 \cdot 10^{-7}\ \text{1/h}, \quad \sigma_0 = 100\ \text{MPa}, \quad n = 12$$

The explicit time integration method with the automatic time step size control was applied. The time step procedure was terminated after reaching the steady stress state. Figure 4.13 illustrates the results based on the ABAQUS finite element code and the approximate solutions of Eqs (4.2.42). We observe that the stress distributions agree well for all assumed values of the normalized pressure.

4.3 Rotating Components

Rotating components like rotor, turbine blade, flywheel, etc. are subjected to stress states due to high centrifugal forces. During long-term operation at high temperature, inelastic deformations may case significant changes of geometry, and as a result lead to structural failures. For example, creep of a turbine blade during operation at high temperature may cause the tips of the turbine blades to severely rub into the non-rotating shroud.

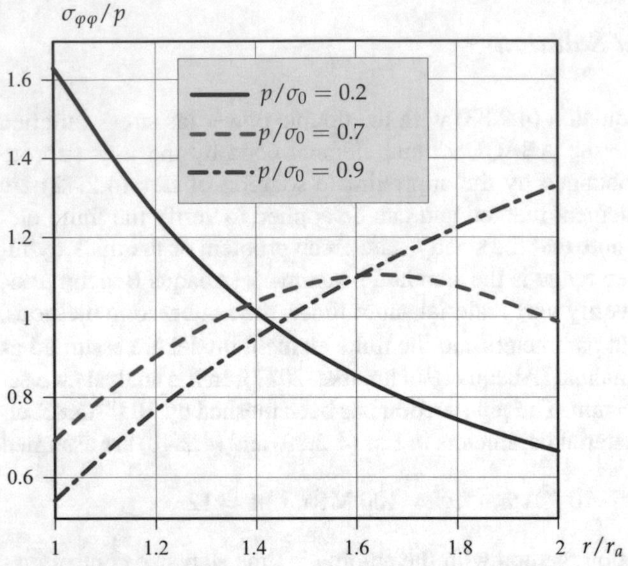

Fig. 4.10 Normalized hoop stress vs. normalized radial coordinate for $\zeta = 0.5$ and different values of pressure

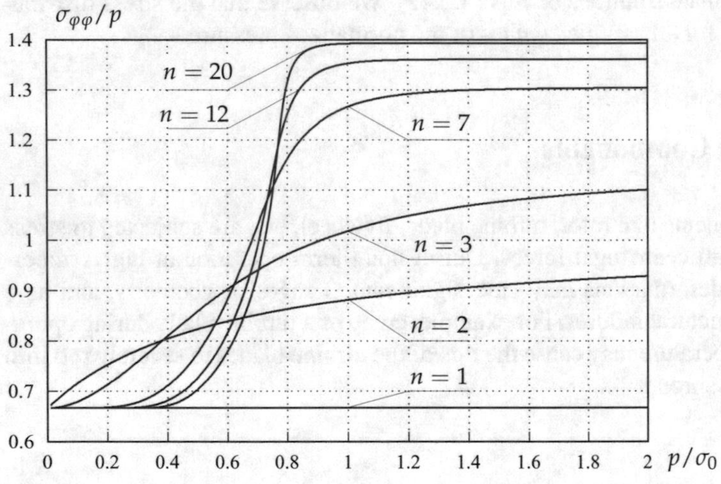

Fig. 4.11 Normalized hoop stress vs. normalized pressure (outer surface, $\zeta = 0.5$) for different values of creep exponent

Numerical studies of turbine rotors in a non-stationary thermal environment and centrifugal forces are presented in Benaarbia et al (2018); Kostenko et al (2013); Naumenko et al (2011); Wang et al (2016). The aim of this section is to present some elementary solutions for rotating components including a uniform rod and a uniform disc in a steady-state creep regime.

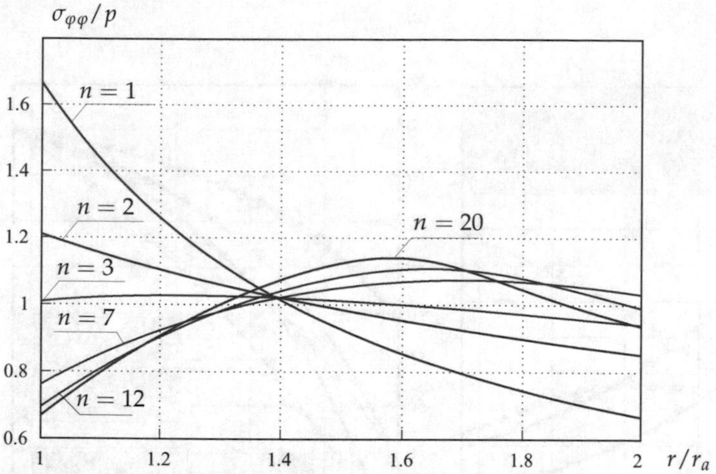

Fig. 4.12 Normalized hoop stress vs. normalized radial coordinate ($p/\sigma_0 = 0.7$, $\zeta = 0.5$) for different values of creep exponent

4.3.1 Rotating Rod

To illustrate basic features of the stress and deformation state in rotating components, let us consider a uniform rod with a constant cross section, as shown in Fig. 4.14a. Assuming steady operation with the constant angular velocity of rotation ω, we can apply the static equilibrium condition to compute the internal forces in the rod. Indeed with the free-body diagram for the part of the rod shown in Fig. 4.14b, the equilibrium condition reads

$$N(r_2) - N(r_1) + f_C = 0, \tag{4.3.51}$$

where N is the internal force in the rod, r_1 and r_2 are coordinates of bottom and top cross sections, and f_C is the total centrifugal force acting on the considered part of the rod. For the elementary section of the rod $\mathrm{d}r$ the mass is $\mathrm{d}m = \rho A \mathrm{d}r$, where ρ is the density, A is the cross section area and r is the radial coordinate. The centrifugal force acting on this section is $\mathrm{d}f_C = \omega^2 r \mathrm{d}m$. The equilibrium condition (4.3.51) can be formulated as follows

$$N(r_2) - N(r_1) + \int_{r_1}^{r_2} \rho A \omega^2 r \mathrm{d}r = 0, \tag{4.3.52}$$

With the fundamental theorem of the integral calculus

$$N(r_2) - N(r_1) = \int_{r_1}^{r_2} N' \mathrm{d}r, \quad (\ldots)' = \frac{\mathrm{d}(\ldots)}{\mathrm{d}r},$$

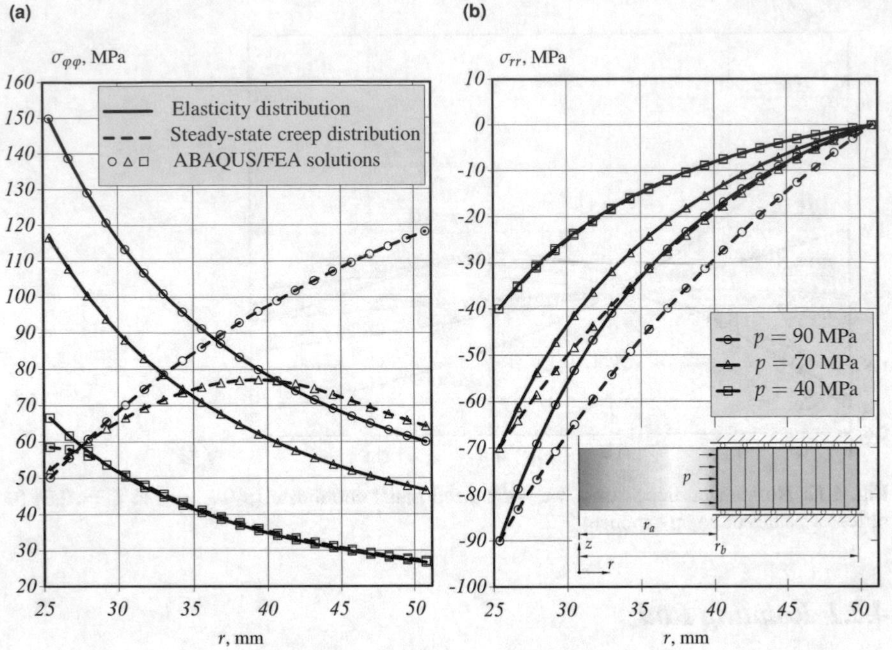

Fig. 4.13 Normalized stresses vs. radial coordinate for $\zeta = 0.5$ using the presented method and ABAQUS finite element code. **(a)** Hoop stress, **(b)** radial stress

Eq. (4.3.52) takes the form

$$\int_{r_1}^{r_2} (N' + \rho A \omega^2 r) \mathrm{d}r = 0 \tag{4.3.53}$$

Equation (4.3.53) is valid for any part of the rod. Since r_1 and r_2 are arbitrary coordinates, the integral (4.3.53) is only zero if

$$N' + \rho A \omega^2 r = 0 \tag{4.3.54}$$

For the sake of brevity let us assume that the rod is homogeneous with a constant cross section area. Then the ordinary differential equation (4.3.54) can be integrated providing the internal force as a function of the radial coordinate. With the boundary condition $N(b) = 0$ we obtain

$$N(r) = \frac{1}{2} \rho A \omega^2 (b^2 - r^2) \tag{4.3.55}$$

The maximum internal force in the root of the rod is

$$N_{\max} = N(r_a) = \frac{1}{2} \rho A \omega^2 (b^2 - a^2) = \omega^2 m \frac{a+b}{2},$$

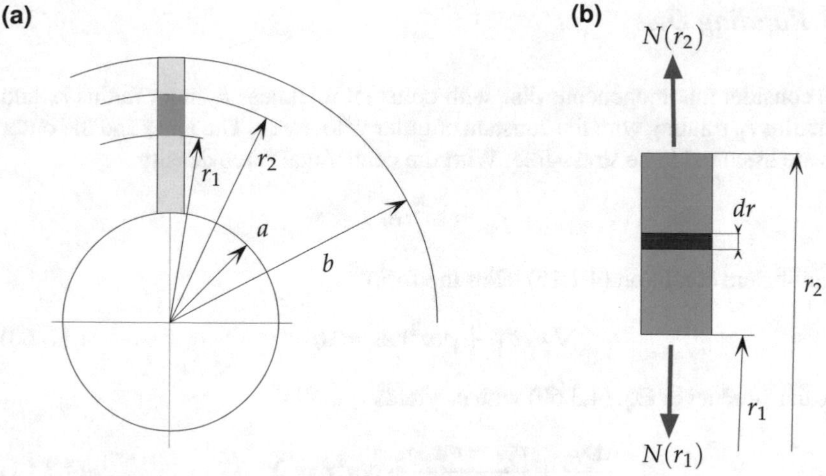

Fig. 4.14 Rotating rod. **(a)** Geometry, **(b)** free-body diagram

where m is the mass of the rod. Assuming steady-state creep let us apply the constitutive equation (4.2.33). For the uni-axial stress state we have

$$\dot{\varepsilon} = \dot{\varepsilon}_0 f\left(\frac{\sigma}{\sigma_0}\right) \tag{4.3.56}$$

Applying the strain-displacement relation $\varepsilon = u'$ we can compute the rate of the maximum displacement of the rod as follows

$$\dot{u}_{\max} = \dot{u}(b) = \int_a^b \dot{\varepsilon} \mathrm{d}r \tag{4.3.57}$$

With Eqs (4.3.55) and (4.3.56) we obtain

$$\dot{u}_{\max} = \dot{\varepsilon}_0 \int_a^b f\left(\frac{\rho\omega^2(b^2 - r^2)}{2\sigma_0}\right) \mathrm{d}r \tag{4.3.58}$$

As an example consider the power law stress function $f(x) = x^n$. In this case

$$\dot{u}_{\max} = \dot{\varepsilon}_0 \left(\frac{\rho\omega^2}{2\sigma_0}\right)^n \int_a^b \left(b^2 - r^2\right)^n \mathrm{d}r \tag{4.3.59}$$

4.3.2 Rotating Disc

Let us consider a homogeneous disc with constant thickness h, inner radius r_a and outer radius r_b rotating with the constant angular velocity ω. The inner and the outer radius are assumed to be stress-free. With the centrifugal force density

$$\bar{f}_{\rm p} = \omega^2 r \boldsymbol{e}_r$$

the equilibrium condition (4.1.15) takes the form

$$\nabla_{\rm p} \cdot \boldsymbol{\sigma}_{\rm p} + \rho\omega^2 r\, \boldsymbol{e}_r = \boldsymbol{0}, \tag{4.3.60}$$

The scalar product of Eq. (4.3.60) with $\boldsymbol{e}_r$ yields

$$\frac{{\rm d}\sigma_{rr}}{{\rm d}r} + \frac{\sigma_{rr} - \sigma_{\varphi\varphi}}{r} + \rho\omega^2 r = 0 \tag{4.3.61}$$

In the case of a statically determined rod, the stress is computed from the equilibrium conditions, does not depend on the creep strain, and consequently remains constant during the creep process. For the rotating disc the equilibrium condition (4.3.61) is not sufficient to compute the stresses. In what follows we analyze the stress distributions in the steady creep state. To this end the constitutive equations and the compatibility condition must be applied.

For the plane stress state of the rotating disc the von Mises-Odqvist flow rule of steady-state creep (4.2.33) provides the strain rates as follows

$$\begin{aligned} \dot{\varepsilon}_{rr} &= \frac{1}{2}\dot{\varepsilon}_0 f\left(\frac{\sigma_{\rm vM}}{\sigma_0}\right)\frac{2\sigma_{rr} - \sigma_{\varphi\varphi}}{\sigma_{\rm vM}}, \\ \dot{\varepsilon}_{\varphi\varphi} &= \frac{1}{2}\dot{\varepsilon}_0 f\left(\frac{\sigma_{\rm vM}}{\sigma_0}\right)\frac{2\sigma_{\varphi\varphi} - \sigma_{rr}}{\sigma_{\rm vM}}, \end{aligned} \tag{4.3.62}$$

where the von Mises equivalent stress takes the form

$$\sigma_{\rm vM}^2 = \sigma_{rr}^2 - \sigma_{rr}\sigma_{\varphi\varphi} + \sigma_{\varphi\varphi}^2 \tag{4.3.63}$$

For axisymmetric problems the compatibility condition (4.1.14) simplifies to

$$\nabla_{\rm p} \cdot (\boldsymbol{\varepsilon}_{\rm p} - {\rm tr}\, \boldsymbol{\varepsilon}_{\rm p} \boldsymbol{P}) = \boldsymbol{0} \tag{4.3.64}$$

The time derivative of Eq. (4.3.64) provides the compatibility condition for the strain rate tensor

$$\nabla_{\rm p} \cdot (\dot{\boldsymbol{\varepsilon}}_{\rm p} - {\rm tr}\, \dot{\boldsymbol{\varepsilon}}_{\rm p} \boldsymbol{P}) = \boldsymbol{0} \tag{4.3.65}$$

For components of the strain rate tensor Eq. (4.3.65) reads

$$\frac{{\rm d}\dot{\varepsilon}_{\varphi\varphi}}{{\rm d}r} + \frac{\dot{\varepsilon}_{\varphi\varphi} - \dot{\varepsilon}_{rr}}{r} = 0 \tag{4.3.66}$$

Inserting the components of the strain rate tensor defined by the constitutive Eqs (4.3.62) into the compatibility condition (4.3.66), the non-linear equation with respect to the components of the stress tensor can be derived. Together with the equilibrium condition (4.3.61) the compatibility condition can be solved providing the radial and the hoop stress as functions of the radial coordinate in the steady-state creep regime.

Let us introduce the following normalized variables

$$x = \frac{r}{r_b}, \quad Y = \frac{\sigma_{rr}}{\sigma_0}, \quad Z = \frac{\sigma_{\varphi\varphi}}{\sigma_0}, \quad S = \frac{\sigma_{\mathrm{vM}}}{\sigma_0} \tag{4.3.67}$$

The equilibrium condition (4.3.61) takes the form

$$Y' + \frac{Y - Z}{x} + \Omega x = 0, \tag{4.3.68}$$

where

$$(\ldots)' = \frac{\mathrm{d}(\ldots)}{\mathrm{d}x}, \quad \Omega = \frac{\rho\omega^2 r_b^2}{\sigma_0}$$

Inserting Eqs (4.3.62) into Eq. (4.3.66) we obtain after some transformations the following non-linear differential equation

$$Y'[(2Z - Y)(2Y - Z)g(S) - 1] + Z'[(2Z - Y)^2 g(S) + 2] - 3\frac{Y - Z}{x} = 0, \tag{4.3.69}$$

where

$$2g(S) = \frac{1}{Sf(S)}\frac{\mathrm{d}f}{\mathrm{d}S} - \frac{1}{S^2}, \quad S^2 = Y^2 - YZ + Z^2$$

For the power law stress function $f(S) = S^n$ it follows

$$g(S) = \frac{n-1}{2S^2} \tag{4.3.70}$$

while for the double power law stress function $g(S) = S + S^n$, $n > 1$ we obtain

$$g(S) = \frac{n-1}{2S^2}\frac{S^{n-1}}{1 + S^{n-1}} \tag{4.3.71}$$

With Eqs (4.3.68) and (4.3.69) the following system of nonlinear first order differential equations can be derived

$$\begin{aligned} Y' &= \frac{Z - Y}{x} - \Omega x \\ Z' &= \frac{(Y - Z)[3 + \Phi(Y,Z)] + \Omega x^2 \Phi(Y,Z)}{\Psi(Y,Z)x} \end{aligned} \tag{4.3.72}$$

with

$$\Phi(Y,Z) = (2Z-Y)(2Y-Z)g(S) - 1, \quad \Psi(Y,Z) = (2Z-Y)^2 g(S) + 2$$

Equations (4.3.72) must be supplemented by the boundary conditions for the normalized radial stress $Y(r_a/r_b) = 0$ and $Y(1) = 0$. The numerical solution can be performed according to the following steps

1. Apply a standard numerical solution algorithm for the initial-value problem, such as Euler or Runge-Kutte methods to Eqs (4.3.72) for $r_a/r_b \leq x \leq 1$ with the initial conditions $Y(r_a/r_b) = 0$ and $Z(r_a/r_b) = Z_a$, see Subsect. 1.3.3 and 1.4.7.
2. Apply a standard procedure, such as the fixed point iteration, to find a root of a non-linear equation to compute Z_a from the condition $Y(1) = 0$. With the initial guess Z_a^0, for example, $Z_a^0 = 1$, find the final value of Z_a^k by iterative solution of Eqs (4.3.72) such that $Y^k(1) = 0$ is satisfied.
3. With the obtained value of the normalized hoop stress at inner radius, solve the initial-value problem to compute the stresses as functions of the normalized coordinate.

Figure 4.15 illustrates the normalized hoop stress at inner radius of a rotating disc with $r_a/r_b = 0.25$ as a function of the creep exponent. The values are obtained after iterative solution of Eqs (4.3.72) with initial guess $Z_a^0 = 1$. The normalized hoop stress attains the maximum value for $n = 1$ for linear viscous or linear-elastic behavior. With increase of n, the hoop stress decreases towards the asymptotic value corresponding to rigid plasticity. Figure 4.16 illustrates the normalized radial and hoop stresses as functions of the normalized radial coordinate. A significant stress redistribution can be observed by comparing solutions for initial elastic state, $n = 1$ and steady creep state for $n = 5$. The hoop stress relaxes at the inner radius and

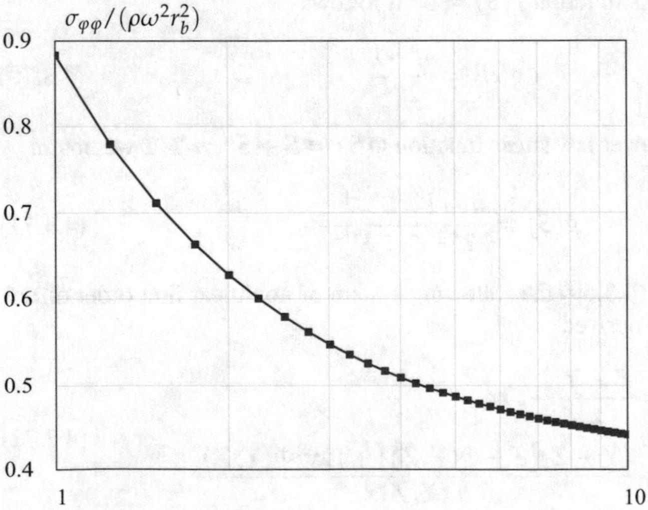

Fig. 4.15 Normalized hoop stress at inner radius of a rotating disc as a function of creep exponent for $r_a/r_b = 0.25$

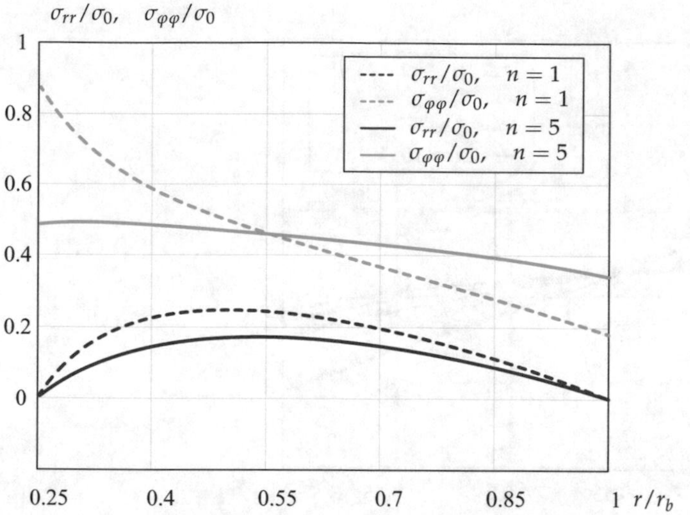

Fig. 4.16 Normalized stresses in a rotating disc as function of normalized radial coordinate for two values of creep exponent, $r_a/r_b = 0.25$ and $\Omega = 1$

increases at the outer radius of the disc. The radial stresses are zero at the inner and outer radius of the disc according to the boundary conditions, and attain the maximum values at the core of the disc. Figure 4.17 shows the normalized hoop stresses as a function of the normalized radial coordinate for different values of the creep exponent. With an increase of n, the hoop stress decreases at the inner radius (see also Fig. 4.15) and increases at the outer radius. With increase of n, the solutions approach an asymptotic distribution according to the assumption of rigid plasticity. Based on the known solutions for the stresses, the radial displacement rate can be computed from the rate of the radial hoop strain as follows

$$\dot{u}_r = r\dot{\varepsilon}_{\varphi\varphi} \tag{4.3.73}$$

With the constitutive Eq. (4.3.62)$_2$ and the normalized variables (4.3.67) we obtain

$$\dot{\delta} = \frac{1}{2}xf(S)\frac{2Z - Y}{S}, \quad \dot{\delta} = \frac{\dot{u}_r}{\dot{\varepsilon}_0 r_b} \tag{4.3.74}$$

4.4 Plate with a Circular Hole

Figure 4.18a shows a rectangular plate with a circular hole subjected to a remote stress σ_{rem}. Far from the hole, the stress state is uniform and bi-axial. The equivalent problem is to find the stress distribution in a circular plate subjected to a constant radial stress, Fig. 4.18a under biaxial loading. In the case of linear elasticity, the Lamé solution provides the maximum value of the hoop stress at the hole (Hahn,

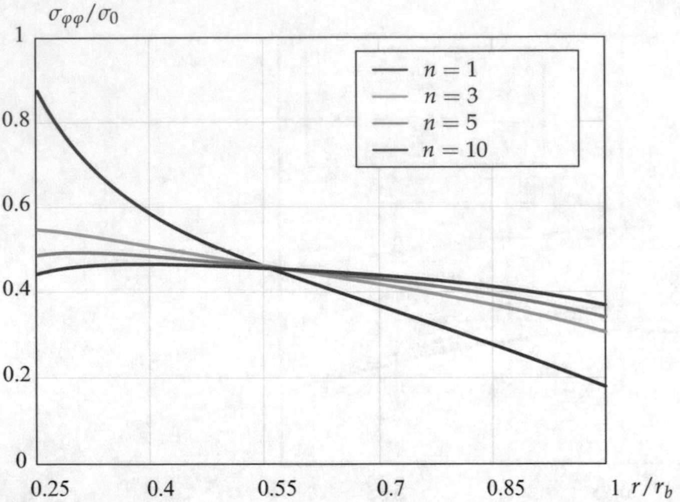

Fig. 4.17 Normalized hoop stress in a rotating disc as function of normalized radial coordinate for different values of creep exponent, $r_a/r_b = 0.25$ and $\Omega = 1$

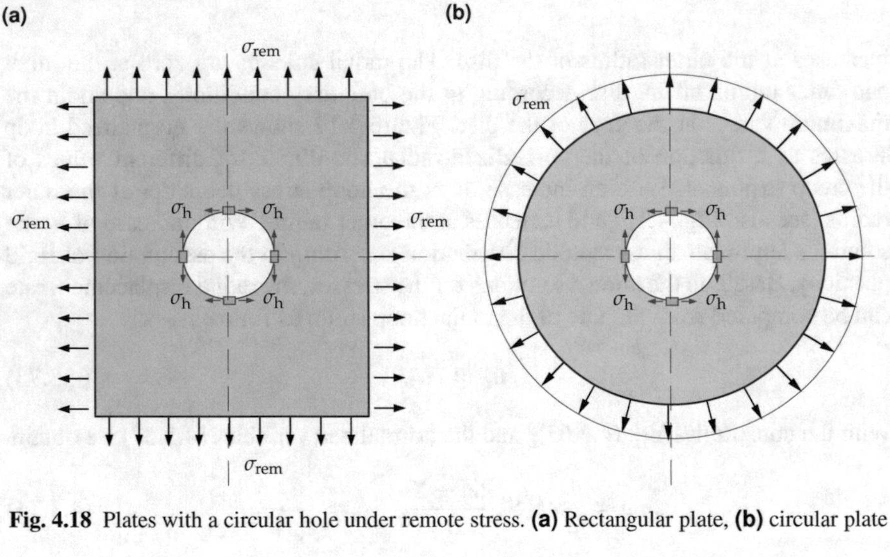

Fig. 4.18 Plates with a circular hole under remote stress. **(a)** Rectangular plate, **(b)** circular plate

1985; Lurie, 2005; Timoshenko and Goodier, 1951)

$$\sigma_{\rm h} = \sigma_{\varphi\varphi_{\max}} = 2\sigma_{\rm rem}$$

with the stress concentration factor 2. For an inelastic analysis, two following features should be considered. First, the stress distributions differ for plane strain and plane stress state assumptions and depend on the material parameters, for example the creep exponent, if power law creep is assumed. Second, the hoop stress in the hole σ_{h} is not necessarily the maximum stress in the plate. Closed form solutions for steady-sate creep assuming power law stress function are presented in Boyle and Spence (1983); Malinin (1981); Rabotnov (1969). Below we recall these classical solutions and present additional results for stress-regime dependent creep.

4.4.1 Plane Stress Solutions

For the plane stress assumptions, we can apply the governing equations derived in Subsect. 4.3.2 for the rotating disc. With the normalized variables

$$\eta = \frac{r}{r_a}, \quad Y = \frac{\sigma_{rr}}{\sigma_0}, \quad Z = \frac{\sigma_{\varphi\varphi}}{\sigma_0}, \quad S = \frac{\sigma_{\mathrm{vM}}}{\sigma_0} \tag{4.4.75}$$

and by setting the angular velocity of rotation to zero, the equilibrium condition (4.3.61) takes the form

$$Y' + \frac{Y - Z}{\eta} = 0, \quad (\ldots)' = \frac{\mathrm{d}(\ldots)}{\mathrm{d}\eta} \tag{4.4.76}$$

Inserting the constitutive equations (4.3.62) into the compatibility condition (4.3.66) and taking into account (4.4.75) we obtain

$$Y'[(2Z - Y)(2Y - Z)g(S) - 1] + Z'[(2Z - Y)^2 g(S) + 2] - 3\frac{Y - Z}{\eta} = 0, \tag{4.4.77}$$

where

$$2g(S) = \frac{1}{Sf(S)}\frac{\mathrm{d}f}{\mathrm{d}S} - \frac{1}{S^2}, \quad S^2 = Y^2 - YZ + Z^2$$

With Eqs (4.4.76) and (4.4.77) the following system of nonlinear first order differential equations can be derived

$$\begin{aligned} Y' &= \frac{Z - Y}{\eta} \\ Z' &= \frac{(Y - Z)}{\eta}\frac{\Phi(Y, Z)}{\Psi(Y, Z)} \end{aligned} \tag{4.4.78}$$

with

$$\Phi(Y, Z) = (2Z - Y)(2Y - Z)g(S) + 2, \quad \Psi(Y, Z) = (2Z - Y)^2 g(S) + 2$$

Equations (4.4.78) must be supplemented by the following boundary conditions

$$Y(1) = 0, \quad Y(r_b/r_a) = Y_{\mathrm{rem}}, \quad Y_{\mathrm{rem}} = \frac{\sigma_{\mathrm{rem}}}{\sigma_0} \tag{4.4.79}$$

For the power law stress function $f(S) = S^n$ we obtain

$$g(S) = \frac{n-1}{2S^2} \tag{4.4.80}$$

With the new variable $\xi = \ln \eta$ and Eq. (4.4.80) Eqs (4.4.78) take the form

$$\begin{aligned} \frac{dY}{d\xi} &= Z - Y \\ \frac{dZ}{d\xi} &= (Y - Z)\frac{(2Z - Y)(2Y - Z)(n-1) + 4S^2}{(2Z - Y)^2(n-1) + 4S^2} \end{aligned} \tag{4.4.81}$$

Equations (4.4.81) can be solved in a closed analytical form. Indeed by substitution

$$Y = \frac{2}{\sqrt{3}} S \cos \Psi, \quad Z = \frac{2}{\sqrt{3}} S \cos\left(\Psi - \frac{\pi}{3}\right),$$

where $\Psi(\xi)$ is a new variable, Eqs (4.4.81) can be put into a form, which can be solved by separation of variables. However, the resulting expressions for stress components are rather cumbersome. For detailed derivations we refer to Boyle and Spence (1983); Malinin (1981); Rabotnov (1969). Here we present only the following expression for the hoop stress at the hole radius

$$\sigma_{\mathrm{h}} = \sigma_{\varphi\varphi}(0) = \left(\frac{n+3}{2n}\right)^{\frac{n+3}{n^2+3}} \exp\left[\frac{\pi}{\sqrt{3}}\left(\frac{n-1}{n^2+3}\right)\right] \sigma_{\mathrm{rem}} \tag{4.4.82}$$

Another approach is to apply the numerical procedure discussed in Subsect. 4.3.2. According to this, Eqs (4.4.78) are treated as the initial value problems with the initial conditions $Y(0) = 0$ and $Z(0) = Z_0$. The unknown value Z_0 is determined from the condition $Y[\ln(r_b/r_a)] = Y_{\mathrm{rem}}$ during the iteration procedure.

Figure 4.19 presents the numerical solution for the hoop stress at the hole as the function of the creep exponent. For the comparison, the closed form solution (4.4.82) is plotted, illustrating that the results agree very well. Figure 4.20 shows the distribution of the normalized stress components over the normalized radial coordinate for $n = 1$ and $n = 5$. The significant relaxation of the hoop stress from the initial elastic solution $n = 1$ towards the steady creep state with $n = 5$ can be observed. Figure 4.21 shows the normalized hoop stresses as a function of the normalized radial coordinate for different values of the creep exponent. With an increase of n the hoop stress decreases substantially at the radius of the hole. For $n > 3$, the location of the maximum hoop stress is not at the hole radius but shifts to the ligament ahead of the hole.

Let as apply the double power law stress function

$$f(S) = S + S^n, \quad n > 1 \tag{4.4.83}$$

In this case we obtain

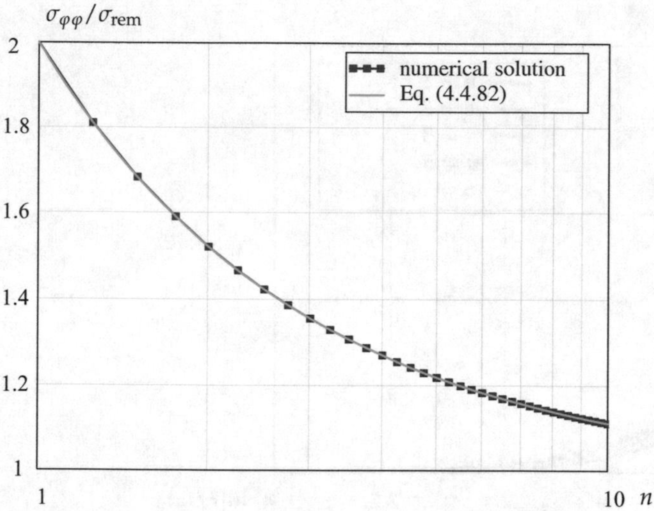

Fig. 4.19 Normalized hoop stress at inner radius of a plate with a circular hole

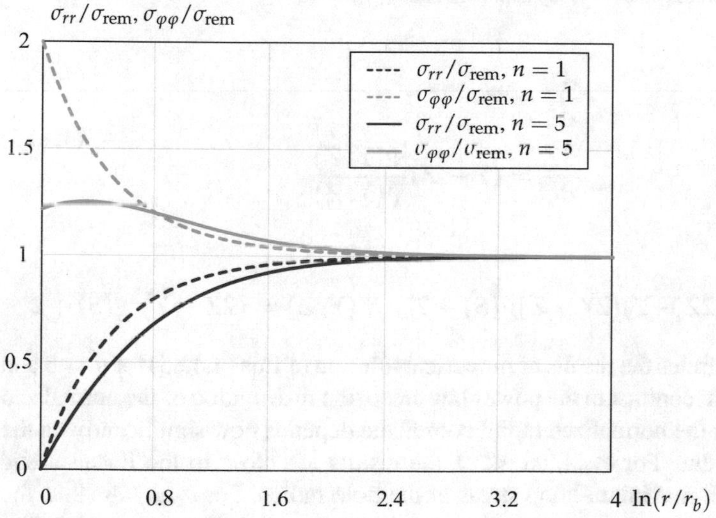

Fig. 4.20 Normalized stresses in a plate with a circular hole as function of normalized radial coordinate for two values of creep exponent and $r_a/r_b = 10^{-3}$

$$g(S) = \frac{n-1}{S^2}\frac{S^{n-1}}{1+S^{n-1}} \tag{4.4.84}$$

Instead of Eqs (4.4.81), we have to solve the following equations

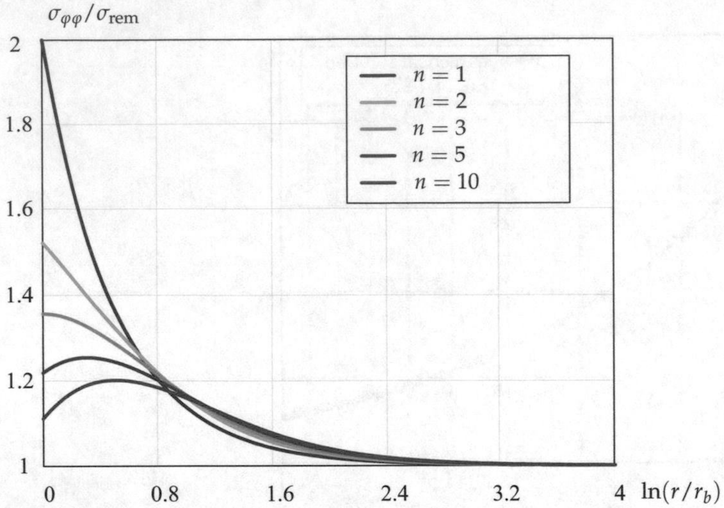

Fig. 4.21 Normalized hoop stress in a plate with a circular hole as function of normalized radial coordinate for different values of creep exponent and $r_a/r_b = 10^{-3}$

$$\begin{aligned}\frac{\mathrm{d}Y}{\mathrm{d}\xi} &= Z - Y\\ \frac{\mathrm{d}Z}{\mathrm{d}\xi} &= (Y - Z)\frac{\Phi(Y,Z)}{\Psi(Y,Z)}\end{aligned} \tag{4.4.85}$$

with

$$\Phi(Y,Z) = (2Z - Y)(2Y - Z)g(S) + 2, \quad \Psi(Y,Z) = (2Z - Y)^2 g(S) + 2$$

Figure 4.22 illustrates the results of numerical solution of Eqs (4.4.85) for $n = 5$ and $r_a/r_b = 10^{-3}$. In contrast to the power law creep, the distribution of the normalized hoop stress over the normalized radial coordinate depends now significantly on the remote stress value. For $\sigma_{\text{rem}}/\sigma_0 \ll 1$ the results are close to the linear-elastic solution with the maximum hoop stress at the hole radius. For $\sigma_{\text{rem}}/\sigma_0 \gg 1$ the results are close to the power law creep solution with $n = 5$. In this case the location of the maximum of the hoop stress is at the ligament ahead of the hole.

4.4.2 Plane Strain Solutions

For moderately thick plates, the solutions presented in Subsect. 4.4.2 may be inaccurate since the out-of-plane stress σ_{33} is neglected. One remedy is the application of the plane strain assumption. In this case we can use the governing equations presented in Subsect. 4.2.1 for a thick cylinder. Indeed the stress tensor is defined by the general solution given by Eqs (4.2.35) – (4.2.39). The integration constants C

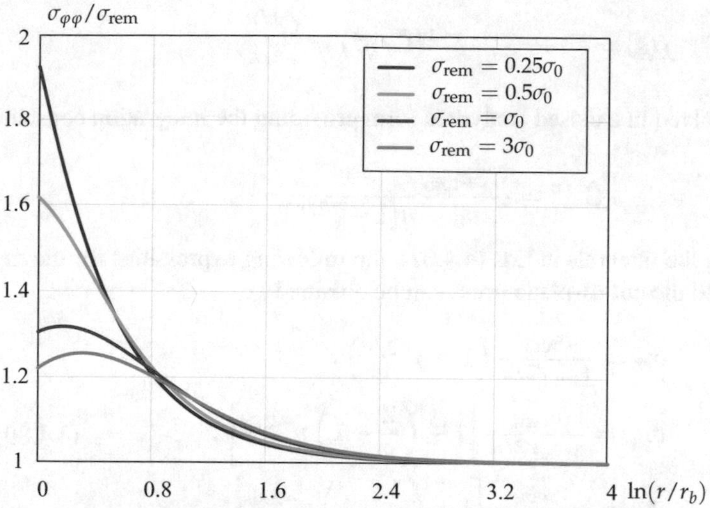

Fig. 4.22 Normalized hoop stress in a plate with a circular hole as function of normalized radial coordinate for different values of remote stress, double power law with $n = 5$ and $r_a/r_b = 10^{-3}$

and D must be computed from the following boundary conditions

$$\sigma_{rr}(r_a) = 0, \quad \sigma_{rr}(r_b) = \sigma_{\text{rem}} \tag{4.4.86}$$

With the normalized variables (4.2.41) and the boundary condition $(4.4.86)_1$, Eqs. (4.2.38) and (4.2.39) provide the following expressions for the stress components

$$\begin{aligned}
\sigma_{rr}(\eta, \tilde{C}) &= \frac{2\sigma_0}{\sqrt{3}} \int_1^{\eta} \frac{1}{\eta} f^{-1}(\tilde{C}/\eta^2)\mathrm{d}\eta, \\
\sigma_{\varphi\varphi}(\eta, \tilde{C}) &= \frac{2\sigma_0}{\sqrt{3}} \left[\int_1^{\eta} \frac{1}{\eta} f^{-1}(\tilde{C}/\eta^2)\mathrm{d}\eta + f^{-1}(\tilde{C}/\eta^2) \right] \\
\sigma_{\text{vM}}(\eta, \tilde{C}) &= \sigma_0 f^{-1}(\tilde{C}/\eta^2), \quad \sigma_{33} = \sigma_{\text{m}} = \frac{1}{2}\left(\sigma_{rr} + \sigma_{\varphi\varphi}\right)
\end{aligned} \tag{4.4.87}$$

With the boundary condition $(4.4.86)_2$, Eq. $(4.4.87)_1$ yields

$$\int_1^{1/\zeta} \frac{2}{\eta} f^{-1}(\tilde{C}/\eta^2)\mathrm{d}\eta = \frac{\sqrt{3}\sigma_{\text{rem}}}{\sigma_0} \tag{4.4.88}$$

The non-linear Eq. (4.4.88) can be solved providing the integration constant $\tilde{C}$. With the the power law stress function

$$f(\tilde{s}) = \tilde{s}^n \quad \Rightarrow \quad f^{-1}(\tilde{C}/\eta^2) = \frac{\tilde{C}^{1/n}}{\eta^{2/n}},$$

Eq. (4.4.88) is solved in a closed analytical form providing the integration constant $\tilde{C}$ as follows

$$\tilde{C}^{1/n} = \frac{\sqrt{3}\sigma_{\text{rem}}}{\sigma_0} \frac{1}{n(1 - \zeta^{2/n})}$$

After computing the integrals in Eqs (4.4.87), the following expressions for the radial, the hoop and the out-of-plane stress can be obtained

$$\begin{aligned}
\sigma_{rr} &= \frac{\sigma_{\text{rem}}}{1 - \zeta^{2/n}} \left(1 - \eta^{-2/n}\right), \\
\sigma_{\varphi\varphi} &= \frac{\sigma_{\text{rem}}}{1 - \zeta^{2/n}} \left[1 + \left(\frac{2}{n} - 1\right) \eta^{-2/n}\right], \\
\sigma_{33} &= \frac{\sigma_{\text{rem}}}{1 - \zeta^{2/n}} \left[1 + \left(\frac{1}{n} - 1\right) \eta^{-2/n}\right]
\end{aligned} \tag{4.4.89}$$

For plates with $r_b \gg r_a$ we can set $\zeta \to 0$ and the stress components can be computed as follows

$$\begin{aligned}
\sigma_{rr} &= \sigma_{\text{rem}} \left(1 - \eta^{-2/n}\right), \\
\sigma_{\varphi\varphi} &= \sigma_{\text{rem}} \left[1 + \left(\frac{2}{n} - 1\right) \eta^{-2/n}\right], \\
\sigma_{33} &= \sigma_{\text{rem}} \left[1 + \left(\frac{1}{n} - 1\right) \eta^{-2/n}\right]
\end{aligned} \tag{4.4.90}$$

For $n = 1$ the classical Lamé solution for a linear-elastic stress distribution can be obtained

$$\sigma_{rr} = \sigma_{\text{rem}} \left(1 - \eta^{-2}\right), \quad \sigma_{\varphi\varphi} = \sigma_{\text{rem}} \left(1 + \eta^{-2}\right) \tag{4.4.91}$$

The out-of plane stress is constant throughout the plate. Furthermore $\sigma_{33} = \sigma_{\text{rem}}$ is valud due to the assumed incompressibility.

References

Abaqus Benchmarks (2017) Benchmarks Manual

Altenbach H, Gorash Y, Naumenko K (2008) Steady-state creep of a pressurized thick cylinder in both the linear and the power law ranges. Acta Mechanica 195:263 – 274

Altenbach H, Altenbach J, Naumenko K (2016) Ebene Flächentragwerke. Springer, Berlin

Benaarbia A, Rae Y, Sun W (2018) Unified viscoplasticity modelling and its application to fatigue-creep behaviour of gas turbine rotor. International Journal of Mechanical Sciences 136:36–49

Betten J (2008) Creep Mechanics, 3rd edn. Springer, Berlin

Boyle JT (2012) The creep behavior of simple structures with a stress range-dependent constitutive model. Archive of Applied Mechanics 82(4):495 – 514

Boyle JT, Spence J (1983) Stress Analysis for Creep. Butterworth, London

Hahn HG (1985) Elastizitätstheorie. B.G. Teubner, Stuttgart

Hult JA (1966) Creep in Engineering Structures. Blaisdell Publishing Company, Waltham

Kostenko Y, Almstedt H, Naumenko K, Linn S, Scholz A (2013) Robust methods for creep fatigue analysis of power plant components under cyclic transient thermal loading. In: ASME Turbo Expo 2013: Turbine Technical Conference and Exposition, American Society of Mechanical Engineers, pp V05BT25A040 – V05BT25A040

Lurie A (2005) Theory of Elasticity. Foundations of Engineering Mechanics, Springer

Malinin NN (1975) Prikladnaya teoriya plastichnosti i polzuchesti (Applied Theory of Plasticity and Creep, in Russ.). Mashinostroenie, Moskva

Malinin NN (1981) Raschet na polzuchest' konstrukcionnykh elementov (Creep Calculations of Structural Elements, in Russ.). Mashinostroenie, Moskva

Naumenko K, Altenbach H (2016) Modeling High Temperature Materials Behavior for Structural Analysis: Part I: Continuum Mechanics Foundations and Constitutive Models, Advanced Structured Materials, vol 28. Springer

Naumenko K, Kostenko Y (2009) Structural analysis of a power plant component using a stress-range-dependent creep-damage constitutive model. Materials Science and Engineering: A 510:169–174

Naumenko K, Altenbach H, Gorash Y (2009) Creep analysis with a stress range dependent constitutive model. Archive of Applied Mechanics 79:619 – 630

Naumenko K, Kutschke A, Kostenko Y, Rudolf T (2011) Multi-axial thermo-mechanical analysis of power plant components from 9-12%Cr steels at high temperature. Engineering Fracture Mechanics 78:1657 – 1668

Odqvist FKG (1974) Mathematical Theory of Creep and Creep Rupture. Oxford University Press, Oxford

Rabotnov YN (1969) Creep Problems in Structural Members. North-Holland, Amsterdam

Skrzypek JJ (1993) Plasticity and Creep. CRC Press, Boca Raton

Timoshenko SP, Goodier JN (1951) Theory of Elasticity. McGraw-Hill, New York

Wang W, Buhl P, Klenk A, Liu Y (2016) The effect of in-service steam temperature transients on the damage behavior of a steam turbine rotor. International Journal of Fatigue 87:471–483

Chapter 5
Plates and Shells

Thin and moderately thick shell structures are designed as structural components in many engineering applications because of light weight and high load-carrying capacity. In many cases they are subjected to high temperature environment and mechanical loadings, such that inelastic material behavior must be taken into account. Examples of high-temperature shell components include pressure vessels, boiler tubes, steam transfer lines, thin coatings, etc. A steam transfer line under long-term operation considering creep-damage material behavior is discussed in Naumenko and Altenbach (2016, Chapt. 1).

Chapter 5 presents examples of inelastic structural analysis of plates and shells. Section 5.1 gives an overview of modeling approaches including various theories of plates and shells as well as various constitutive models of inelastic material behavior. Governing equations of the first order shear deformation theory of plates are presented in Sect. 5.2. An emphasis is placed on the direct formulation of inelastic constitutive laws. Section 5.3 illustrates examples of steady-state creep analysis of circular plates. Advanced constitutive models with internal state variables, such as the damage parameter require the use of advanced plate theories to consider edge effects. Section 5.4 illustrates an example of a rectangular plate with different types of boundary conditions. The results based on the plate theory are compared with the results according to the three-dimensional theory. Section 5.5 presents governing equations and the solution procedure for the creep behavior of a thin-walled pipe subjected to the internal pressure and the bending moment.

5.1 Approaches to the Analysis of Plates and Shells

At high temperature the load carrying capacity and the lifetime of plates and shells are limited by the development of inelastic strains and damage processes. The failure modes under creep-damage conditions may include unacceptable changes of the components shape, creep buckling and loss of material strength, e.g. Roche et al (1992). The first two modes are associated with excessive inelastic deformations and

K. Naumenko and H. Altenbach, *Modeling High Temperature Materials Behavior for Structural Analysis*, Advanced Structured Materials 112,
https://doi.org/10.1007/978-3-030-20381-8_5

stress redistributions. Local changes in shape of the component may lead to the loss of functionality of the whole structure. Creep buckling may occur if external loading leads to compressive stresses. A thin-walled structure designed against spontaneous elastic buckling may fail after a certain critical time as a consequence of stress redistribution. In Miyazaki and Hagihara (2015), creep buckling is categorized into two types. The first one is the case where the characteristic deformation of a mechanical system tends to increase and becomes infinite after a certain critical time. The second type is the creep buckling due to kinetic instability, in which the critical time can be determined by examining the shape of total potential energy in the vicinity of a quasi-static equilibrium state. Bifurcation buckling and snap-through buckling during creep deformation are examples. The degradation of material strength as a result of creep cavitation, thermal ageing, oxidation, etc. may lead to damage evolution, initiation and growth of macro-cracks until final collapse of the structure. Examples include creep failures of pipes and pipe bends (Jones, 2004; Le May et al, 1994; Psyllaki et al, 2009).

To discuss approaches for the inelastic analysis of plate and shell structures, let us categorize the published results according to the problem statement, the type of constitutive model and the type of structural mechanics model. Problems for inelastic analysis of thin and moderately thick plates and shells are summarized in Table 5.1. Constitutive equations describing inelastic material processes have been introduced in Naumenko and Altenbach (2016, Chapt. 5). Table 5.2 provides an overview of several creep constitutive models applied to the analysis of plates and shells. The corresponding structural mechanics models are given in Table 5.3. The overviews presented in Tables 5.1 – 5.3 lead to the conclusion that the type and the order of complexity of the applied structural mechanics models are connected with the problem statement and with the type of the material behavior description.

The early works were primarily concerned with the analysis of steady-state creep in plates and shells. The creep behavior was assumed to exhibit only primary and secondary creep stages and the Norton-Bailey-Odqvist creep constitutive equation, sometimes extended by strain or time hardening functions was applied. The structural mechanics models were those of the Kirchhoff plate and the Kirchhoff-Love shell. In Lin (1962), the biharmonic equation describing a deflection surface of the Kirchhoff plate taking into account the given distribution of creep strains has been derived. It is shown that the deflection of the plate can be computed by applying additional fictitious lateral loads on the plate face and additional fictitious moments on the plate edges. In many cases this equation can be solved by special numerical methods, e.g. the finite difference method (Podgorny et al, 1984) or direct variational methods (Boyle and Spence, 1983; Altenbach and Naumenko, 1997). The results obtained by special methods are useful in verifying general purpose finite element codes and creep material subroutines. Creep equations for axisymmetrically loaded shells of revolution were derived in Penny (1964) by the use of the Kirchhoff-Love hypotheses. The influence of creep is expressed in terms of fictitious in-plane forces and bending moments. In Byrne and Mackenzie (1966); Murakami and Suzuki (1971, 1973) stress redistributions from the reference state of elastic de-

Table 5.1 Examples for inelastic structural analysis of plates and shells

References	Type of Problem
Byrne and Mackenzie (1966); Murakami and Suzuki (1971, 1973)	Shells of revolution, steady-state creep
Altenbach et al (1996); Combescure and Jullien (2017); Betten and Borrmann (1987); Betten and Butters (1990); Miyazaki and Hagihara (2015)	Shells of revolution, thin plates, finite deflections, finite rotations, creep buckling
Altenbach et al (2000a); Burlakov et al (1977); Burlakov et al (1981); Breslavsky et al (2014)	Shells of revolution, dynamic creep
Altenbach et al (1997b,a,c); Altenbach and Naumenko (1997)	Shells of revolution, shallow shells, creep-damage
Takezono and Fujoka (1981); Takezono et al (1988); Kashkoli et al (2017)	Moderately thick and layered shells, steady-state creep
Bodnar and Chrzanowski (2001); Ganczarski and Skrzypek (2000)	Plates, thermo-mechanical coupling, creep-damage
Altenbach et al (2001, 2004)	Moderately thick plates, creep-damage
Altenbach et al (2002); Bialkiewicz and Kuna (1996); Ganczarski and Skrzypek (2004)	Moderately thick plates damage induced anisotropy
Fessler and Hyde (1994); Galishin et al (2017); Koundy et al (1997); Krieg (1999)	Moderately thick shells, creep-damage

formation towards the steady creep state were analyzed for axisymmetrically loaded shells of revolution by means of the finite difference method.

Classical models of Kirchhoff plate or Kirchhoff-Love shell are based on geometrically linear equations. Since the development of creep strains may lead to significant changes of the components shape, geometrically nonlinear terms should be taken into account in the kinematical equations and as well as in the equilibrium conditions. For elastic plates, the governing equations (finite deflection model) were originally proposed by von Kármán (1911). Geometrically nonlinear equations for creep in membranes and plates have been derived by Odqvist (1962). Problems of long-term stability and strength have required the use of refined geometrically nonlinear structural mechanics models. Creep buckling analysis of cylindrical shells under internal pressure and compressive force has been performed in Betten and Butters (1990); Miyazaki (1987, 1988, see also references citet therein). The governing equations correspond to the Kirchhoff-Love type shell with geometrical nonlinearities in the von Kármán's sense. In Altenbach et al (1996, 1997b); Altenbach and Naumenko (1997); Altenbach et al (1997c) a geometrically nonlinear theory is applied to the creep-damage analysis of rectangular plates and cylindrical shells. The results show that the effect of geometrical nonlinearity may be associated with

Table 5.2 Constitutive models applied to analysis of plates and shells in creep range

	Constitutive Model for Creep Stages			
	Primary	Secondary	Tertiary	
References	Time or Strain Hardening	Power or sinh law	Scalar damage variable	Tensor damage variable
Murakami and Suzuki (1971, 1973);		x		
Kashkoli et al (2017)		x		
Altenbach et al (1996);	x	x		
Betten and Borrmann (1987);	x	x		
Betten and Butters (1990) ;	x	x		
Takezono and Fujoka (1981)	x	x		
Altenbach et al (2000a);		x	x	
Altenbach et al (2001, 2004);		x	x	
Burlakov et al (1977, 1981);		x	x	
Altenbach et al (1997b);		x	x	
Galishin et al (2017);		x	x	
Altenbach et al (1997a,c);	x	x	x	
Bodnar and Chrzanowski (2001);	x	x	x	
Ganczarski and Skrzypek (2000);	x	x	x	
Fessler and Hyde (1994);	x	x	x	
Koundy et al (1997); Krieg (1999)	x	x	x	
Altenbach et al (2002);		x		x
Bialkiewicz and Kuna (1996);		x		x
Ganczarski and Skrzypek (2004)		x		x

"structural hardening", i.e. an increase in the structural resistance to time-dependent deformations. Furthermore, even in the case of moderate bending, the classical geometrically linear theory leads to a significant underestimation of the life-time and overestimation of the deformation.

A first order shear deformation shell theory (FSDT) has been firstly applied in Takezono and Fujoka (1981) to analyze primary and secondary creep of simply supported cylindrical shells under internal pressure. The initial-boundary value problem is solved by the finite difference method. Time dependent distributions of displacements and stress resultants are compared with those according to the Kirchhoff-Love type theory. It is demonstrated that the results agree well only for thin shells. In the case of moderately thick shells the difference between the results is significant and increases with time. Reissner type plate equations were applied in Ganczarski and Skrzypek (2000, 2004) for the creep-damage analysis of a simply supported circular plate considering thermo-mechanical couplings. The derived plate equations as well as the equations of the three-dimensional theory were solved by means of the finite difference method. The results show that in the tertiary creep range the thickness

Table 5.3 Structural mechanics models of plates and shells (FSDT – first order shear deformation theory)

	Structural Mechanics Model				
References	Kirchhoff-Love Type	FSDT	Geometrical Non-linearities	3-D Models	Experimental Analysis
Byrne and Mackenzie (1966);	x				
Murakami and Suzuki (1973)	x				
Altenbach et al (1996);	x		x		
Altenbach et al (1997b);	x		x		
Betten and Borrmann (1987);	x		x		
Betten and Butters (1990)	x		x		
Altenbach et al (2000a);	x				x
Burlakov et al (1977, 1981)	x				x
Takezono and Fujoka (1981);		x			
Takezono et al (1988);		x			
Kashkoli et al (2017)		x			
Bodnar and Chrzanowski (2001);		x			
Ganczarski and Skrzypek (2000)		x			
Altenbach et al (2001, 2004);		x		x	
Altenbach et al (2002);		x		x	
Bialkiewicz and Kuna (1996);		x		x	
Ganczarski and Skrzypek (2004)		x		x	
Fessler and Hyde (1994);				x	x
Koundy et al (1997);				x	x
Krieg (1999)				x	x

distribution of the transverse shear stress differs from the parabolic one. Similar effects have been illustrated for beams in Sect. 3.3.

It is worth noting that unlike the Kirchhoff-Love type theories, the first order shear deformation theories have been found more convenient for the finite element implementations due to C^0 continuity (Zienkiewicz and Taylor, 1991). They are implemented as standard elements in commercial finite element codes. Examples of creep-damage analysis of plates and shells by the use of ANSYS code are presented in Altenbach et al (2000b, 2001).

Numerous refined finite element techniques were designed to solve nonlinear problems of shells. For reviews we refer to Wriggers (2008); Yang et al (2000). One feature of the refined theories of plates and shells is that they describe additional edge zone effects, except special types of boundary conditions (e.g. simple support). The use of the finite element or the finite difference method to solve refined equations of plates and shells requires advanced numerical techniques to represent the rapidly varying behavior in the edge zones. Several closed form and approximate analytical solutions of the first order shear deformation plate equations in the linear elastic range illustrate edge zone effects for different types of boundary conditions, e.g. Eisenträger et al (2015a); Naumenko et al (2001); Zhilin and Ivanova (1995). Similar solutions in the case of creep-damage in plates and shells are not available. Further investigations should be made to formulate corresponding benchmark problems and to assess the validity of different available shell and solid type finite elements in problems of creep mechanics.

In recent years, sandwich and laminate shells and plates have become important components for lightweight and civil engineering applications. For example, curved laminated panels with glass skin layers and a core layer from polyvinyl butyral are widely used in the civil engineering and automotive industry (Chen et al, 2016; Naumenko and Eremeyev, 2017). Crystalline or thin film photovoltaic modules are laminates with front glass layer, solar cell layer embedded in a polymeric encapsulant and back glass or polymeric layer (Paggi et al, 2011; Schulze et al, 2012; Weps et al, 2013). Lightweight designs of photovoltaic modules replace front and/or back glasses by polymer or polymer composite layers (Weps et al, 2013). For in-building integrated photovoltaics as well as for new designs achieving the installation of a module on the roof of a car, doubly curved panels are used. Polymers like polyvinyl butyral exhibit time-dependent deformation even at room temperature. Therefore, for the analysis of glass and photovoltaic laminates, creep deformation of the core layer should be taken into account.

Recently, higher-order shell theories were developed and applied to the analysis of laminated structures. A widely used approach is the zig-zag theory, where the displacements are approximated by piecewise functions with respect to the thickness coordinate such that the compatibility between the layers is fulfilled. Applying these approximations the governing equations of the three-dimensional elasticity theory are reduced to two-dimensional shell equations by the use of variational methods or asymptotic techniques (Carrera, 2003; Filippi et al, 2018).

Within the layer-wise theory (LWT), balance equations and constitutive models are derived for each ply independently. With constitutive assumptions for interaction forces and compatibility conditions between the layers, a model for the laminate can be developed. One advantage of LWT is that the load transfer between the layers can be analyzed explicitly since forces of interactions are directly accessible, while within zig-zag-type approximations Lagrange multipliers appear to be reactions to the constraints. Furthermore, within LWT additional assumptions can be made with regard to kinematics and/or dynamics of individual layers leading to robust governing equations. For example, for curved laminated beams with core layer from soft polymers, the assumption is made that glass skin layers deform according to the Bernoulli-Euler beam theory, i. e. they are shear-rigid. In Naumenko and Eremeyev (2014); Eisenträger et al (2015b) closed-form analytical solutions are presented for rectangular plates with various boundary conditions. In Eisenträger et al (2015b) a finite element is developed and utilized inside Abaqus for the analysis of laminates based on the LWT. The element possesses nine degrees of freedom: two mean in-plane displacements, two relative in-plane displacements, the deflection, two mean cross-section rotations, and two relative cross-section rotations.

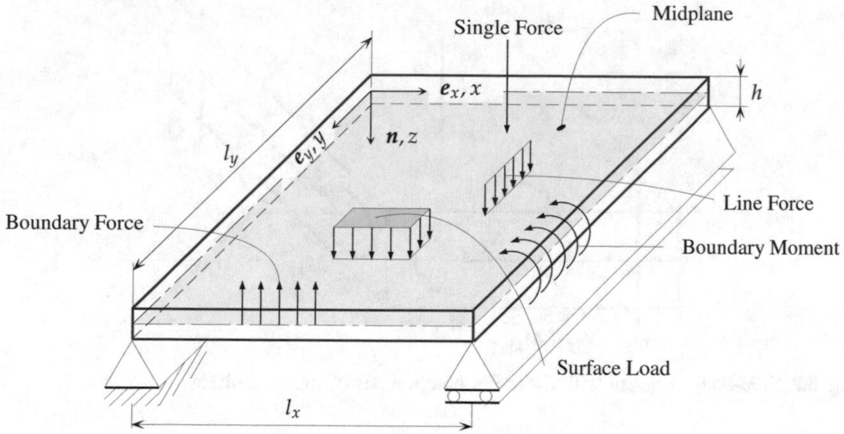

Fig. 5.1 Simply supported rectangular plate

5.2 Governing Equations of the FSDT Plate Theory

In this section we present governing equations of the FSDT plate theory. The classical Kirchhoff-type plate equations will be considered as a special case. For the sake of simplicity, we assume infinitesimal strains and normal rotation such that the geometrically linear theory can be applied. Below we specify the in-plane base vectors by $\boldsymbol{i}_1$ and $\boldsymbol{i}_2$, the corresponding coordinates by x_1 and x_2. The Greek indices take values 1 and 2 and the Einstein summation convention over repeated indices will be used.

Figure 5.1 shows a sketch of a rectangular homogeneous plate. The Cartesian base vectors $\boldsymbol{e}_x = \boldsymbol{i}_1$, $\boldsymbol{e}_y = \boldsymbol{i}_2$, $\boldsymbol{e}_z = \boldsymbol{n}$ and the corresponding coordinates $x = x_1$, $y = x_2$ and z are used to specify the position vectors in the reference state. l_x and l_y designate the length and the width of the plate while h denotes the plate thicknesses. The origin for the z coordinate is placed in the midplane of the plate as shown in Fig. 5.1, such that $-h/2 \leq z \leq h/2$. In this section we present basic equations of the first order shear deformation theory. They include the equilibrium conditions, the constitutive equations and the compatibility conditions for strains and curvatures.

5.2.1 Equilibrium conditions

Figure 5.2 illustrates the free-body diagram for a plate section with Cartesian components of stress resultant vectors and second order tensors. They include the in-plane force tensor $\boldsymbol{N} = N_{\alpha\beta}\boldsymbol{i}_\alpha \otimes \boldsymbol{i}_\beta$, the shear force vector $\boldsymbol{Q} = Q_\alpha \boldsymbol{i}_\alpha$ and the bending/twisting moment tensor $\boldsymbol{\mathcal{M}} = -M_{\alpha\beta}\boldsymbol{i}_\alpha \otimes \boldsymbol{i}_\beta \times \boldsymbol{n}$. The stress resultants can

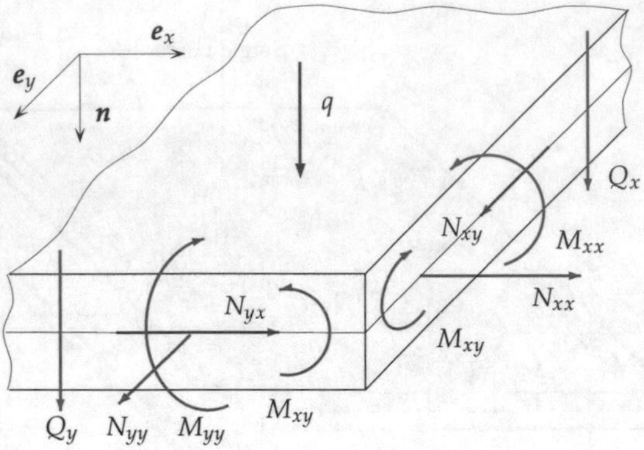

Fig. 5.2 Free-body-diagram with Cartesian components of stress resultants

be obtained by through-the-thickness integration of the stress tensor $\boldsymbol{\sigma}$ as follows (Naumenko and Eremeyev, 2017)

$$\boldsymbol{T} = \langle \boldsymbol{P} \cdot \boldsymbol{\sigma} \rangle = \boldsymbol{N} + \boldsymbol{Q} \otimes \boldsymbol{n}, \tag{5.2.1}$$

$$\boldsymbol{\mathcal{M}} = -\langle z\boldsymbol{P} \cdot \boldsymbol{\sigma} \times \boldsymbol{n} \rangle, \tag{5.2.2}$$

$$N_{\alpha\beta} = \langle \sigma_{\alpha\beta} \rangle, \quad Q_\alpha = \langle \sigma_{\alpha 3} \rangle, \quad M_{\alpha\beta} = \langle z\sigma_{\alpha\beta} \rangle,$$

$$\boldsymbol{P} = \boldsymbol{i}_\alpha \otimes \boldsymbol{i}_\alpha, \quad \langle \ldots \rangle = \int\limits_{-h/2}^{h/2} (\ldots) \mathrm{d}z,$$

where $\boldsymbol{T}$ is the force tensor. The equilibrium conditions for the plate can either be derived from the equilibrium conditions for the stress tensor and Eqs (5.2.1) and (5.2.2), e.g. Lebedev et al (2010); Libai and Simmonds (1998) or by applying a direct approach, i.e. by considering an infinitesimal plate element with stress resultants (Altenbach et al, 2005). The equilibrium conditions take the form

$$\nabla_{\mathrm{p}} \cdot \boldsymbol{T} + q\boldsymbol{n} = \boldsymbol{0}, \quad \nabla_{\mathrm{p}} \cdot \boldsymbol{\mathcal{M}} + \boldsymbol{T}_\times = \boldsymbol{0}, \tag{5.2.3}$$

where

$$\nabla_{\mathrm{p}} = \boldsymbol{i}_\alpha \frac{\partial}{\partial x_\alpha}$$

is the nabla (Hamilton) operator and $\boldsymbol{T}_\times$ denotes the Gibbs cross or vectorial invariant of the force tensor (Naumenko and Altenbach, 2016, Appendix A). With Eq. (5.2.1) and $(5.2.3)_1$ the equilibrium conditions for the in-plane force tensor and the shear force vector can be derived as follows

$$\nabla_{\mathrm{p}} \cdot \boldsymbol{N} = \boldsymbol{0}, \quad \nabla_{\mathrm{p}} \cdot \boldsymbol{Q} + q = 0 \tag{5.2.4}$$

Instead of axial moment tensor $\boldsymbol{\mathcal{M}}$, the polar moment tensor $\boldsymbol{M}$ can be introduced as follows

$$\boldsymbol{M} = \boldsymbol{\mathcal{M}} \times \boldsymbol{n} = M_{\alpha\beta} \boldsymbol{i}_\alpha \otimes \boldsymbol{i}_\beta, \quad \boldsymbol{M} = \langle z\boldsymbol{P} \cdot \boldsymbol{\sigma} \cdot \boldsymbol{P} \rangle \tag{5.2.5}$$

The equilibrium condition $(5.2.3)_2$ takes the following form

$$\nabla_{\mathrm{p}} \cdot \boldsymbol{M} - \boldsymbol{Q} = \boldsymbol{0}, \tag{5.2.6}$$

5.2.2 *Constitutive equations*

For the sake of brevity let us neglect the thermal expansion. The constitutive equations can be derived by applying two approaches. With the direct approach, the stress resultants are related to the corresponding kinematic quantities. The constitutive functions and material parameters are determined either from tests on thin-walled structures or by comparing solutions of the plate equations for specific loadings with the corresponding solutions of the three-dimensional continuum mechanics. The second approach is to derive constitutive equations from three-dimensional continuum mechanics applying specific through-the-thickness approximations for displacement and/or stress fields. In the case of linear elasticity, both approaches are well established. The corresponding results are presented in Subsubsect. 5.2.2.1. In the case of inelastic material behavior, direct constitutive equations are available only for special cases of loading. Examples will be presented in Subsubsect. 5.2.2.2.

5.2.2.1 Linear-Elastic Behavior

In this case, linear dependencies between the stress resultants and elastic deformation measures can be established. For a homogeneous plate, the following uncoupled equations for the in-plane force tensor, the bending/twisting moment tensor, and the shear force vector can be formulated

$$\boldsymbol{N} = {}^{4}\boldsymbol{C} \cdot\cdot (\boldsymbol{\varepsilon}_{\mathrm{p}} - \boldsymbol{\varepsilon}_{\mathrm{p}}^{\mathrm{pl}}), \quad \boldsymbol{Q} = \boldsymbol{\Gamma} \cdot (\boldsymbol{\gamma} - \boldsymbol{\gamma}^{\mathrm{pl}}), \tag{5.2.7}$$

$$\boldsymbol{M} = {}^{4}\boldsymbol{D} \cdot\cdot (\boldsymbol{\chi} - \boldsymbol{\chi}^{\mathrm{pl}}), \tag{5.2.8}$$

where

$$\boldsymbol{\varepsilon}_{\mathrm{p}} \equiv \frac{1}{2}\left[\nabla_{\mathrm{p}}\boldsymbol{u} + (\nabla_{\mathrm{p}}\boldsymbol{u})^T\right] = \varepsilon_{\mathrm{p}_{\alpha\beta}} \boldsymbol{i}_\alpha \otimes \boldsymbol{i}_\beta, \tag{5.2.9}$$

with

$$\boldsymbol{\gamma} \equiv \nabla_{\mathrm{p}} w + \boldsymbol{\varphi} = \gamma_\alpha \boldsymbol{i}_\alpha, \tag{5.2.10}$$

$$\boldsymbol{\chi} \equiv \frac{1}{2}\left[\nabla_{\mathrm{p}}\boldsymbol{\varphi} + (\nabla_{\mathrm{p}}\boldsymbol{\varphi})^T\right] = \chi_{\alpha\beta} \boldsymbol{i}_\alpha \otimes \boldsymbol{i}_\beta \tag{5.2.11}$$

Here $\boldsymbol{\varepsilon}_{\mathrm{p}}$ is the in-plane strain tensor, $\boldsymbol{u}_{\mathrm{p}}$ is the in-plane displacement vector, $\boldsymbol{\gamma}$ is the transverse shear strain vector, $\boldsymbol{\varphi}$ is the normal rotation vector, w is the deflection, and $\boldsymbol{\chi}$ is the curvature change tensor. $\boldsymbol{C}$ and $\boldsymbol{D}$ are fourth-rank in-plane and out-of-plane stiffness tensors, while $\boldsymbol{\Gamma}$ is the second-order transverse shear stiffness tensor. For isotropic and homogeneous material behavior these tensors can be represented as follows

$$\boldsymbol{C} = \frac{E\nu h}{1-\nu^2}\boldsymbol{P}\otimes\boldsymbol{P} + \frac{Eh}{2(1+\nu)}(\boldsymbol{i}_\alpha\otimes\boldsymbol{P}\otimes\boldsymbol{i}_\alpha + \boldsymbol{i}_\alpha\otimes\boldsymbol{i}_\beta\otimes\boldsymbol{i}_\alpha\otimes\boldsymbol{i}_\beta) \tag{5.2.12}$$

$$\boldsymbol{D} = \frac{E\nu h^3}{12(1-\nu^2)}\boldsymbol{P}\otimes\boldsymbol{P} + \frac{Eh^3}{24(1+\nu)}(\boldsymbol{i}_\alpha\otimes\boldsymbol{P}\otimes\boldsymbol{i}_\alpha + \boldsymbol{i}_\alpha\otimes\boldsymbol{i}_\beta\otimes\boldsymbol{i}_\alpha\otimes\boldsymbol{i}_\beta) \tag{5.2.13}$$

The shear stiffness tensor has the following form

$$\boldsymbol{\Gamma} = \Gamma\boldsymbol{P}, \quad \Gamma = kGh \tag{5.2.14}$$

k is the shear correction factor. With Eqs (5.2.12) – (5.2.14) the constitutive equations (5.2.7) and (5.2.8) take the following form

$$\begin{aligned} \boldsymbol{N} &= \frac{Eh}{1+\nu}\left[\boldsymbol{\varepsilon}_{\mathrm{P}} - \boldsymbol{\varepsilon}_{\mathrm{P}}^{\mathrm{pl}} + \frac{\nu}{1-\nu}\mathrm{tr}\left(\boldsymbol{\varepsilon}_{\mathrm{P}} - \boldsymbol{\varepsilon}_{\mathrm{P}}^{\mathrm{pl}}\right)\boldsymbol{P}\right], \\ \boldsymbol{Q} &= kGh\left(\boldsymbol{\gamma} - \boldsymbol{\gamma}^{\mathrm{pl}}\right), \\ \boldsymbol{M} &= \frac{Eh^3}{12(1+\nu)}\left[\boldsymbol{\chi} - \boldsymbol{\chi}^{\mathrm{pl}} + \frac{\nu}{1-\nu}\mathrm{tr}\left(\boldsymbol{\chi} - \boldsymbol{\chi}^{\mathrm{pl}}\right)\boldsymbol{P}\right] \end{aligned} \tag{5.2.15}$$

5.2.2.2 Steady-State Flow

Let us assume a pure bending state of the plate such that $\boldsymbol{N} = \boldsymbol{0}$ and $\boldsymbol{Q} = \boldsymbol{0}$. For example, such a state can be realized by applying constant moments on the boundaries of a thin rectangular plate leading to cylindrical, parabolic, hyperbolic or torsional bending surface (Altenbach et al, 2016). Furthermore, in the case of small bending of homogeneous thin plates, the in-plane forces and transverse shear strains are negligible. In this case the inelastic curvature rate tensor is related to the moment tensor. By analogy to the three-dimensional theory (Naumenko and Altenbach, 2016, Sect. 5.4.), we can assume that the inelastic potential W exists, such that the inelastic curvature rate tensor is determined by the flow rule as follows

$$\dot{\boldsymbol{\chi}}^{\mathrm{pl}} = \frac{\partial W}{\partial \boldsymbol{M}} \tag{5.2.16}$$

The inelastic potential is a scalar function of the moment tensor. For isotropic materials, W depends on two invariants, for example the principal invariants

$$J_{1\boldsymbol{M}} = \mathrm{tr}\,\boldsymbol{M}, \quad J_{2\boldsymbol{M}} = \frac{1}{2}\left[(\mathrm{tr}\,\boldsymbol{M})^2 - \mathrm{tr}\,\boldsymbol{M}^2\right]$$

By analogy to the three-dimensional theory, we may introduce an equivalent moment as a quadratic function of invariants as follows

$$m_{\text{eq}}^2 = J_{1\boldsymbol{M}}^2 - \beta J_{2\boldsymbol{M}}^2, \tag{5.2.17}$$

where β is a material parameter. With $W(\boldsymbol{M}) = W(m_{\text{eq}})$, Eqs (5.2.16) and (5.2.17) provide

$$\dot{\boldsymbol{\chi}}^{\text{pl}} = \frac{1}{2}\dot{\chi}_{\text{eq}}^{\text{pl}} \frac{\beta \boldsymbol{M} + (2-\beta) J_{1\boldsymbol{M}} \boldsymbol{P}}{m_{\text{eq}}}, \quad \dot{\chi}_{\text{eq}}^{\text{pl}} = \frac{\mathrm{d}W}{\mathrm{d}m_{\text{eq}}} \tag{5.2.18}$$

To identify the parameter β and the constitutive function $\dot{\chi}_{\text{eq}}^{\text{pl}}(m_{\text{eq}})$, creep bending tests on a thin plate should be performed. As an example, consider uniaxial bending of a rectangular plate with $\boldsymbol{M} = M\boldsymbol{e}_x \otimes \boldsymbol{e}_x$, where M is the magnitude of the applied moment. In this case Eq. (5.2.18) yields

$$\dot{\boldsymbol{\chi}}^{\text{pl}} = \dot{\chi}_{\text{eq}}^{\text{pl}} \left[\boldsymbol{e}_x \otimes \boldsymbol{e}_x + \left(1 - \frac{\beta}{2}\right) \boldsymbol{e}_y \otimes \boldsymbol{e}_y \right] \tag{5.2.19}$$

By measurements of the curvature rates in two directions for different values of the bending moment M, the function $\dot{\chi}_{\text{eq}}^{\text{pl}}(M)$ and the constant β can be determined.

Alternatively, three-dimensional equations can be solved and the results compared to the results of the plate theory in order to identify the material parameters. Indeed, in the case of pure bending the velocity vector and the strain rate tensor have the form

$$\boldsymbol{v}(x, y, z) = \boldsymbol{\varphi}(x, y) z, \quad \Rightarrow \quad \dot{\boldsymbol{\varepsilon}}(x, yz) - \dot{\boldsymbol{\chi}}(x, y) z$$

Assume that the von Mises-Odqvist flow rule can be applied for the inelastic flow. For the plane stress state, we can use Eqs (4.1.29)

$$\begin{aligned} \dot{\boldsymbol{\varepsilon}}_{\text{p}}^{\text{pl}} &= \frac{3}{2} \frac{\dot{\varepsilon}_{\text{vM}}^{\text{pl}}}{\sigma_{\text{vM}}} \left(\boldsymbol{\sigma}_{\text{p}} - \frac{1}{3} \text{tr}\, \boldsymbol{\sigma}_{\text{p}} \boldsymbol{P} \right), \\ \dot{\varepsilon}_{\text{vM}}^{\text{pl}} &= \dot{\varepsilon}_0 f\left(\frac{\sigma_{\text{vM}}}{\sigma_0} \right), \quad \sigma_{\text{vM}}^2 = \frac{3}{2} \text{tr}\, \boldsymbol{\sigma}_{\text{p}}^2 - \frac{1}{2} (\text{tr}\, \boldsymbol{\sigma}_{\text{p}})^2 \end{aligned} \tag{5.2.20}$$

or in the inverse form

$$\begin{aligned} \boldsymbol{\sigma}_{\text{p}} &= \frac{2}{3} \frac{\sigma_{\text{vM}}}{\dot{\varepsilon}_{\text{vM}}^{\text{pl}}} \left(\dot{\boldsymbol{\varepsilon}}_{\text{p}}^{\text{pl}} + \text{tr}\, \dot{\boldsymbol{\varepsilon}}_{\text{p}}^{\text{pl}} \boldsymbol{P} \right), \\ \sigma_{\text{vM}} &= \sigma_0 f^{-1} \left(\frac{\dot{\varepsilon}_{\text{vM}}^{\text{pl}}}{\dot{\varepsilon}_0} \right), \quad \dot{\varepsilon}_{\text{vM}}^{\text{pl}^2} = \frac{2}{3} \text{tr} \left(\dot{\boldsymbol{\varepsilon}}_{\text{p}}^{\text{pl}} \right)^2 + \frac{2}{3} \left(\text{tr}\, \dot{\boldsymbol{\varepsilon}}_{\text{p}}^{\text{pl}} \right)^2 \end{aligned} \tag{5.2.21}$$

In a steady creep state we can set

$$\dot{\boldsymbol{\varepsilon}}_{\text{p}}^{\text{pl}} = \dot{\boldsymbol{\varepsilon}}_{\text{p}} = \dot{\boldsymbol{\chi}}^{\text{pl}}(x, y) z = \dot{\boldsymbol{\chi}}(x, y) z \tag{5.2.22}$$

Inserting Eq. (5.2.22) into Eq. (5.2.21) we obtain

$$\boldsymbol{\sigma}_{\mathrm{P}} = \frac{2}{3}\sigma_0 f^{-1}\left(\frac{z\dot{\chi}^{\mathrm{pl}}_{\mathrm{vM}}}{\dot{\varepsilon}_0}\right)\frac{\dot{\boldsymbol{\chi}}^{\mathrm{pl}}_{\mathrm{P}} + \mathrm{tr}\,\dot{\boldsymbol{\chi}}^{\mathrm{pl}}_{\mathrm{P}}\boldsymbol{P}}{\dot{\chi}^{\mathrm{pl}}_{\mathrm{vM}}}, \tag{5.2.23}$$

where

$$\dot{\chi}^{\mathrm{pl}^2}_{\mathrm{vM}} = \frac{2}{3}\mathrm{tr}\left(\dot{\boldsymbol{\chi}}^{\mathrm{pl}}_{\mathrm{P}}\right)^2 + \frac{2}{3}\left(\mathrm{tr}\,\dot{\boldsymbol{\chi}}^{\mathrm{pl}}_{\mathrm{P}}\right)^2$$

With Eq. (5.2.5) we obtain

$$\boldsymbol{M} = \frac{2}{3}g_M(\dot{\chi}^{\mathrm{pl}}_{\mathrm{vM}})\frac{\dot{\boldsymbol{\chi}}^{\mathrm{pl}}_{\mathrm{P}} + \mathrm{tr}\,\dot{\boldsymbol{\chi}}^{\mathrm{pl}}_{\mathrm{P}}\boldsymbol{P}}{\dot{\chi}^{\mathrm{pl}}_{\mathrm{vM}}}, \tag{5.2.24}$$

$$g_M(\dot{\chi}^{\mathrm{pl}}_{\mathrm{vM}}) = \left\langle \sigma_0 z f^{-1}\left(\frac{z\dot{\chi}^{\mathrm{pl}}_{\mathrm{vM}}}{\dot{\varepsilon}_0}\right)\right\rangle \tag{5.2.25}$$

The inverse form of (5.2.24) is

$$\begin{aligned}\dot{\boldsymbol{\chi}}^{\mathrm{pl}} &= \frac{3}{2}\dot{\chi}^{\mathrm{pl}}_{\mathrm{vM}}\frac{\boldsymbol{M} - \mathrm{tr}\,\boldsymbol{M}\boldsymbol{P}}{m_{\mathrm{vM}}},\\ \dot{\chi}^{\mathrm{pl}}_{\mathrm{vM}} &= f_M(m_{\mathrm{vM}}), \quad f_M(m_{\mathrm{vM}}) = g_M^{-1}(m_{\mathrm{vM}}),\\ m^2_{\mathrm{vM}} &= \frac{3}{2}\mathrm{tr}\,\boldsymbol{M}^2 - \frac{1}{2}(\mathrm{tr}\,\boldsymbol{M})^2\end{aligned} \tag{5.2.26}$$

Now the direct constitutive equation (5.2.18) can be identified. Indeed, by comparing Eqs (5.2.18) and (5.2.26) we can set

$$\beta = 3, \quad \dot{\chi}^{\mathrm{pl}}_{\mathrm{eq}} = \dot{\chi}^{\mathrm{pl}}_{\mathrm{vM}}, \quad m_{\mathrm{eq}} = m_{\mathrm{vM}}$$

For the power law stress function, the integral (5.2.25) can be evaluated in a closed analytical form. As a result we obtain

$$g_M(\dot{\chi}^{\mathrm{pl}}_{\mathrm{vM}}) = D_n\frac{\sigma_0}{\dot{\varepsilon}_0^{1/n}}\left(\dot{\chi}^{\mathrm{pl}}_{\mathrm{vM}}\right)^{1/n}, \quad D_n = \frac{2n}{2n+1}\left(\frac{h}{2}\right)^{2+1/n} \tag{5.2.27}$$

or in the inverse form

$$f_M(m_{\mathrm{vM}}) = \frac{\dot{\varepsilon}_0}{\sigma_0^n}\frac{m_{\mathrm{vM}}}{D_n^n} \tag{5.2.28}$$

The constitutive equation for the equivalent inelastic curvature rate $(5.2.26)_2$ can also be formulated as follows

$$\dot{\chi}^{\mathrm{pl}}_{\mathrm{vM}} = \dot{\chi}_0 f\left(\frac{m_{\mathrm{vM}}}{m_0}\right), \quad f(x) = x^n, \quad \dot{\chi}_0 = \frac{2\dot{\varepsilon}_0}{h}, \quad m_0 = \frac{2n}{2n+1}\frac{h}{2}\sigma_0 \tag{5.2.29}$$

The presented direct approach to derive constitutive equations can be generalized for combined loadings by $\boldsymbol{M}$, $\boldsymbol{N}$, and $\boldsymbol{Q}$. In this case, the inelastic potential is a function of five invariants including two considered invariants of the moment tensor, two invariants of the in-plane force tensor, and the magnitude of the shear force vector. Examples of direct constitutive equations for combined bending and in-plane states are presented in Boyle and Spence (1983); Rabotnov (1969).

For advanced constitutive modeling, the direct approach would require the introduction of internal state variables, for example backstress-type tensors corresponding to bending, in-plane and transverse shear loadings. Such constitutive and evolution equations are useful for thin-walled components, for which the material does not exist in a "three-dimensional state" and the usual material testing procedures are not applicable. Examples include thin films (Bagheri et al, 2019; Nase et al, 2016), coatings (Nordmann et al, 2018), etc.

Alternatively, constitutive equations for the stress resultants can be derived from three-dimensional continuum mechanics by means of through-the thickness integration of the Cauchy stress tensor with Eqs (5.2.1) and (5.2.2). This requires a step-by-step numerical procedure to update inelastic strains and internal state variables. For beams this approach is discussed in Subsect. 3.1.2 and Sect. 3.3.

5.2.3 Shear-rigid Plates

In the Kirchhoff plate theory, the transverse shear strains are neglected. This can be accomplished by assuming that the plate is transverse shear-rigid. Indeed by setting the transverse shear stiffness to infinity and by taking into account that the shear forces must be finite, the elastic transverse shear strain vector in the constitutive equation (5.2.7) must be set to zero. Furthermore, the transverse shear creep strain rate is assumed to be negligible. Therefore

$$\boldsymbol{\gamma} = 0, \quad \boldsymbol{\gamma}^{\mathrm{pl}} = 0$$

According to Eq. (5.2.10)

$$\boldsymbol{\varphi} = -\boldsymbol{\nabla}_{\mathrm{P}} w \tag{5.2.30}$$

The cross-section rotations are not independent anymore, but related to the deflection function. The shear force vector is not defined by the constitutive equation and can be determined from the equilibrium condition. From Eq. (5.2.6) it follows

$$\boldsymbol{Q} = \boldsymbol{\nabla}_{\mathrm{P}} \cdot \boldsymbol{M} \tag{5.2.31}$$

After inserting the above expression for $\boldsymbol{Q}$ into Eq, $(5.2.4)_2$ we obtain

$$\boldsymbol{\nabla}_{\mathrm{P}} \cdot (\boldsymbol{\nabla}_{\mathrm{P}} \cdot \boldsymbol{M}) + q = 0 \tag{5.2.32}$$

The divergence operator applied to Eq. (5.2.11) yields

$$\boldsymbol{\nabla}_{\mathrm{P}} \cdot \boldsymbol{\chi} = \frac{1}{2}\left(\Delta_{\mathrm{P}} \boldsymbol{\varphi} + \boldsymbol{\nabla}_{\mathrm{P}} \boldsymbol{\nabla}_{\mathrm{P}} \cdot \boldsymbol{\varphi}\right), \tag{5.2.33}$$

Taking the divergence of Eq. (5.2.33) we obtain

$$\nabla_{\mathrm{p}} \cdot (\nabla_{\mathrm{p}} \cdot \boldsymbol{\chi}) = \Delta_{\mathrm{p}} \nabla_{\mathrm{p}} \cdot \boldsymbol{\varphi} \tag{5.2.34}$$

On the other hand, by taking the trace of Eq. (5.2.11) we obtain

$$\operatorname{tr} \boldsymbol{\chi} = \nabla_{\mathrm{p}} \cdot \boldsymbol{\varphi} \tag{5.2.35}$$

From Eqs (5.2.34) and (5.2.35) the following compatibility condition for the curvature tensor $\boldsymbol{\chi}$ can be derived

$$\nabla_{\mathrm{p}} \cdot (\nabla_{\mathrm{p}} \cdot \boldsymbol{\chi}) = \Delta_{\mathrm{p}} \operatorname{tr} \boldsymbol{\chi} \tag{5.2.36}$$

For isotropic plates, the constitutive equations can be given by Eqs $(5.2.15)_2$ and (5.2.26) as follows

$$\begin{aligned} \boldsymbol{M} &= \frac{Eh^3}{12(1+\nu)} \left[\boldsymbol{\chi} - \boldsymbol{\chi}^{\mathrm{pl}} + \frac{\nu}{1-\nu} \operatorname{tr} \left(\boldsymbol{\chi} - \boldsymbol{\chi}^{\mathrm{pl}} \right) \boldsymbol{P} \right] \\ \dot{\boldsymbol{\chi}}^{\mathrm{pl}} &= \frac{3}{2} \dot{\chi}^{\mathrm{pl}}_{\mathrm{vM}} \frac{\boldsymbol{M} - \operatorname{tr} \boldsymbol{MP}}{m_{\mathrm{vM}}}, \\ \dot{\chi}^{\mathrm{pl}}_{\mathrm{vM}} &= \dot{\chi}_0 f\left(\frac{m_{\mathrm{vM}}}{m_0} \right), \quad m^2_{\mathrm{vM}} = \frac{3}{2} \operatorname{tr} \boldsymbol{M}^2 - \frac{1}{2} (\operatorname{tr} \boldsymbol{M})^2, \end{aligned} \tag{5.2.37}$$

where $\dot{\chi}_0$ and m_0 are material parameters and f is the constitutive function, for example the power law. Equations (5.2.30) – (5.2.37) together with the initial condition for the inelastic part of the curvature tensor $\boldsymbol{\chi}^{\mathrm{pl}}$ and the boundary conditions can be applied to analyze creep bending of plates of various shapes. Different types of boundary conditions are discussed in detail in Altenbach et al (2016); Timoshenko and Woinowsky-Krieger (1959).

5.3 Creep Bending of Circular Plates

Figure (5.3) illustrates examples for circular and annular plates with various types of boundary conditions. In the case of axisymmetrical loading, the moment stress tensor and the shear force vector have the following representations

$$\boldsymbol{M} = M_{rr} \boldsymbol{e}_r \otimes \boldsymbol{e}_r + M_{\varphi\varphi} \boldsymbol{e}_\varphi \otimes \boldsymbol{e}_\varphi, \quad \boldsymbol{Q} = Q_r \boldsymbol{e}_r$$

The equilibrium conditions (5.2.4) take the form

$$\frac{1}{r} \frac{\mathrm{d}(Q_r r)}{\mathrm{d}r} + q = 0, \qquad \frac{\mathrm{d}M_{rr}}{\mathrm{d}r} + \frac{M_{rr} - M_{\varphi\varphi}}{r} - Q_r = 0 \tag{5.3.38}$$

For the analysis of steady-state creep, the constitutive equations (5.2.37) provide the rate of the curvature tensor. For the components we obtain

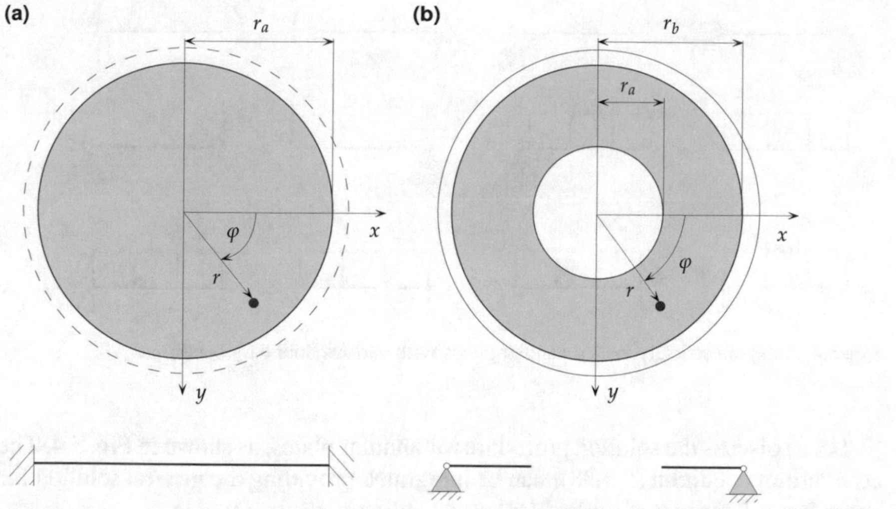

Fig. 5.3 Circular plates. **(a)** clamped circular plate, **(b)** annular plate with outer radius simply supported and inner radius free

$$\dot{\chi}_{rr} = \frac{1}{2}\dot{\chi}_0 f\left(\frac{m_{\mathrm{vM}}}{m_0}\right)\frac{2M_{rr} - M_{\varphi\varphi}}{m_{\mathrm{vM}}},$$
$$\dot{\chi}_{\varphi\varphi} = \frac{1}{2}\dot{\chi}_0 f\left(\frac{m_{\mathrm{vM}}}{m_0}\right)\frac{2M_{\varphi\varphi} - M_{rr}}{m_{\mathrm{vM}}}, \qquad (5.3.39)$$

where the von Mises equivalent moment takes the form

$$m_{\mathrm{vM}}^2 = M_{rr}^2 - M_{rr}M_{\varphi\varphi} + M_{\varphi\varphi}^2 \qquad (5.3.40)$$

The time derivative of Eq. (5.2.36) provides the compatibility condition for the curvature rate tensor. For axisymmetrical cases it simplifies to

$$\frac{\mathrm{d}\dot{\chi}_{\varphi\varphi}}{\mathrm{d}r} + \frac{\dot{\chi}_{\varphi\varphi} - \dot{\chi}_{rr}}{r} = 0 \qquad (5.3.41)$$

Inserting the components of the curvature rate tensor defined by the constitutive Eqs (5.3.39) into the compatibility condition (5.3.41), the nonlinear equation with respect to the components of the moment tensor can be derived. Together with the equilibrium condition (5.3.38) the compatibility condition can be solved providing the radial and hoop stress as functions of the radial coordinate in the steady-state creep regime. To compute the deflection and the cross-section rotation rates Eqs (5.2.30) and (5.2.11) can be applied. With respect to the rates we obtain

$$\frac{\mathrm{d}^2\dot{w}}{\mathrm{d}r^2} = -\dot{\chi}_{rr}, \quad \frac{1}{r}\frac{\mathrm{d}\dot{w}}{\mathrm{d}r} = -\dot{\chi}_{\varphi\varphi}, \quad \dot{\varphi}_r = -\frac{\mathrm{d}\dot{w}}{\mathrm{d}r} \qquad (5.3.42)$$

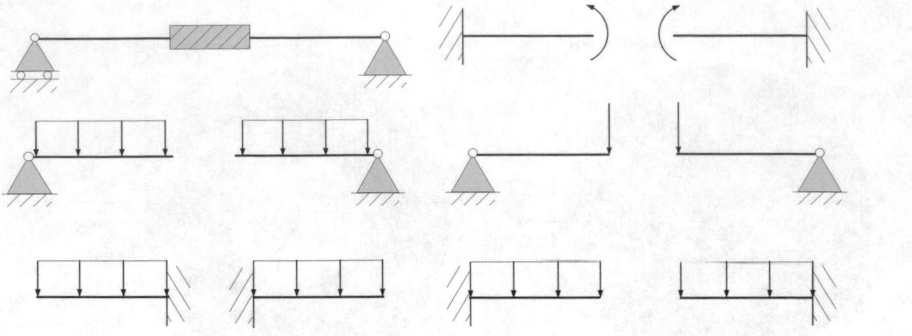

Fig. 5.4 Axisymmetrically loaded annular plates with various boundary conditions

Let us discuss the solution procedure for annular plates, as shown in Fig. 5.4. The equilibrium condition $(5.3.38)_1$ can be integrated providing the general solution for shear force. For the uniformly distributed loading q, the solution is

$$Q_r r = -\frac{qr^2}{2} + C, \tag{5.3.43}$$

where C is an integration constant. As an example we assume that the inner radius is free. With $Q_r = 0$ for $r = r_a$ we obtain

$$Q_r r = \frac{q}{2}\left(r_a^2 - r^2\right), \tag{5.3.44}$$

The equilibrium condition $(5.3.38)_2$ takes the following form

$$\frac{\mathrm{d}M_{rr}}{\mathrm{d}r} + \frac{M_{rr} - M_{\varphi\varphi}}{r} + \frac{q}{2r}\left(r^2 - r_a^2\right) = 0 \tag{5.3.45}$$

Let us introduce the following normalized variables

$$x = 1 - \frac{r}{r_b}, \quad X = \frac{\dot{w}}{r_b^2 \dot{\chi}_0}, \quad Y = \frac{M_{rr}}{m_0}, \quad Z = \frac{M_{\varphi\varphi}}{m_0}, \quad \mu = \frac{m_{\mathrm{vM}}}{m_0} \tag{5.3.46}$$

The equilibrium condition (5.3.45) can be rewritten as follows

$$Y' - \frac{Y - Z}{1 - x} - \frac{\tilde{q}}{2}\frac{(1-x)^2 - \zeta^2}{1 - x} = 0, \tag{5.3.47}$$

where

$$(\ldots)' = \frac{\mathrm{d}(\ldots)}{\mathrm{d}x}, \quad \tilde{q} = \frac{qr_b^2}{m_0}, \quad \zeta = \frac{r_a}{r_b} \tag{5.3.48}$$

Inserting Eqs (5.3.39) into Eq. (5.3.41) we obtain after some transformations the following nonlinear differential equation

$$Y'[(2Z-Y)(2Y-Z)g(\mu)-1]+Z'[(2Z-Y)^2g(\mu)+2]+3\frac{Y-Z}{1-x}=0, \tag{5.3.49}$$

where

$$2g(\mu)=\frac{1}{\mu f(\mu)}\frac{\mathrm{d}f}{\mathrm{d}\mu}-\frac{1}{\mu^2}, \quad \mu^2=Y^2-YZ+Z^2$$

For the power law stress function $f(\mu)=\mu^n$ it follows

$$g(\mu)=\frac{n-1}{2\mu^2} \tag{5.3.50}$$

while for the double power law stress function $g(\mu)=\mu+\mu^n$, $n>1$ we obtain

$$g(\mu)=\frac{n-1}{2\mu^2}\frac{\mu^{n-1}}{1+\mu^{n-1}} \tag{5.3.51}$$

Let us note that Eq. (5.3.49) formally coincides with Eq. (4.3.69) derived for the rotating disc. From Eqs (5.3.42) and (5.3.39) we obtain the following differential equation for the normalized deflection

$$X'=\frac{1}{2}(1-x)f(\mu)\frac{2Z-Y}{\mu}, \tag{5.3.52}$$

With Eqs (5.3.47), (5.3.49) and (5.3.52), the following system of nonlinear first-order differential equations can be derived

$$\begin{aligned} X'&=(1-x)(2Z-Y)\Omega(Y,Z),\\ Y'&=\frac{Y-Z}{1-x}+\frac{\bar{q}}{2}\frac{(1-x)^2-\zeta^2}{1-x}=0,\\ Z'&=\frac{(Z-Y)[3+\Phi(Y,Z)]-\frac{\bar{q}}{2}[(1-x)^2-\zeta^2]\,\Phi(Y,Z)}{\Psi(Y,Z)(1-x)} \end{aligned} \tag{5.3.53}$$

with

$$\begin{aligned} \Omega(Y,Z)&=\frac{1}{2}\frac{f(\mu)}{\mu},\\ \Phi(Y,Z)&=(2Z-Y)(2Y-Z)g(\mu)-1,\\ \Psi(Y,Z)&=(2Z-Y)^2g(\mu)+2 \end{aligned}$$

For the solution we have to specify three boundary conditions. As an example consider the annular plate simply supported at outer radius and free at the inner radius, as shown in Fig. 5.3b. In this case we have to set $X(0)=0$, $Y(0)=0$ and $Y(1-\zeta)=0$. Furthermore, let us assume power law creep such that the stress function $f(x)=x^n$ and the constants $\dot{\chi}_0$ and m_0 are defined by Eqs (5.2.29). For the numerical solution of Eqs (5.3.53) we apply the approach presented in Subsect. 4.3.2. Accordingly, Eqs (5.3.53) are treated as the initial value problems with ini-

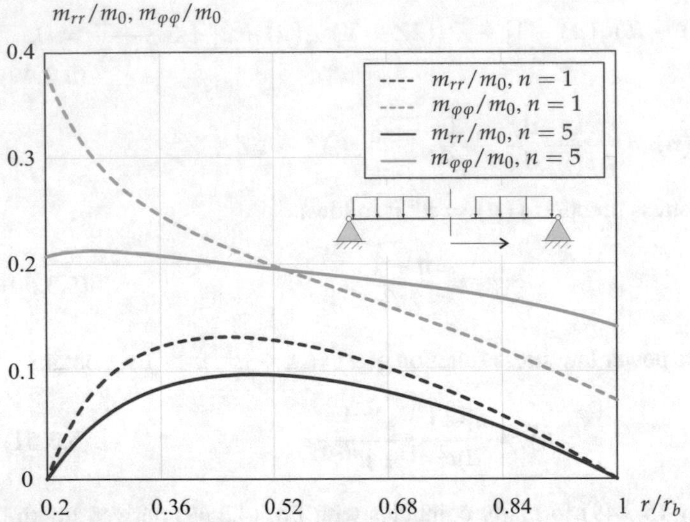

Fig. 5.5 Normalized moments in a plate with a circular hole as function of normalized radial coordinate for two values of creep exponent, $\tilde{q} = 1$ and $r_a/r_b = 0.2$

tial conditions $X(0) = 0$, $Y(0) = 0$ and $Z(0) = Z_0$. The unknown value of the normalized circumferential bending moment at the inner radius of the plate Z_0 is determined from the condition $Y(1 - \zeta) = 0$ applying the iteration procedure.

Figure 5.5 shows the distributions of the normalized bending moments along the normalized radial coordinate for two values of creep exponent. For the normalized uniformly distributed load we assume $\tilde{q} = 1$. In this case according to Eq. (5.3.48)$_2$ we may set the normalized factor $m_0 = qr_b^2$. For $n = 1$, the classical distributions of the components of the moment tensor are obtained, as known from the linear theory of plates (Altenbach et al, 2016; Timoshenko and Woinowsky-Krieger, 1959). The circumferential moment attains the maximum at the inner radius of the plate. As a consequence of the creep process, significant redistributions of the moments can be observed. The circumferential moment relaxes at the inner radius and increases at the outer radius of the plate, as shown in Fig. 5.5 for $n = 5$. The effect of creep on the redistribution of the bending moments depends on the value of the creep exponent, as illustrated in Fig. 5.6. For $n > 2$ the maximum circumferential moment is not at the inner radius of the plate but shifts to the ligament ahead of the hole.

Figure 5.7 shows the distribution of the normalized deflection rate along the normalized radial coordinate for different values of the creep exponent. To normalize the data the maximum value of the deflection rate $\dot{w}_{\max}$ for the given value of the creep exponent is applied. We observe that with an increase of the creep exponent, the distribution approaches a linear function.

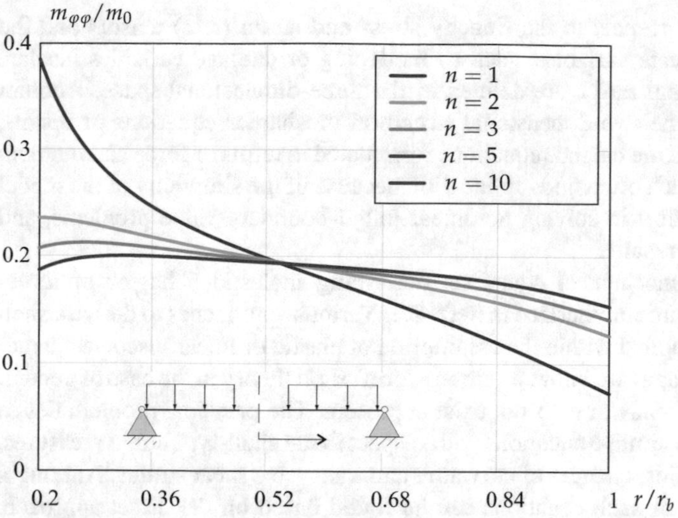

Fig. 5.6 Normalized circumferential moment in a plate with a circular hole as function of normalized radial coordinate for different values of creep exponent ($\tilde{q} = 1$ and $r_a/r_b = 0.2$)

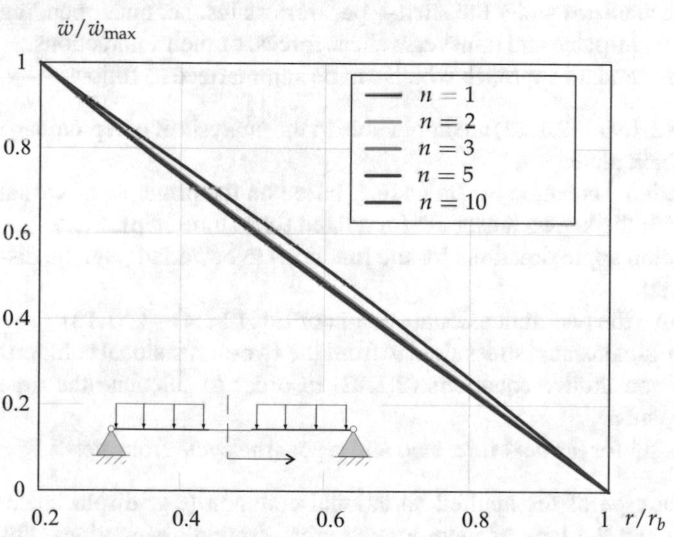

Fig. 5.7 Normalized deflection rate in a plate with a circular hole as function of normalized radial coordinate for different values of creep exponent ($\tilde{q} = 1$ and $r_a/r_b = 0.2$)

5.4 Edge Effects in a Moderately Thick Rectangular Plate

An important step in the inelastic analysis of plates and shells is to select a suitable structural mechanics model. One way is the "three-dimensional approach" which is based on the three-dimensional equations of continuum mechanics. This approach seems more preferable since many unified constitutive models of inelastic analysis are developed with respect to the Cauchy stress and strain (rate) tensors and the proposed internal state variables, such as hardening or damage variables (scalars or tensors of different rank), are defined in the three-dimensional space. Another way is the use of the two-dimensional structural mechanics equations of beams, plates and shells and the balance equations formulated in terms of force and moment tensors. This approach often finds application because of the simplicity of the model creation, smaller effort in solving nonlinear initial-boundary value problems, and easily interpretable results.

The governing mechanical equations describing inelastic behavior of three-dimensional solids are summarized in Sect. 2.1. Various approaches to derive a shell theory have been applied within the assumption of elastic or linear viscoelastic material behavior. As far as we know, a "closed-form" shell theory in the case of general nonlinear inelastic behavior does not exist at present. The principal problem lies in establishing the constitutive equations with respect to the shell-type strain measures, i.e. the in-plane strains, changes of curvature and transverse shear strains. Although, a general structure of such equations can be found based on the direct approach, e.g. Altenbach et al (2005); Altenbach and Zhilin (2004), the open question is the introduction of appropriate hardening and damage measures as well as the identification of damage mechanisms under the shell-type stress states, i.e. under bending and twisting moments, in-plane and transverse shear forces, or their interactions.

Here we apply the standard approach which can be summarized as follows:

1. Assume that Eqs (2.1.4) – (2.1.12) are applicable to the analysis of creep-damage in a moderately thick plate.
2. Formulate a variational equation of statics (e.g. based on the principle of virtual displacements) with the known tensor $\boldsymbol{\varepsilon}^{\mathrm{pl}}$ for a fixed time (time step).
3. Specify cross-section approximations for the functions to be varied (e.g. the displacement vector $\boldsymbol{u}$).
4. Formulate and solve the two-dimensional version of Eqs (2.1.4) – (2.1.12).
5. Recover the three-dimensional stress field $\boldsymbol{\sigma}$ from the two-dimensional solution.
6. Insert $\boldsymbol{\sigma}$ into the constitutive equations (2.1.12) in order to calculate the time increments of $\boldsymbol{\varepsilon}^{\mathrm{pl}}$ and ω.
7. Update the tensor $\boldsymbol{\varepsilon}^{\mathrm{pl}}$ for the next time step and repeat the cycle from step 2.

Depending on the type of the applied variational equation (e.g. displacement type or mixed type) and the type of incorporated cross-section assumptions, different two-dimensional versions of Eqs (2.1.4) – (2.1.12) with a different order of complexity can be obtained (i.e. models with forces and moments or models with higher order stress resultants). In the case of linear elastic plates, a huge number of such kind of plate theories has been proposed, e.g. in Lo et al (1977); Reddy (1984). Note that the steps 2 and 3 can be performed numerically applying, e.g. the Galerkin method to Eqs (2.1.4) – (2.1.12). Various types of finite elements which were developed for the inelastic analysis of shells are reviewed in Wriggers (2008). Let us note that when studying inelastic behavior coupled with damage, the type of assumed cross-section approximations may have a significant influence on the result. For example, if we use a mixed type variational equation and approximate both the

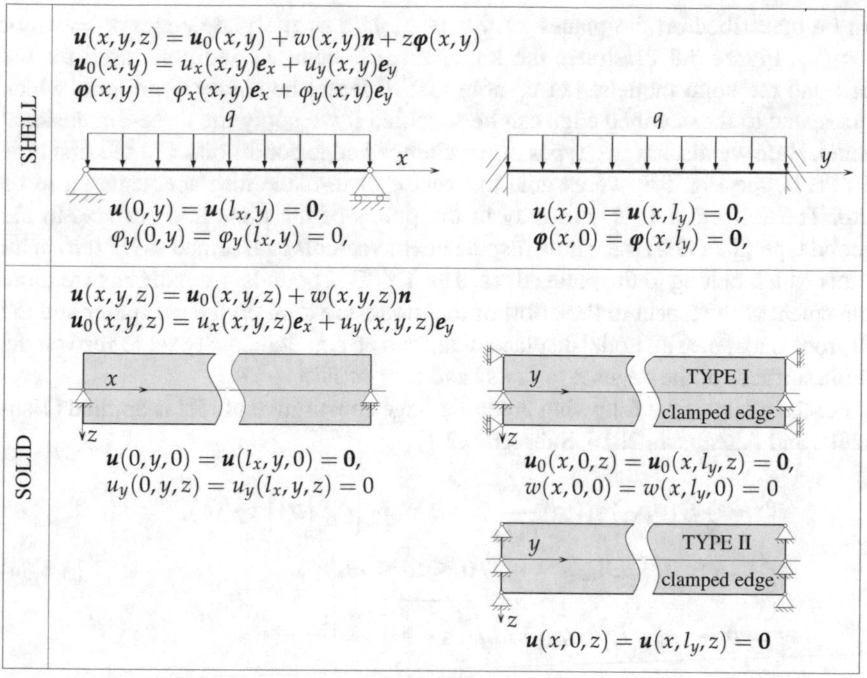

Fig. 5.8 Types of kinematical boundary conditions

displacements and stresses, a parabolic through-the-thickness approximation for the transverse shear stress or a linear approximation for the in-plane stresses is in general not suitable for creep-damage estimations (Altenbach and Naumenko, 2002). In what follows we compare finite element solutions based on the three-dimensional approach and a two-dimensional plate model and discuss the possibilities and limitations of each approach in connection with the creep-damage analysis.

Consider a square plate with $l_x = l_y = 1000$ mm, $h = 100$ mm, loaded by a pressure $q = 2$ MPa uniformly distributed on the top surface. The edges $x = 0$ and $x = l_x$ are simply supported (hard hinged support) and the edges $y = 0$ and $y = l_y$ are clamped. According to the first order shear deformation plate model we can specify the vectors of "plate displacements" $\boldsymbol{u}_p(x,y) = \boldsymbol{u}_0(x,y) + w(x,y)\boldsymbol{n}$, $\boldsymbol{u}_0 \cdot \boldsymbol{n} = 0$ and cross-section rotations $\boldsymbol{\varphi}(x,y)$ on the lines $x = const$ or $y = const$, Fig. 5.8. Applying such a model and assuming infinitesimal cross-section rotations, the displacement vector $\boldsymbol{u}(x,y,z)$ is usually formulated as follows

$$\boldsymbol{u}(x,y,z) \approx \boldsymbol{u}_p(x,y) + z\boldsymbol{\varphi}(x,y)$$

In the case of the three-dimensional model, the displacement vector

$$\boldsymbol{u}(x,y,z) = \boldsymbol{u}_0(x,y,z) + w(x,y,z)\boldsymbol{n}$$

can be prescribed on the planes x_c, y, z or x, y_c, z of the plate edges $x = x_c$ or $y = y_c$. Figure 5.8 illustrates the kinematical boundary conditions used for the shell and the solid models. Let us note that different boundary conditions which correspond to the clamped edge can be specified if we apply the three-dimensional model. Here we discuss two types of the clamped edge conditions. For the first type (TYPE I), see Fig. 5.8, we assume the vector of in-plane displacements $\boldsymbol{u}_0$ to be zero. The deflection w is zero only in the points of the plate mid-surface. In the second type (TYPE II) the whole displacement vector $\boldsymbol{u}$ is assumed to be zero in all points which belong to the plate edges. The TYPE II boundary conditions are more convenient with respect to the effort in the model creation on the computer and the preprocessing since all nodal displacements can be simultaneously set to zero on the whole surfaces of the edges $x = const$ and $y = const$.

For the analysis the following creep-damage constitutive model is applied (Naumenko and Altenbach, 2016, Subsect. 6.2.1)

$$\begin{aligned}
&\dot{\boldsymbol{\varepsilon}}^{\mathrm{pl}} = \frac{3}{2} f_1(\sigma_{\mathrm{vM}}) g_1(\omega) \frac{\boldsymbol{s}}{\sigma_{\mathrm{vM}}}, \quad \dot{\omega} = f_2\left[\sigma_{\mathrm{eq}}^{\omega}(\boldsymbol{\sigma})\right] g_2(\omega), \\
&\boldsymbol{\varepsilon}^{\mathrm{pl}}|_{t=0} = \boldsymbol{0}, \quad \omega|_{t=0} = 0, \quad 0 \leq \omega \leq \omega_*, \\
&\boldsymbol{s} = \boldsymbol{\sigma} - \frac{1}{3} \mathrm{tr}\, \boldsymbol{\sigma} \boldsymbol{I}, \quad \sigma_{\mathrm{vM}} = \sqrt{\frac{3}{2} \boldsymbol{s} \cdot\cdot \boldsymbol{s}}
\end{aligned} \tag{5.4.54}$$

Here $\boldsymbol{\varepsilon}^{\mathrm{pl}}$ is the creep strain tensor, $\boldsymbol{\sigma}$ is the stress tensor, ω is the scalar-valued damage parameter and $\sigma_{\mathrm{eq}}^{\omega}$ is the damage equivalent stress. The constitutive functions and material parameters are taken for a type 316 stainless steel at 650°C from Liu et al (1994) as follows

$$\begin{aligned}
f_1(\sigma) &= a\sigma^n, \quad g_1(\omega) = (1-\omega)^{-n}, \\
f_2(\sigma) &= b\sigma^k, \quad g_2(\omega) = (1-\omega)^{-k}
\end{aligned} \tag{5.4.55}$$

$$\begin{aligned}
&a = 2.13 \cdot 10^{-13}\,\mathrm{MPa}^{-n}/\mathrm{h}, \quad b = 9.1 \cdot 10^{-10}\,\mathrm{MPa}^{-k}/\mathrm{h}, \\
&n = 3.5, \quad k = 2.8
\end{aligned} \tag{5.4.56}$$

It is assumed that the damage evolution is controlled by the maximum tensile stress. Therefore, the damage equivalent stress takes the form

$$\sigma_{\mathrm{eq}}^{\omega}(\boldsymbol{\sigma}) = \frac{\sigma_I + |\sigma_I|}{2},$$

where σ_I is the first principal stress. The elastic material behavior is characterized by the following values of the Young's modulus E and the Poisson's ratio ν

$$E = 1.44 \cdot 10^5\,\mathrm{MPa}, \quad \nu = 0.314 \tag{5.4.57}$$

The analysis has been performed using the ANSYS finite element code after incorporating the material model (3.2.55) with the help of a user-defined creep-damage material subroutine. In Subsect. 3.1.4 we discussed various examples for

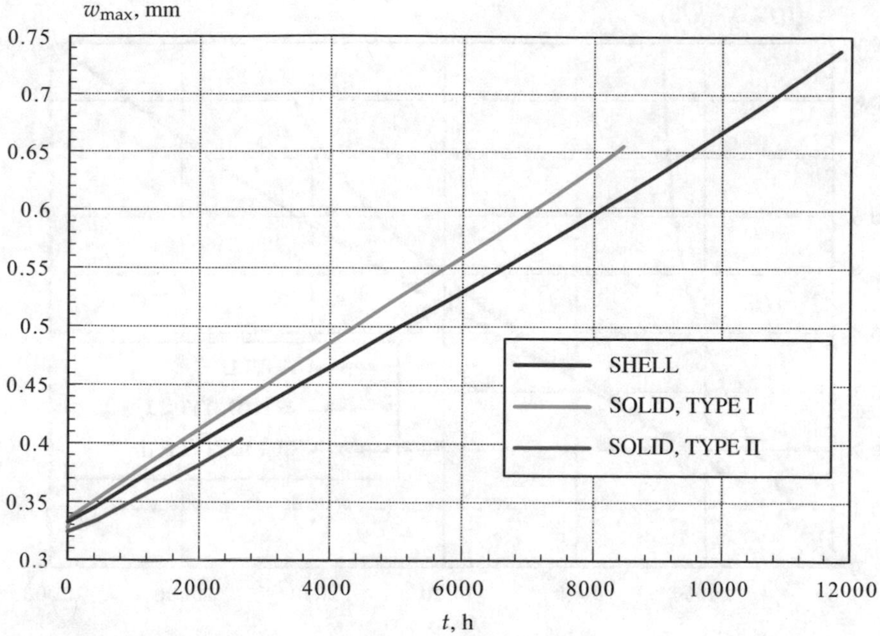

Fig. 5.9 Time variations of the maximum deflection

beams, which verify the developed subroutine. The finite elements available in the ANSYS code for plasticity and creep analysis were applied as follows: the 20-nodes solid element SOLID 95 and the 4-nodes shell element SHELL 43. 30 $\times$ 15 elements were used for a half of the plate in the case of the shell model and 30 $\times$ 15 $\times$ 3 elements in the case of the solid model. The meshes have been justified based on the elasticity solutions and the steady-state creep solutions neglecting damage. With these meshes, the reference stress distributions as well as the distributions of the von Mises stresses in the steady creep state were approximately the same for both the solid and the shell elements and did not change anymore by further re-meshing. The automatical time stepping feature with a minimum time step of 0.1 h has been applied. The time step based calculations were performed up to $\omega = \omega_* = 0.9$, where ω_* is the selected critical value of the damage parameter. Figures 5.9 and 5.10 illustrate the results of the computations, where the maximum deflection and the maximum value of the damage parameter are plotted as functions of time. From Fig. 5.9 we observe that the starting values of maximum deflection as well as the starting rates of the deflection growth due to creep are approximately the same for the shell and the two solid models. Consequently the type of the elements (shell or solid) and the type of the applied boundary conditions in the case of the solid elements has a small influence on the description of the steady-state creep process. However, the three used models lead to quite different life time predictions. The difference can be seen in Fig. 5.10. The shell model overestimates the time to failure, while the results based on the solid model depend significantly on the type of the

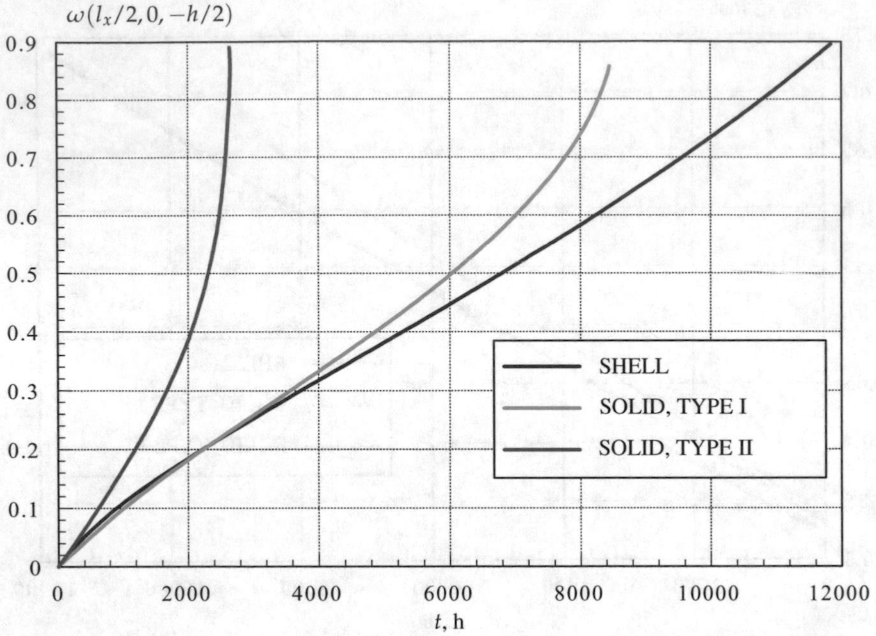

Fig. 5.10 Time variations of the damage parameter

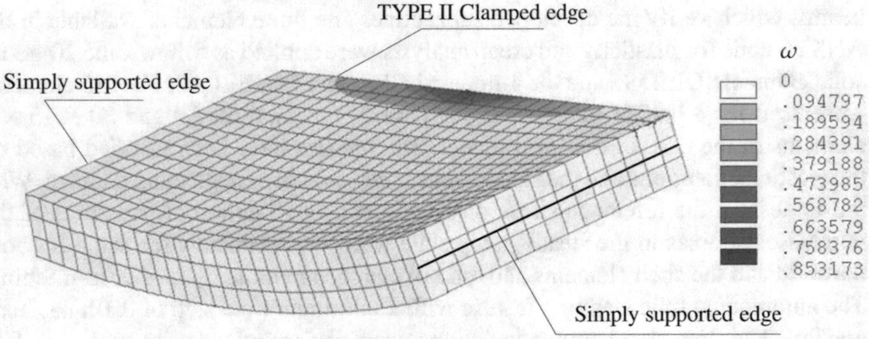

Fig. 5.11 Deformed shape of a half of the plate and distribution of the damage parameter in the zone of a clamped edge (SOLID elements, TYPE II boundary conditions, last time step)

clamped edge boundary conditions. In the case of the TYPE II clamped edge much more accelerated damage growth is obtained. The corresponding time to failure is approximately four times shorter compared to those based on the TYPE I clamped edge. All considered models predict the zone of maximum damage to be in the midpoint of the clamped edge on the plate top surface, as shown in Fig. 5.11.

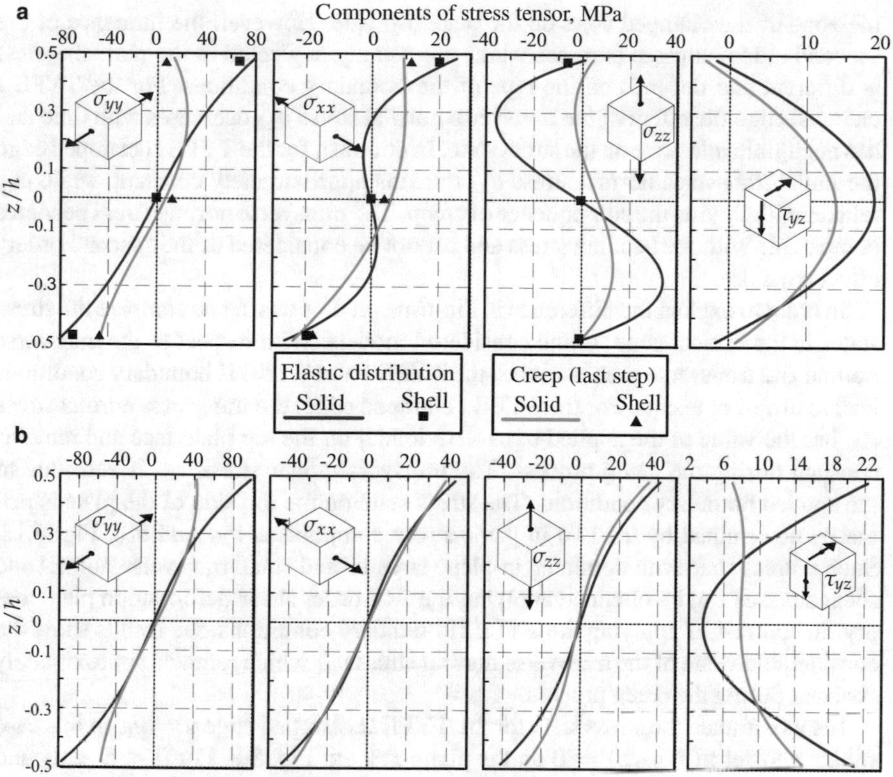

Fig. 5.12 Local stress state in a midpoint of the clamped edge vs. thickness coordinate (last time step). **a** TYPE I clamped edge, **b** TYPE II clamped edge

The creep response of a structure is connected with the time-dependent stress redistributions. If the applied load and the boundary conditions are assumed to be constant and the effect of tertiary creep is ignored, then an asymptotic stress state exists, which is known as the state of stationary or steady creep (see Sect. 1.2). If tertiary creep is considered, then stresses change with time up to the critical damage state. It is obvious that the damage growth and the tertiary creep behavior of the considered plate are controlled by the local stress state in the vicinity of the clamped edges. Figure 5.12 illustrates the stress states in the midpoint of the clamped edge with the coordinates $x = l_x/2, y = 0$. Four components of the stress tensor (the two remaining components are zero due to symmetry conditions) are plotted as functions of the normalized thickness coordinate. The starting elastic distributions (solid lines) as well as the creep solutions at the last time step (dotted lines) are presented. The maximum starting stresses obtained by the use of the three considered models are the normal in-plane stresses σ_{yy} and σ_{xx} (the stresses which result in the maximum bending and twisting moments in the clamped edges), Fig. 5.12. These in-plane stresses remain dominant during the whole creep process for the used shell and solid elements. Therefore, all the applied models predict the damage evolution in the zone of the clamped edge on the plate top side. However, the influence of the "second order" stresses (stresses which are usually neglected in the plate theories) is different and depends on the type of the boundary conditions. For the TYPE I clamped edge the effect of the transverse normal stress σ_{zz} decreases with time and has negligible influence on the stress state. In contrast, for the TYPE II clamped edge the initial transverse normal stress σ_{zz} remains approximately constant, while σ_{yy} relaxes with time as the consequence of creep. The transverse normal stress becomes comparable with the bending stress and cannot be considered as the "second order" effect anymore.

In order to explain the difference in life-time predictions, let us compare the stress states in the critical zone for the considered models. With respect to the transverse normal and transverse shear stresses, the TYPE I and TYPE II boundary conditions lead to different results. For the TYPE I clamped edge, the transverse normal stress σ_{zz} has the value of the applied transverse load q on the top plate face and remains constant during the creep process. The transverse shear stress τ_{xz} is zero due to the applied boundary conditions. The stress state on the top side of the plate is primarily determined by the two in-plane stress components σ_{xx} and σ_{yy}, Fig. 5.12. Such a stress state with dominant in-plane stresses and small transverse normal and shear stresses can be obtained applying the first order shear deformation plate theory. In contrast, if applying the TYPE II boundary conditions, the results show the considerable value of the transverse normal stress σ_{zz} which remains approximately constant during the creep process.

Let us estimate the stress state for the TYPE II clamped edge $y = y_c$. In this case we have to set $\boldsymbol{u}(x, y_c, z) = \boldsymbol{0}$ on the plane x, y_c, z, Fig. 5.8. For $0 < x < l_x$ and $-h/2 < z < h/2$ we can write

$$\frac{\partial \boldsymbol{u}}{\partial x} = \frac{\partial \boldsymbol{u}}{\partial z} = \boldsymbol{0} \quad \Rightarrow \quad \nabla \boldsymbol{u}(x, y_c, z) = \boldsymbol{e}_y \otimes \frac{\partial \boldsymbol{u}}{\partial y},$$
$$\operatorname{tr} \boldsymbol{\varepsilon}(x, y_c, z) = \nabla \cdot \boldsymbol{u} = \frac{\partial u_y}{\partial y} \tag{5.4.58}$$

In addition, we can set $\boldsymbol{e}_x \cdot \boldsymbol{u}(l_x/2, y, z) = 0$ due to the symmetry condition. The starting elastic stress state at $t = 0$ can be obtained from the constitutive equations (2.1.9) by setting $\boldsymbol{\varepsilon}^{\mathrm{pl}} = \boldsymbol{0}$

$$\sigma_{\mathrm{m}}|_{t=0} = \frac{1}{3}\frac{1+\nu}{1-2\nu}\sigma_0, \qquad \sigma_0 = 2G\frac{\partial u_y}{\partial y}|_{t=0}, \qquad \tau_0 = G\frac{\partial w}{\partial y}|_{t=0},$$
$$\boldsymbol{s}|_{t=0} = \frac{1}{3}\sigma_0\left[2\boldsymbol{e}_y \otimes \boldsymbol{e}_y - (\boldsymbol{I} - \boldsymbol{e}_y \otimes \boldsymbol{e}_y)\right] + \tau_0(\boldsymbol{e}_y \otimes \boldsymbol{n} + \boldsymbol{n} \otimes \boldsymbol{e}_y),$$
$$\boldsymbol{\sigma}|_{t=0} = \frac{1-\nu}{1-2\nu}\sigma_0\left[\boldsymbol{e}_y \otimes \boldsymbol{e}_y + \frac{\nu}{1-\nu}(\boldsymbol{I} - \boldsymbol{e}_y \otimes \boldsymbol{e}_y)\right] + \tau_0(\boldsymbol{e}_y \otimes \boldsymbol{n} + \boldsymbol{n} \otimes \boldsymbol{e}_y) \tag{5.4.59}$$

From the last equation in (5.4.59) we observe that

$$\sigma_{zz} = \sigma_{xx} = \sigma_{yy}\nu/(1-\nu)$$

This well known result of the theory of linear isotropic elasticity agrees with the obtained finite element solution for $\nu = 0.314$, Fig. 5.12b (solid lines).

Let us estimate the stress redistribution in the TYPE II clamped edge as a consequence of creep. For this purpose, we neglect the damage evolution by setting $\omega = 0$ in (3.2.55). Since the boundary conditions and the applied pressure are independent of time, we can estimate the type of the stress state under stationary state creep by setting $\dot{\boldsymbol{\varepsilon}} \approx \dot{\boldsymbol{\varepsilon}}^{\text{pl}}$, $\dot{\varepsilon}_V \approx 0$ or

$$\frac{1}{2}\left(\boldsymbol{e}_y \otimes \frac{\partial \dot{\boldsymbol{u}}}{\partial y} + \frac{\partial \dot{\boldsymbol{u}}}{\partial y} \otimes \boldsymbol{e}_y\right) \approx \dot{\boldsymbol{\varepsilon}}^{\text{pl}} = \frac{3}{2} a \sigma_{\text{vM}}^{n-1} \boldsymbol{s}, \qquad \nabla \cdot \dot{\boldsymbol{u}} \approx 0 \tag{5.4.60}$$

Consequently

$$\frac{1}{2}\frac{\partial \dot{w}}{\partial y}(\boldsymbol{e}_y \otimes \boldsymbol{n} + \boldsymbol{n} \otimes \boldsymbol{e}_y) \approx \frac{3}{2} a \sigma_{\text{vM}}^{n-1} \boldsymbol{s} \tag{5.4.61}$$

From Eq. (5.4.61) we observe that the stress deviator in the steady-state creep has the form $\boldsymbol{s} \approx \tau(\boldsymbol{e}_y \otimes \boldsymbol{n} + \boldsymbol{n} \otimes \boldsymbol{e}_y)$ and is completely determined by the transverse shear stress. The mean stress σ_{m} cannot be determined from the constitutive equation, it must be computed from the equilibrium conditions (2.1.6). The stress state in the zone of the clamped edge $(l_x/2, y, z)$ is then of the type

$$\boldsymbol{\sigma} \approx \sigma_{\text{m}} \boldsymbol{I} + \tau(\boldsymbol{e}_y \otimes \boldsymbol{n} + \boldsymbol{n} \otimes \boldsymbol{e}_y)$$

We observe that $\sigma_{zz} \approx \sigma_{yy} \approx \sigma_{xx} \approx \sigma_{\text{m}}$ after the transient stress redistribution. This estimation agrees again with the obtained finite element solution, Fig. 5.12b (dotted lines). The transverse normal stress is approximately equal to the in-plane stresses and cannot be neglected.

Let us compare the finite element results for the mean stress and the von Mises equivalent stress. Figure 5.13 shows the corresponding time variations in the element A of the solid model for the TYPE I and TYPE II boundary conditions. We observe that the TYPE II boundary condition leads to a lower starting value of the von Mises stress and a higher starting value of the mean stress when compared with those for the TYPE I boundary condition. In addition, for the TYPE II clamped edge we observe that the mean stress rapidly decreases within the short transition time and after that remains constant, while the von Mises stress relaxes during the whole creep process. With the relaxation of σ_{vM}, the stress state tends to $\boldsymbol{\sigma} = \sigma_{\text{m}} \boldsymbol{I}$. The relatively high constant value of σ_{m} is the reason for the obtained increase of damage and much shorter time to fracture in the case of the TYPE II clamped edge (see Fig. 5.10). The above effect of the mean stress has a local character and is observed only in the neighborhood of the edge. As Fig. 5.14 shows, the value of the transverse normal stress decreases rapidly with increased distance from the boundary.

We discussed the possibilities of creep-damage behavior modeling in moderately thick structural elements. The selected constitutive model of creep is based on the assumption that the secondary creep strain rate is determined by the deviatoric part of the stress tensor and the von Mises equivalent stress, while the increase of the

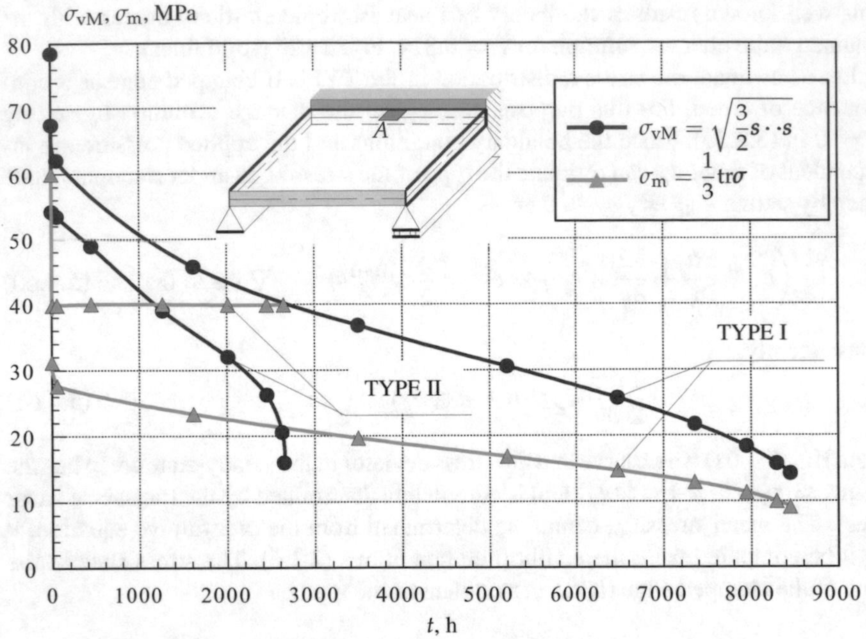

Fig. 5.13 Time variations of the von Mises equivalent stress and the hydrostatic stress in element A of the clamped edge

creep rate in the tertiary range is due to isotropic damage evolution which is controlled by the mean stress, the first principal stress and the von Mises equivalent stress. The use of this model in connection with long-term predictions of structural elements has motivated a numerical comparative study of two approaches: the three-dimensional approach and the approach based on the first order shear deformation type plate theory. The finite element results as well as some simplified estimates have shown that the approaches based on standard solid and shell finite elements provide quite different predictions. The model based on the shell elements overestimates the fracture time. The reason for the obtained differences is the local stress response in the zone of the clamped edge. In the case of linear isotropic elasticity, the transverse normal and shear stresses in the zone of the clamped edge can be assumed to be second order quantities in comparison to the dominant in-plane stresses. In the case of steady-state creep, the transverse normal and shear stresses are comparable with the in-plane stresses due to the stress redistribution. If studying the creep behavior coupled with damage, the influence of these factors cannot be ignored.

If a shell or a plate theory is considered to be an approximate version of the three-dimensional equations (2.1.4) – (2.1.12), then we can conclude that "more accurate" cross-section approximations for the transverse normal and shear stresses have to be used in the case of creep. In this sense it is more reliable to solve the three-

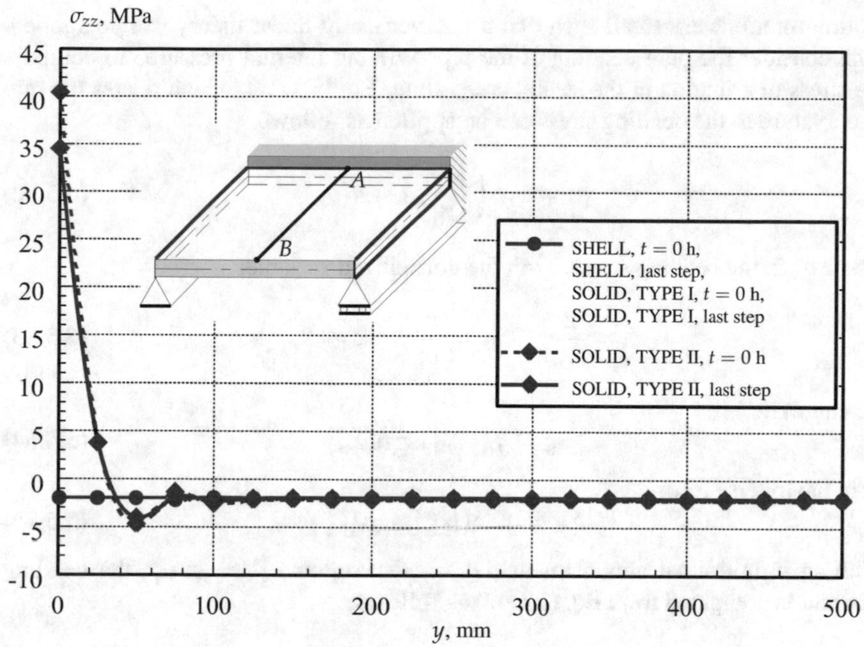

Fig. 5.14 Distribution of the transverse normal stress σ_{zz} in elements along the line AB

dimensional equations (2.1.4) – (2.1.12) which are "free" from ad hoc assumptions for the displacements and stresses.

5.5 Thin-walled Tube under Internal Pressure and Bending Moment

Figure 5.15 shows a long thin-walled tube with the mean radius of the cross section R_m and the wall thickness h. For the analysis we assume that strains and cross-

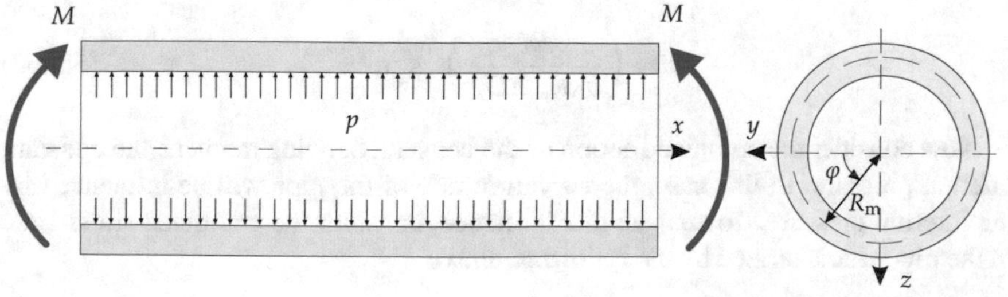

Fig. 5.15 Thin-walled tube under internal pressure and bending moment

section rotations are small such that a geometrically-linear theory can be applied. First consider the pure bending of the pipe without internal pressure. To compute the stress distribution in the steady creep state, Eq. (3.1.27), which relates the rate of curvature to the bending stress can be applied as follows

$$\dot{\chi} z = \dot{\varepsilon}_0 f\left(\frac{|\sigma_x|}{\sigma_0}\right) \operatorname{sgn}(\sigma_\mathrm{b}), \tag{5.5.62}$$

where σ_x is the bending stress. With the normalized variables

$$\xi = \frac{z}{R_\mathrm{m}}, \quad \Sigma_x = \frac{\sigma_x}{\sigma_0}, \quad \dot{\kappa} = R_\mathrm{m}\frac{\dot{\chi}}{\dot{\varepsilon}_0} \tag{5.5.63}$$

we can write

$$\dot{\kappa}\xi = f(|\Sigma_x|)\operatorname{sgn}(\Sigma_x) \tag{5.5.64}$$

or in the inverse form

$$\Sigma_x = f^{-1}(|\dot{\kappa}\xi|)\operatorname{sgn}(\dot{\kappa}\xi) \tag{5.5.65}$$

With an infinitesimal area element $\mathrm{d}A = R_m h \mathrm{d}\varphi$ and $\xi = \sin\varphi$, the bending moment is computed from Eq. (3.1.42) as follows

$$M = \int_A \sigma_x z \mathrm{d}A = \sigma_0 R_\mathrm{m}^2 h \int_0^{2\pi} f^{-1}(|\dot{\kappa}\sin\varphi|)\operatorname{sgn}(\dot{\kappa}\sin\varphi)\sin\varphi \mathrm{d}\varphi \tag{5.5.66}$$

For the given function of stress, Eq. (5.5.66) can be integrated numerically providing the relationship between the bending moment and the rate of the normalized curvature in the steady creep state. With the power law stress function $f^{-1}(x) = x^{1/n}$, the integral in Eq. (5.5.66) can be computed as follows

$$M = \sigma_0 R_\mathrm{m}^2 h D_n |\dot{\kappa}|^{1/n} \operatorname{sgn}\dot{\kappa}, \tag{5.5.67}$$

where

$$D_n = \int_0^{2\pi} |\sin\varphi|^{\frac{n+1}{n}} \mathrm{d}\varphi$$

The inverse form of Eq. (5.5.67) provides the rate of the normalized curvature

$$\dot{\kappa} = \left(\frac{M}{\sigma_0 R_\mathrm{m}^2 h D_n}\right)^n \operatorname{sgn}M, \tag{5.5.68}$$

Now consider the combined action of the constant bending moment and constant internal pressure. In this case, the curvature rate of the pipe will be influenced by the internal pressure. To analyze this influence, consider the combined stress state in the pipe characterized by the following tensor

$$\boldsymbol{\sigma} = \sigma_x \boldsymbol{e}_x \otimes \boldsymbol{e}_x + \frac{pR_\mathrm{m}}{h} \boldsymbol{e}_\varphi \otimes \boldsymbol{e}_\varphi, \tag{5.5.69}$$

where the unit vectors $\boldsymbol{e}_x$ and $\boldsymbol{e}_\varphi$ designate the axial and the circumferential direction, respectively.

To specify the strain rate tensor we assume that any cross section of the pipe remains plane during the deformation. In this case, the rate of the axial strain is a linear function of the z coordinate such that

$$\dot{\varepsilon}_x = \dot{\varepsilon}_\mathrm{c} + \dot{\chi} z, \tag{5.5.70}$$

where ε_c is the axial strain of the centerline, or the mean axial strain of the pipe.

For the plane stress state (5.5.69) we can apply the constitutive equation (5.2.21). For the steady creep state the plane part of the strain rate tensor is

$$\dot{\boldsymbol{\varepsilon}}^\mathrm{pl}_\mathrm{p} = \dot{\boldsymbol{\varepsilon}}_\mathrm{p} = \dot{\varepsilon}_x \boldsymbol{e}_x \otimes \boldsymbol{e}_x + \dot{\varepsilon}_\varphi \boldsymbol{e}_\varphi \otimes \boldsymbol{e}_\varphi \tag{5.5.71}$$

Inserting Eq. (5.5.71) into Eq. (5.2.21) we obtain

$$\begin{gathered}
\sigma_x = \frac{2}{3} \frac{\sigma_\mathrm{vM}}{\dot{\varepsilon}_\mathrm{vM}} \left(2\dot{\varepsilon}_x + \dot{\varepsilon}_\varphi\right), \quad \sigma_\varphi = \frac{2}{3} \frac{\sigma_\mathrm{vM}}{\dot{\varepsilon}_\mathrm{vM}} \left(2\dot{\varepsilon}_\varphi + \dot{\varepsilon}_x\right) = \frac{pR_\mathrm{m}}{h}, \\
\sigma_\mathrm{vM} = \sigma_0 f^{-1}\left(\frac{\dot{\varepsilon}_\mathrm{vM}}{\dot{\varepsilon}_0}\right), \quad \dot{\varepsilon}^2_\mathrm{vM} = \frac{4}{3}\left(\dot{\varepsilon}^2_x + \dot{\varepsilon}_x \dot{\varepsilon}_\varphi + \dot{\varepsilon}^2_\varphi\right)
\end{gathered} \tag{5.5.72}$$

With the normalized variables in Eq. (5.5.63) we obtain

$$\Sigma_x = \frac{2}{3} \frac{f^{-1}(\dot{\epsilon}_\mathrm{vM})}{\dot{\epsilon}_\mathrm{vM}} \left(2\dot{\epsilon}_x + \dot{\epsilon}_\varphi\right), \quad 2\dot{\epsilon}_\varphi = \frac{3}{2} \frac{\dot{\epsilon}_\mathrm{vM}}{f^{-1}(\dot{\epsilon}_\mathrm{vM})} \tilde{p} - \dot{\epsilon}_x, \tag{5.5.73}$$

where

$$\tilde{p} = \frac{pR_\mathrm{m}}{h\sigma_0}, \quad \dot{\epsilon}_x = \frac{\dot{\varepsilon}_x}{\dot{\varepsilon}_0}, \quad \dot{\epsilon}_\varphi = \frac{\dot{\varepsilon}_\varphi}{\dot{\varepsilon}_0}, \quad \dot{\epsilon}_\mathrm{vM} = \frac{\dot{\varepsilon}_\mathrm{vM}}{\dot{\varepsilon}_0}$$

Inserting Eq. $(5.5.73)_2$ into Eq. $(5.5.73)_1$ we obtain

$$\Sigma_x = f^{-1}\left(\dot{\epsilon}_\mathrm{vM}\right) \frac{\dot{\epsilon}_\mathrm{c} + \dot{\kappa}\xi}{\dot{\epsilon}_\mathrm{vM}} + \frac{\tilde{p}}{2}, \quad \dot{\epsilon}_\mathrm{c} = \frac{\dot{\varepsilon}_\mathrm{c}}{\dot{\varepsilon}_0} \tag{5.5.74}$$

From Eqs $(5.5.72)_4$ and $(5.5.73)_2$ we obtain

$$\dot{\epsilon}_\mathrm{vM} \left[1 - \frac{3}{4}\left(\frac{\tilde{p}}{f^{-1}\left(\dot{\epsilon}_\mathrm{vM}\right)}\right)^2\right]^{1/2} = |\dot{\epsilon}_\mathrm{c} + \dot{\kappa}\xi| \tag{5.5.75}$$

For the given value of internal pressure, the nonlinear equation (5.5.75) can be solved providing the normalized von Mises equivalent strain rate as a function of $|\dot{\epsilon}_\mathrm{c} + \dot{\kappa}\xi|$. Then the normal stress can be computed from Eq. $(5.5.74)_1$ as follows

$$\Sigma_x(\xi, \tilde{p}) = g(|\dot{\varepsilon}_c + \dot{\kappa}\xi|, \tilde{p})(\dot{\varepsilon}_c + \dot{\kappa}\xi) + \frac{\tilde{p}}{2}, \quad g = \frac{f^{-1}(\dot{\varepsilon}_{vM})}{\dot{\varepsilon}_{vM}} \tag{5.5.76}$$

The resultant axial force and the bending moment are computed as follows

$$\begin{aligned} \tilde{N} - 2\pi\frac{\tilde{p}}{2} &= \int_0^{2\pi} g(|\dot{\varepsilon}_c + \dot{\kappa}\sin\varphi|, \tilde{p})(\dot{\varepsilon}_c + \dot{\kappa}\sin\varphi)\mathrm{d}\varphi, \\ \tilde{M} &= \int_0^{2\pi} g(|\dot{\varepsilon}_c + \dot{\kappa}\sin\varphi|, \tilde{p})(\dot{\varepsilon}_c + \dot{\kappa}\sin\varphi)\sin\varphi\mathrm{d}\varphi, \end{aligned} \tag{5.5.77}$$

where

$$\tilde{N} = \frac{N}{\sigma_0 R_m h}, \quad \tilde{M} = \frac{M}{\sigma_0 R_m^2 h}$$

By specifying the value for the axial force N or mean axial strain rate $\dot{\varepsilon}_c$, Eqs (5.5.77) can be solved providing the relation between the bending moment and the curvature rate for the given value of internal pressure. As an example consider the power law creep $f(x) = x^n$, $f^{-1}(x) = x^{1/n}$. Furthermore assume that the mean axial strain rate is zero, which is a good approximation for a long pipe. This case was analyzed in Spence (1973); Boyle and Spence (1983). Equation (5.5.75) takes the form

$$\dot{\varepsilon}_{vM}\left[1 - \frac{3}{4}\left(\frac{\tilde{p}}{(\dot{\varepsilon}_{vM})^{1/n}}\right)^2\right]^{1/2} = |\dot{\kappa}\sin\varphi| \tag{5.5.78}$$

The normalized resultants defined by Eqs (5.5.77) can be computed as follows

$$\begin{aligned} \tilde{N} - 2\pi\frac{\tilde{p}}{2} &= \int_0^{2\pi} \dot{\varepsilon}_{vM}^{\frac{1-n}{n}}(|\dot{\kappa}\sin\varphi|, \tilde{p})\dot{\kappa}\sin\varphi\mathrm{d}\varphi, \\ \tilde{M} &= \int_0^{2\pi} \dot{\varepsilon}_{vM}^{\frac{1-n}{n}}(|\dot{\kappa}\sin\varphi|, \tilde{p})\dot{\kappa}\sin^2\varphi\mathrm{d}\varphi, \end{aligned} \tag{5.5.79}$$

In order to find the relationship between the bending moment and the curvature rate, two numerical procedures can be applied. First, a standard procedure to find the root of a nonlinear equation can be applied to solve Eq. (5.5.78). Then, by varying $\dot{\kappa}$ in the range $[0, \dot{\kappa}_{max}]$, the integrals (5.5.79) can be evaluated by applying a standard numerical quadrature rule. Let us introduce the following variables

$$Y = \frac{\dot{\varepsilon}_{vM}}{(\sqrt{3/4}\tilde{p})^n}, \quad x = \frac{\dot{\kappa}\sin\varphi}{(\sqrt{3/4}\tilde{p})^n}$$

Then the nonlinear Eq. (5.5.78) takes the form

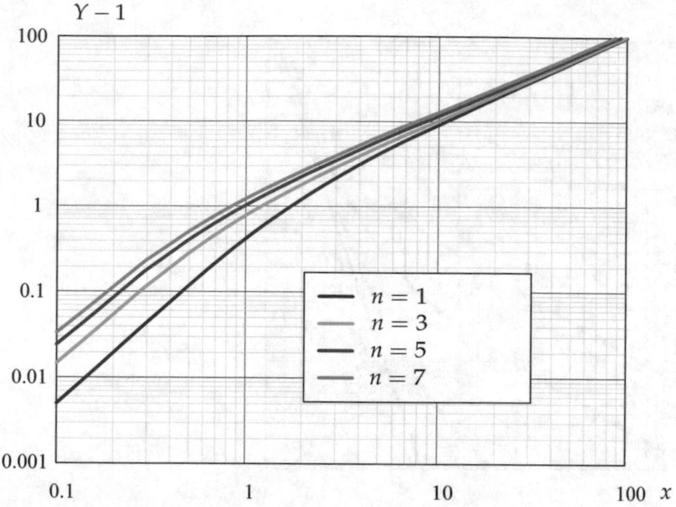

Fig. 5.16 Numerical solutions to Eq. (5.5.80) for different values of creep exponent

$$Y\sqrt{1 - Y^{-\frac{2}{n}}} = |x| \tag{5.5.80}$$

Note that for $x = 0$ Eq. (5.5.80) has two solutions: $Y = 0$ (no deformation in the absence of loading) and $Y = 1$ (equivalent strain rate for loading by internal pressure without bending). For the numerical solution we assume that $Y = 1$ is given for $x = 0$, i.e. the pipe is subjected to internal pressure in the reference state. Figure 5.16 illustrates the results of the numerical solution to Eq. (5.5.80) for different values of the creep exponent n. For large values of x, the results for all values of the creep exponent approach the pure bending solution $Y = |x|$. With the known function $Y(x)$, the bending moment $(5.5.79)_2$ can be computed as follows

$$\tilde{M} = \sqrt{\frac{3}{4}}\tilde{p} \int_0^{2\pi} Y^{\frac{1-n}{n}}(|x|) x \sin\varphi \mathrm{d}\varphi, \quad x = \frac{\dot{\kappa}\sin\varphi}{(\sqrt{3/4}\tilde{p})^n} \tag{5.5.81}$$

Figure 5.17 illustrates the results of numerical evaluation of the integral (5.5.81). The normalized rate of curvature is shown as a function of the normalized bending moment for $n = 5$ and different values of the internal pressure. We observe that the internal pressure leads to a significant increase of the curvature rate. Furthermore, contrary to the pure bending of the pipe, a loading-range dependent creep is observed if internal pressure is additionally applied. For large values of the bending moment, the solution approaches the power law creep regime in the pure bending, while for moderate and low values of the bending moment the creep-bending exponent decreases.

In Spence (1973); Boyle and Spence (1983) a flexibility factor is introduced as the ratio of the curvature rate under combined loading to the curvature rate under

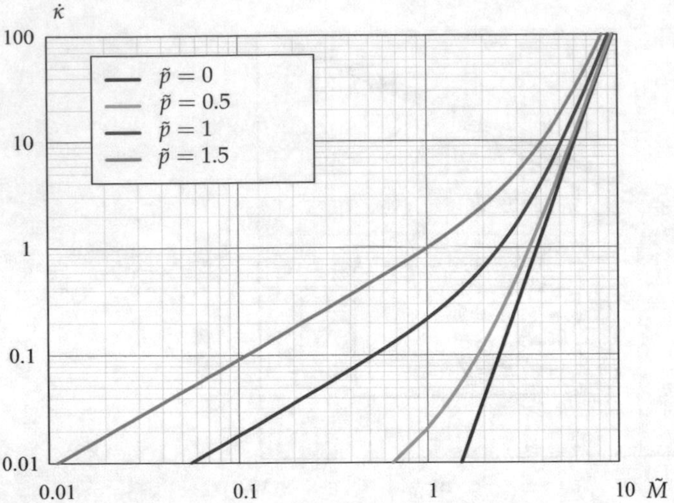

Fig. 5.17 Normalized rate of curvature vs normalized bending moment for different values of internal pressure for $n = 5$

pure bending with the same moment

$$K = \frac{\dot{\kappa}(\tilde{M}, \tilde{p})}{\dot{\kappa}_0(\tilde{M})}, \tag{5.5.82}$$

where $\dot{\kappa}_0$ is defined by Eq. (5.5.68). Inserting Eq. (5.5.82) into Eq. (5.5.81) we obtain

$$P^{-1} = \sqrt{\frac{3}{4}} \int\limits_0^{2\pi} \Upsilon^{\frac{1-n}{n}}(|x|) x \sin\varphi \, d\varphi, \quad x = \frac{K}{P} \frac{\sin\varphi}{(\sqrt{3/4} D_n)^n}, \tag{5.5.83}$$

where

$$P = \frac{\tilde{p}}{\tilde{M}} = \frac{p R_m^3}{M}$$

is the loading parameter. Figure 5.18 illustrates the results of numerical solution to Eq. (5.5.83) for different values of the creep exponent. We observe that the internal pressure has a significant influence on the flexibility parameter. The effect of the internal pressure increases as n increases. Note that in the case of elasticity or linear-viscous creep with $n = 1$ the pressure does not influence the rate of curvature.

References

Altenbach H, Naumenko K (1997) Creep bending of thin-walled shells and plates by consideration of finite deflections. Computational Mechanics 19:490 – 495

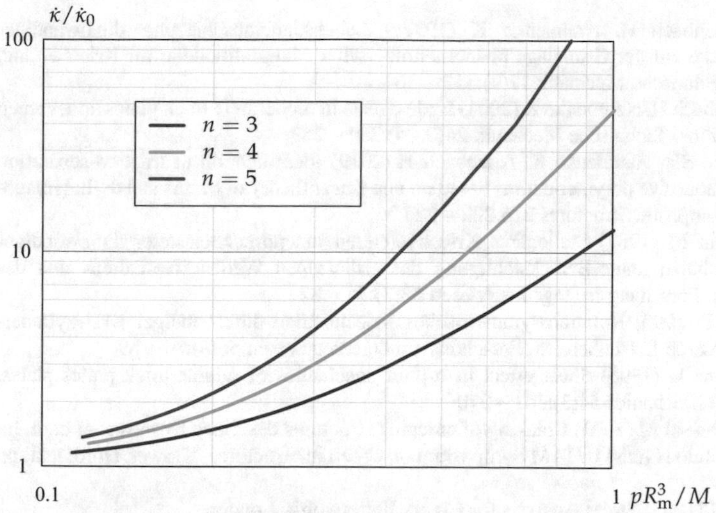

Fig. 5.18 Flexibility parameter vs loading parameter for different values of creep exponent

Altenbach H, Naumenko K (2002) Shear correction factors in creep-damage analysis of beams, plates and shells. JSME International Journal Series A, Solid Mechanics and Material Engineering 45:77 – 83

Altenbach H, Zhilin PA (2004) The theory of simple elastic shells. In: Kienzler R, Altenbach H, Ott I (eds) Theories of Plates and Shells. Critical Review and New Applications, Springer, Berlin, pp 1 – 12

Altenbach H, Morachkovsky O, Naumenko K, Sichov A (1996) Zum Kriechen dünner Rotationsschalen unter Einbeziehung geometrischer Nichtlinearität sowie der Asymmetrie der Werkstoffeigenschaften. Forschung im Ingenieurwesen 62(6):47 – 57

Altenbach H, Altenbach J, Naumenko K (1997a) On the prediction of creep damage by bending of thin-walled structures. Mechanics of Time Dependent Materials 1:181 – 193

Altenbach H, Morachkovsky O, Naumenko K, Sychov A (1997b) Geometrically nonlinear bending of thin-walled shells and plates under creep-damage conditions. Archive of Applied Mechanics 67:339 – 352

Altenbach H, Breslavsky D, Morachkovsky O, Naumenko K (2000a) Cyclic creep damage in thin-walled structures. The Journal of Strain Analysis for Engineering Design 35(1):1 – 11

Altenbach H, Kolarow G, Morachkovsky O, Naumenko K (2000b) On the accuracy of creep-damage predictions in thinwalled structures using the finite element method. Computational Mechanics 25:87 – 98

Altenbach H, Kushnevsky V, Naumenko K (2001) On the use of solid- and shell-type finite elements in creep-damage predictions of thinwalled structures. Archive of Applied Mechanics 71:164 – 181

Altenbach H, Huang C, Naumenko K (2002) Creep damage predictions in thin-walled structures by use of isotropic and anisotropic damage models. The Journal of Strain Analysis for Engineering Design 37(3):265 – 275

Altenbach H, Naumenko K, Zhilin PA (2005) A direct approach to the formulation of constitutive equations for rods and shells. In: Pietraszkiewicz W, Szymczak C (eds) Shell Structures: Theory and Applications, Taylor & Francis, Leiden, pp 87 – 90

Altenbach H, Altenbach J, Naumenko K (2016) Ebene Flächentragwerke. Springer, Berlin

Altenbach J, Altenbach H, Naumenko K (1997c) Lebensdauerabschätzung dünnwandiger Flächentragwerke auf der Grundlage phänomenologischer Materialmodelle für Kriechen und Schädigung. Technische Mechanik 17(4):353 – 364

Altenbach J, Altenbach H, Naumenko K (2004) Egde effects in moderately thick plates under creep damage conditions. Technische Mechanik 24(3 - 4):254 – 263

Bagheri B, Schulze SH, Naumenko K, Altenbach H (2019) Identification of traction-separation curves for self-adhesive polymeric films based on non-linear theory of beams and digital images of t-peeling. Composite Structures 216:222 – 227

Betten J, Borrmann M (1987) Stationäres Kriechverhalten innendruckbelasteter dünnwandiger Kreiszylinderschalen unter Berücksichtigung des orthotropen Werkstoffverhaltens und des CSD - Effektes. Forschung im Ingenieurwesen 53(3):75 – 82

Betten J, Butters T (1990) Rotationssymmetrisches Kriechbeulen dünnwandiger Kreiszylinderschalen im primären Kriechbereich. Forschung im Ingenieurwesen 56(3):84 – 89

Bialkiewicz J, Kuna H (1996) Shear effect in rupture mechanics of middle-thick plates plates. Engng Fracture Mechanics 54(3):361 – 370

Bodnar A, Chrzanowski M (2001) Cracking of creeping structures described by means of cdm. In: Murakami S, Ohno N (eds) IUTAM Symposium on Creep in Structures, Kluwer, Dordrecht, pp 189 – 196

Boyle JT, Spence J (1983) Stress Analysis for Creep. Butterworth, London

Breslavsky D, Morachkovsky O, Tatarinova O (2014) Creep and damage in shells of revolution under cyclic loading and heating. International Journal of Non-Linear Mechanics 66:87–95

Burlakov AV, Lvov GI, Morachkovsky OK (1977) Polzuchest' tonkikh obolochek (Creep of thin shells, in Russ.). Kharkov State Univ. Publ., Kharkov

Burlakov AV, Lvov GI, Morachkovsky OK (1981) Dlitel'naya prochnost' obolochek (Long-term strength of shells, in Russ.). Vyshcha shkola, Kharkov

Byrne TP, Mackenzie AC (1966) Secondary creep of a cylindrical thin shell subject to axisymmetric loading. J Mech Eng Sci 8(2):215 – 225

Carrera E (2003) Historical review of Zig-Zag theories for multilayered plates and shells. Appl Mech Rev 56(2):287 – 308

Chen S, Zang M, Wang D, Zheng Z, Zhao C (2016) Finite element modelling of impact damage in polyvinyl butyral laminated glass. Composite Structures 138:1–11

Combescure A, Jullien JF (2017) Creep buckling of cylinders under uniform external pressure: Finite element simulation of buckling tests. International Journal of Solids and Structures 124:14–25

Eisenträger J, Naumenko K, Altenbach H, Köppe H (2015a) Application of the first-order shear deformation theory to the analysis of laminated glasses and photovoltaic panels. International Journal of Mechanical Sciences 96:163–171

Eisenträger J, Naumenko K, Altenbach H, Meenen J (2015b) A user-defined finite element for laminated glass panels and photovoltaic modules based on a layer-wise theory. Composite Structures 133:265–277

Fessler H, Hyde TH (1994) The use of model materials to simulate creep behavior. The Journal of Strain Analysis for Engineering Design 29(3):193 – 200

Filippi M, Carrera E, Valvano S (2018) Analysis of multilayered structures embedding viscoelastic layers by higher-order, and zig-zag plate elements. Composites Part B: Engineering 154:77–89

Galishin A, Zolochevskii A, Sklepus S (2017) Feasibility of shell models for determining stress–strain state and creep damage of cylindrical shells. International Applied Mechanics 53(4):398–406

Ganczarski A, Skrzypek J (2000) Damage effect on thermo-mechanical fields in a mid-thick plate. J Theor Appl Mech 38(2):271 – 284

Ganczarski A, Skrzypek J (2004) Anisotropic thermo-creep-damage in 3d thick plate vs. reissner's approach. In: Kienzler R, Altenbach H, Ott I (eds) Theories of Plates and Shells. Critical Review and new Applications, Springer, Berlin, pp 39 – 44

Jones D (2004) Creep failures of overheated boiler, superheater and reformer tubes. Engineering Failure Analysis 11(6):873–893

von Kármán T (1911) Festigkeitsprobleme im Maschinenbau. In: Encyklop. d. math. Wissensch. IV/2, Teubner, Leipzig, pp 311 – 385

Kashkoli M, Tahan KN, Nejad M (2017) Time-dependent creep analysis for life assessment of cylindrical vessels using first order shear deformation theory. Journal of Mechanics 33(4):461–474

Koundy V, Forgeron T, Naour FL (1997) Modeling of multiaxial creep behavior for incoloy 800 tubes under internal pressure. Trans ASME J Pressure Vessel & Technology 119:313 – 318

Krieg R (1999) Reactor Pressure Vessel Under Severe Accident Loading. Final Report of EU-Project Contract FI4S-CT95-0002. Tech. rep., Forschungszentrum Karlsruhe, Karlsruhe

Le May I, da Silveria TL, Cheung-Mak SKP (1994) Uncertainties in the evaluations of high temperature damage in power stations and petrochemical plant. International Journal of Pressure Vessels and Piping 59:335 – 343

Lebedev LP, Cloud MJ, Eremeyev VA (2010) Tensor Analysis with Applications in Mechanics. World Scientific

Libai A, Simmonds JG (1998) The Nonlinear Theory of Elastic Shells. Cambridge University Press, Cambridge

Lin TH (1962) Bending of a plate with nonlinear strain hardening creep. In: Hoff NJ (ed) Creep in Structures, Springer, Berlin, pp 215 – 228

Liu Y, Murakami S, Kageyama Y (1994) Mesh-dependence and stress singularity in finite element analysis of creep crack growth by continuum damage mechanics approach. European Journal of Mechanics A Solids 35(3):147 – 158

Lo KH, Christensen RM, Wu EM (1977) A high – order theory of plate deformation. Part I: Homogeneous plates. Trans ASME J Appl Mech 44(4):663 – 668

Miyazaki N (1987) Creep buckling analyses of circular cylindrical shells under axial compression-bifurcation buckling analysis by the finite element method. Trans ASME J Pressure Vessel & Technol 109:179 – 183

Miyazaki N (1988) Creep buckling analyses of circular cylindrical shell under both axial compression and internal or external pressure. Computers & Struct 28:437 – 441

Miyazaki N, Hagihara S (2015) Creep buckling of shell structures. Mechanical Engineering Reviews 2(2):14–00,522

Murakami S, Suzuki K (1971) On the creep analysis of pressurized circular cylindrical shells. International Journal of Non-Linear Mechanics 6:377 – 392

Murakami S, Suzuki K (1973) Application of the extended newton method to the creep analysis of shells of revolution. Ingenieur-Archiv 42:194 – 207

Nase M, Rennert M, Naumenko K, Eremeyev VA (2016) Identifying traction–separation behavior of self-adhesive polymeric films from in situ digital images under t-peeling. Journal of the Mechanics and Physics of Solids 91:40–55

Naumenko K, Altenbach H (2016) Modeling High Temperature Materials Behavior for Structural Analysis: Part I: Continuum Mechanics Foundations and Constitutive Models, Advanced Structured Materials, vol 28. Springer

Naumenko K, Eremeyev VA (2014) A layer-wise theory for laminated glass and photovoltaic panels. Composite Structures 112:283–291

Naumenko K, Eremeyev VA (2017) A layer-wise theory of shallow shells with thin soft core for laminated glass and photovoltaic applications. Composite Structures 178:434–446

Naumenko K, Altenbach J, Altenbach H, Naumenko VK (2001) Closed and approximate analytical solutions for rectangular Mindlin plates. Acta Mechanica 147:153 – 172

Nordmann J, Thiem P, Cinca N, Naumenko K, Krüger M (2018) Analysis of iron aluminide coated beams under creep conditions in high-temperature four-point bending tests. The Journal of Strain Analysis for Engineering Design 53(4):255–265

Odqvist FKG (1962) Applicability of the elastic analogue to creep problems of plates, membranes and beams. In: Hoff NJ (ed) Creep in Structures, Springer, Berlin, pp 137 – 160

Paggi M, Kajari-Schröder S, Eitner U (2011) Thermomechanical deformations in photovoltaic laminates. The Journal of Strain Analysis for Engineering Design 46(8):772–782

Penny RK (1964) Axisymmetrical bending of the general shell of revolution during creep. J Mech Eng Sci 6:44 – 45

Podgorny AN, Bortovoj VV, Gontarovsky PP, Kolomak VD, Lvov GI, Matyukhin YJ, Morachkovsky OK (1984) Polzuchest' elementov mashinostroitel'nykh konstrykcij (Creep of mashinery structural members, in Russ.). Naukova dumka, Kiev

Psyllaki P, Pantazopoulos G, Lefakis H (2009) Metallurgical evaluation of creep-failed superheater tubes. Engineering Failure Analysis 16(5):1420–1431

Rabotnov YN (1969) Creep Problems in Structural Members. North-Holland, Amsterdam

Reddy JN (1984) A simple higher-order theory for laminated composite plate. Trans ASME J Appl Mech 51:745 – 752

Roche RL, Townley CHA, Regis V, Hübel H (1992) Structural analysis and available knowledge. In: Larson LH (ed) High Temperature Structural Design, Mechanical Engineering Publ., London, pp 161 – 180

Schulze S, Pander M, Naumenko K, Altenbach H (2012) Analysis of laminated glass beams for photovoltaic applications. International Journal of Solids and Structures 49(15 - 16):2027 – 2036

Spence J (1973) Creep of a straight pipe under combined bending and internal pressure. Nuclear Engineering and Design 24(1):88–104

Takezono S, Fujoka S (1981) The creep of moderately thick shells of revolution under axisymmetrical load. In: Ponter ARS, Hayhurst DR (eds) Creep in Structures, Springer-Verlag, Berlin, pp 128 – 143

Takezono S, Migita K, Hirakawa A (1988) Elastic/visco-plastic deformation of multi-layered shells of revolution. JSME, Ser 1 31(3):536 – 544

Timoshenko SP, Woinowsky-Krieger S (1959) Theory of Plates and Shells. McGraw-Hill, New York

Weps M, Naumenko K, Altenbach H (2013) Unsymmetric three-layer laminate with soft core for photovoltaic modules. Composite Structures 105:332–339

Wriggers P (2008) Nonlinear Finite Element Methods. Springer, Berlin, Heidekberg

Yang HTY, Saigal S, Masud A, Kapania RK (2000) A survey of recent shell finite elements. Int J Numer Meth Engng 47:101 – 127

Zhilin PA, Ivanova EA (1995) Modifitsirovannyi funktsional energii v teorii plastin tipa reissnera (a modified energy functional in the reissner type plate theory, in russ.). Izv RAS Mekhanika tverdogo tela 2:120 – 128

Zienkiewicz OC, Taylor RL (1991) The Finite Element Method. McGraw-Hill, London

Index

K. Naumenko and H. Altenbach, *Modeling High Temperature Materials Behavior for Structural Analysis*, Advanced Structured Materials 112,
https://doi.org/10.1007/978-3-030-20381-8

Printed in the United States
By Bookmasters